POLYMER GELS

Fundamentals and Biomedical Applications

POLYMER GELS

Fundamentals and Biomedical Applications

Edited by

Danilo DeRossi

University of Pisa
Pisa, Italy

Kanji Kajiwara

Kyoto Institute of Technology
Kyoto, Japan

Yoshihito Osada

Ibaraki University
Mito-shi, Japan

and

Aizo Yamauchi

HOYA Corporation
Saitama-ken, Japan
Formerly of Research Institute for Polymers and Textiles, AIST
Tsukuba-shi, Japan

PLENUM PRESS • NEW YORK AND LONDON

Library of Congress Cataloging-in-Publication Data

International Symposium on Polymer Gels (1989 : Tsukuba-shi, Japan)
Polymer gels : fundamentals and biomedical applications / edited by Danilo DeRossi ... [et al.].
p. cm.
"Proceedings of an International symposium on Polymer Gels, held September 18-21, 1989, in Tsukuba, Japan"--T.p. verso.
Includes bibliographical references and index.
ISBN-13: 978-1-4684-5894-7 e-ISBN-13: 978-1-4684-5892-3
DOI: 10.1007/978-1-4684-5892-3
1. Gels (Pharmacy)--Congresses. 2. Polymers--Congresses. 3. Drugs--Vehicles--Congresses. I. DeRossi, Danilo. II. Title. [DNLM: 1. Drug Carriers--congresses. 2. Gels--congresses. 3. Polymers--therapeutic use. 4. Technology, Pharmaceutical--congresses. QV 785 I6163p 1989]
RS201.G44I58 1989
615'.19--dc20
DNLM/DLC
for Library of Congress 91-2631
CIP

Proceedings of an International Symposium on Polymer Gels,
held September 18-21, 1989, in Tsukuba, Japan

Softcover reprint of the hardcover 1st edition 1991

A Division of Plenum Publishing Corporation
233 Spring Street, New York, N.Y. 10013

PREFACE

This volume contains a series of papers originally presented at the Symposium on Polymer Gels organized and sponsored by the Research Group on Polymer Gels,The Society of Polymer Science of Japan and co-sponsored by the Science and Technology Agency (STA) and MITI, Japan.

The Symposium took place at Tsukuba Science City on 18th and 19th September, 1989. Recognized experts in their fields were invited to speak and there was a strong attendance from government, academic and industrial research centers. The purpose of the Symposium was to review the state of the art and to present and discuss recent progress in the understanding of the behavioral properties of polymer gels and their application to biomedical, environmental and robotic fields.

Most of the papers and related discussions concentrated on the swelling behavior of hydrogels and chemomechanical systems, both artificial and naturally occurring, in which external stimuli of a physical or chemical nature control energy transformation or signal transduction. The recent great interest in chemomechanical systems based on polymer gels has stimulated considerable effort towards the development of new sensors and actuators, controllable membrane separation processes, and delivery systems in which the functions of sensing, processing and actuation are all built into the polymeric network device. Artificial chemomechanical systems, through the use of environmentally sensitive polymer gels, are emerging as interesting materials for mimicking basic processes previously only confined to the biological world, and commercially viable applications are also foreseen in the not-too-distant future.

The main topics of interest of the Symposium were grouped into the following sessions: (1) Fundamentals, (2) Biologicals, (3) Separation, (4) Sensors and Actuators, and (5) Drug Delivery Systems. This volume is organized into five sections which essentially maintain the same structure as the Symposium.

The five papers which comprise Section 1 deal with characterization of gel structure and studies on gelation, gel swelling and gel mechanical properties.

In Section 2, the properties and excitability of nerve cytoplasm are described in two papers, while three papers are concerned with the preparation and characterization of gels purposely tailored for implantation into the human body or capable of performing catalytic or regulating functions for biotechnological or biomedical applications.

Section 3 comprises four papers which address new materials and techniques for separation and ultrafiltration technology, also addressing the very important aspect of real-time control of membrane properties.

The seven papers of Section 4 are concerned with the study of gel sensors and actuators, with particular emphasis on the development of compliant, "muscle-like" actuators for advanced robotics and prosthetics.

Drug delivery systems based on environmentally sensitive hydrogels are discussed in Section 5. Different factors controlling swelling kinetics and equilibrium are exploited on the way towards reaching the final goal of developing self-regulated delivery systems possessing the intrinsic capabilities of stimulus sensing and actuation under the control of material built-in algorithms.

The Editors of this volume are grateful to all the contributors for their prompt and accurate preparation of manuscripts and to all those who made the Symposium possible and this volume available, including STA, AIST-MITI, the Society of Polymer Science of Japan (SPSJ), the Organizing and Steering Committees of the Symposium, and Plenum Publishing. Finally on behalf of all non-Japanese attendees at the Symposium, we wish to thank Dr Aizo Yamauchi (AIST, SPSJ) and our hosts at Tsukuba for another admirable demonstration of traditional Japanese hospitality.

Danilo De Rossi
University of Pisa
Pisa, Italy
August 1990

CONTENTS

FUNDAMENTALS

BIOLOGICALS

SEPARATION

FUNDAMENTALS

Characterization of Gel Structure by Means of SAXS and SANS

Kanji Kajiwara, Shinzo Kohjiya, Mitsuhiro Shibayama, Hiroshi Urakawa

Kyoto Institute of Technology,
Kyoto, Sakyo-ku, Matsugasaki, 606 Japan

ABSTRACT

The characterization of the gel structure by means of small-angle X-ray scattering (SAXS) and small-angle neutron scattering (SANS) is described by taking the examples of monovalent and bivalent cation κ-carrageenates in aqueous solution, poly(vinyl alcohol)-borate gels and the inorganic/organic hybrid gels from tetraethoxysilane and poly(oxytetramethylene). The SAXS and SANS results were analysed and discussed in terms of the classic gelation thoery as well as the scaling concepts.

1. INTRODUCTION

Structural characterization of polymeric gels faces a serious problem how we can define the structure of gel. Various attempts have been made to specify a gel with simple parameters such as the branching fraction[1] or the correlation length[2]. These theories idealize the gel structure in some extent, so that the analysis in terms of these theories yields inevitably the oversimplified picture of gel structures. For example, in the classic theory, the polymer molecules are represented by complexes of points and lines, and these one-dimensional graphlike molecules are embedded in spaces of higher dimensionality to compute physical properties by discrete combinatorial and/or continuum averaging process. The scaling concept describes the structure of random systems within the framework of fractal geometry, where the universal properties are discussed in terms of scaling laws and characteristic exponents.

Polymeric gels will be classified in two categories with respect to the nature of networks; (i) the chemical gel where the network is formed by covalent links, and (ii) the physical gel which undergoes reversible gelation by externally-induced topological change of chains. In this report, we concern mainly the characterization of gel structures by the scattering method. First we discuss a general scattering theory where we restrict the discussion to the static scattering factor from sol and gel phases. The gel model treated in this discussion is an idealized gel specified by previous theories mentioned above. The main purpose is to demonstrate how we can evaluate the molecular parameters from the scattering experiments in terms of these theories. The results of the scattering experiments are discussed to elucidate the gel structure from the evaluated parameters according to the theories mentioned above. Further attempts will be made by the conventional analysis of

Polymer Gels, Edited by D. DeRossi *et al.*
Plenum Press, New York, 1991

the scattering data to discuss some details of the structures of the gel and crosslinking domains.

We chose two experimental examples of reversible gelation observed in the systems of poly(vinyl alcohol)/borate and cation carrageenate aqueous solutions. Here the chemical structures of the junction points or domains vary with external conditions, and we examine the potential and limit of the model network theories developed earlier in the analysis of the gel structure of poly(vinyl alcohol) and cation carrageenate.

The last example deals with inorganic/organic hybrid gels prepared by the sol-gel process of tetraethoxysilane and organic polymer. This system belongs to the category of chemical gel where the branching proceeds with covalent linking leading to the gel formation. The incompatibility of inorganic and organic domains causes the segregation of these two domains, and the domain structure is expected to change by varying the composition ratio of inorganic and organic components. An attempt will be made to picture the domain structure in the hybrid gel from its SAXS pattern. The inorganic components compose the crosslinking domains, and the elastic property of the hybrid gel is specified by organic chains linking these inorganic domains. The branching mechanism will be speculated from the mechanical property and the structural aspect of the hybrid gels. Here the SAXS is expected to reveal the details of the microscopic structural characteristics of the hybrid gel.

2. THEORETICAL BACKGROUND

The scattered intensity from the gel phase is asymptotically given in terms of the correlation length ξ (the size of the uncorrelated domains compactly arranged in the gel phase) and the fractal dimension D inside the domain by[3]

$$I(q) \sim \frac{c\xi^3}{\left(1+(D+1)\xi^2 q^2/3\right)^{D/2}}. \tag{1}$$

Here q denotes the scattering magnitude given by

$$q = 4\pi \, sin\, \theta/\lambda \tag{2}$$

with θ and λ being the half of the scattering angle and the incident electro-magnetic wavelength, respectively. The assumption involved in the above formulation is that the monomer density correlation in each domain decays as r^{D-d} with a distance r in the d-dimensional space. The uncorrelated domains in sol phase are equivalent to clusters formed by random branching, and the scattered intensity from sol phase is determined by the mass distribution of clusters specified by the critical exponents σ and τ as

$$w(s) = s^{1-\tau} h(\epsilon s^{\sigma}) \tag{3}$$

with a scaling function h and the distance from the gel point $\epsilon(= p_c - p)$ where p denotes the conversion and the subscript c the critical value at the gel point.[4] The correlation length in sol phase is identified with the z-average mean-square radius of gyration of clusters as $D(D+1)\xi^2/2 =< R_G^2 >_z$. The fractal dimension of an isolated Gaussian chain is equal to 2 so that ξ^2 is given directly by the initial slope of the scattering profile as $\xi^2 =< R_G^2 >_z /3$. ξ diverges at the gel point as

$$\xi = \xi_0(p_c - p)^{-1/\sigma D} \tag{4}$$

where the proportional constant ξ_0 is regarded to correspond to the radius of the smallest cluster. When the gel is formed, the system is supposed to consist of a compact arrangement of uncorrelated domains of the size specified by the correlation length ξ. ξ in gel phase scales as eq.(4), and the size distribution of those domains is expected to follow eq.(3) by redefining ϵ as $\epsilon = p - p_c$. Here the domains are formed by the density fluctuation in the gel, which can be observed in the suitable range of $\xi_0 q$. The total scattered intensity from a polydisperse fractal ensemble thus reduces to

$$I(q) \sim \begin{cases} cq^{-D(3-\tau)}, & \text{when } \xi_0 q << 1; \\ cq^{-D}, & \text{when } \xi_0 q >> 1. \end{cases} \tag{5}$$

Thus, the observable scattered intensity scales as

$$I(q) \sim q^{-\mu} \tag{6}$$

where the polydispersity-smeared fractal dimension is given as $\mu = D(3 - \tau)$ or μ can be identified with the fractal dimension D of the particles in the system if the condition $\xi_0 q >> 1$ is satisfied. This condition is, however, seldom satisfied, because in most cases $q < 0.5 \mathring{A}^{-1}$ and ξ_0 is small in the order of $1\mathring{A}$. The chemical characteristics of particles may become significant and smear the fractal dimension given by the scattering exponent eq.(6), when q exceeds $1\mathring{A}^{-1}$. Thus in practice we can evaluate the polydispersity-smeared scattering exponent from the Guinier regime, which is given as $\mu = 2.0$ near the gel point from the accepted values of τ (2.2) and D (2.5) in percolation clusters.[5]

The graph theoretical approach in the frame work of the cascade process provides more systematic description of the scattering behaviour from gel and sol phases.[6] The randomly crosslinked arbitrary primary chain distribution yields the scattering factor $S(q)$ in sol phase as

$$P_w \cdot S(q) = \frac{(1 + p\phi)\overline{f}_w(\phi)}{\{1 - [\overline{f}_w(\phi) - 1]p\phi\}} \tag{7}$$

Here $\overline{f}_w(\phi)$ and $P_w \cdot S(q)$ correspond to the enumeration generating functions for the mean number of distinct paths in a primary chain and a whole ensemble, respectively.[7] When we set

$$\phi = exp(-q^2 b^2/6), \tag{8}$$

then $\overline{f}_w(\phi)$ and $P_w \cdot S(q)$ reduce to the particle scattering factors for linear chains and random vulcanisates with Gaussian subchains, respectively. The weight-average degree of polymerization P_w is given by

$$P_w = (1 + p)\overline{f}_w(1)/[1 - (\overline{f}_w(1) - 1)p] \tag{9}$$

with $\overline{f}_w(1)$ yielding the weight-average degree of polymerization of a primary chain ensemble. A gel point is specified by

$$p_c = 1/(\overline{f}_w(1) - 1). \tag{10}$$

Although the scattered intensity diverses at a gel point, the scattered intensity scales as

$$I(q)(\equiv P_w \cdot S(q)) \sim q^{-2} \tag{11}$$

due to the Gaussian nature of subchains, although the same exponent is predicted from the fractal theory of gel formation. Here again the scattering exponent will be smeared by the chemical characteristics of chains as, for example, the exponent is equal to unity for a completely rigid rod.

3. EXPERIMENTAL

Three examples are chosen to characterize the gel structure by means of small-angle neutron or X-ray scattering (SANS or SAXS).

A first example deals with the sol-gel transition of κ-carrageenan aqueous solution, where the structure of cross-linking domain is analysed from the results of the small-angle X-ray scattering from the sol and gel phase of κ-carrageenan aqueous solution. In the second example, the structure of poly(vinyl alcohol)-borate gels is discussed in terms of the correlation length ξ, a measure of the spatial length of concentration fluctuation, estimated from a modified Zimm plot of small-angle neutron scattered intensities. The inorganic/organic hybrid gels will be investigated in the last example by means of small-angle X-ray scattering in order to specify the network structure. The branching mechanism will be speculated from the network structure thus elucidated.

3.1. Sample Preparation

3.1.1. κ-carrageenate aqueous solutions

κ-carrageenan (Type III) was purchased from SIGMA, and purified by dialysis and subsequent percolation of κ-carrageenan aqueous solution (C_p = 2 %) through an ion exchanger. The resulted acidic carrageenan aqueous solutions were neutralized with KOH, CsOH, $Ca(OH)_2$, or $Ba(OH)_2$ aqueous solution, and then freeze-dried for further use. All reagents were of analytical grade, and water was distilled, deionized through an ion exchanger and distilled again just prior to use. κ-carrageenate aqueous solutions were prepared by dissolving weighted κ-carrageenan into freshly distilled water, and dialysing against sufficient amount of salt solution if necessary. The κ-carrageenan concentration was adjusted to 3 and 6 % by weight in each solution. These solutions were directly subjected to small-angle X-ray scattering measurements.

3.1.2. Poly(vinyl alcohol)-borate gels

Re-saponified poly(vinyl alcohol) (PVA) powder (P_w = 120, supplied by Unitika Chem. Co.) was dissolved in deuterated water at ca. $90°C$ in a flask connected to a reflux condenser and by stirring for several hours. Deuterated water solutions of boric acid and NaOH were prepared separately, and mixed in the PVA solutions, which were then homogenized by heating above the sol-gel transition temperature. Here the composition ratio of the final solutions was kept constant to $10 : 1 : 1$ by weight in terms of the PVA initial solution, 0.2M-$B(OH)_3$ and 1.8N-NaOH, respectively.

3.1.3. Inorganic/organic hybrid gels

The hybrid gels were prepared from the prepolymer triethoxysilyl terminated poly(oxy-tetramethylene) (ET-PTMO) synthesized from poly(oxytetramethylene) glycol, and tetra-ethoxysilane (TEOS) with varying the compositions of ET-PTMO and TEOS. The sol-gel reaction was performed by mixing TEOS and ET-PTMO in an ethanol/THF mixture and adding hydrogen chloride. After one-hour refluxing, the mixture was put in a glass container, which was kept at $50°C$ in open air. The resulted gel was dried first at room temperature, and then at $50°C$ under vacuum to a constant weight. Rubbery gel films of approximately 1 mm thick were subjected to the small-angle X-ray scattering observation with a Kratky compact camera.

3.2. Scattering Measurements

3.2.1. Small-angle X-ray scattering

Small-angle X-ray scattering (SAXS) measurements were performed from the κ-carrageenate aqueous solutions prepared as above at $5°C$ (in the gel state) and $60°C$ (in the sol state) with a SAXES focusing optics installed at the BL10C of the Photon Factory, Tsukuba, Japan.[8] The excess scattered intensity $I(q)$ with respect to respective buffers was Fourier-transformed to the distance distribution function $p(r)$ defined as:

$$p(r) = (2\pi)^2 \int_0^{\infty} I(q) \cdot (qr) \cdot sin(qr)dr. \tag{12}$$

Here q denotes the scattering magnitude given by $4\pi \, sin\,\theta/\lambda$ with θ and λ being the half of the scattering angle and the wavelength of incident X-ray, respectively. The cross-sectional radius of gyration R_c was estimated from the cross-sectional Guinier plot:[9]

$$qI(q) \sim exp(-R_c^2 q^2/2). \tag{13}$$

The scattered intensities were recorded for the time interval of $600sec$ with an equal interval of 512 channels on the one-dimensional position sensitive proportional counter of an effective length 160mm (Rigaku Denki Co.) over the scattering magnitude from $6.61\times10^{-3}\mathring{A}^{-1}$ to $0.33\mathring{A}^{-1}$ (equivalent to the Bragg spacing d_B from $951\mathring{A}$ to $18.8\mathring{A}$). The circulating current of electrons in the storage ring was 100 to $250mA$, and the wavelength of $0.149nm$) was employed. The specimen-to-detector distance was approximately $1900mm$, and an exact scattering magnitude was calibrated by using the diffraction peaks of collagen fiber. The data acquisition was made by the CAMAC system controlled by a MINC11/23 (Digital Equipment Co.). Further analysis of the data was performed with Asahi-Stellar GS1000 (Asahi-Techno Co.).

The small-angle X-ray scattering from inorganic/organic hybrid gels was observed with a Kratky compact camera with an entrance slit of $100\mu m$ (Anton Paar KG) installed in the Phillips X-ray generator Model PW1730 under the applied power of $55kV$ and $45mA$ where the K_β line was eliminated with a nickel filter. A Kratky compact camera was coupled with an electrically programmable step scanner device and a proportional counter with a pulse height discriminator to allow an automatic operation. The system is controlled by a NEC PC-9801. Scattered intensities were recorded at 165 angles for each gel sample, and then deconvoluted with respect to the line profile of the primary beam according to the statistical procedure.[10] The deconvolution and subsequent Fourier-transformation were made with Asahi-Stellar GS1000.

3.2.2. Small-angle neutron scattering

The small-angle neutron scattering experiments were performed on poly(vinyl alcohol)–borate gels at the Research Reactor of the National Institute of Standards and Technology. A flux of cold neutrons was monochromatized with a velocity selector of wavelength $\lambda = 6\mathring{A}$ to be used as an incident beam. A specially-designed copper sample chamber (2 mm thick and 16 mm in diameter with copper and quartz windows for both sides) stores gel specimens for scattering experiments. Scattered intensities were recorded two-dimensionally, and then transformed into one-dimensional data with respect to the scattering vector by averaging concentric-circularly after correcting background noise, fast neutron scattering, air scattering and transmittance. Each measurement lasted for the period till the monitor attained 5×10^7 counts, which approximately took 30 minutes.

3.2.3. Dynamic elastic modulus

The dynamic elastic moduli E' from hybrid gels were evaluated with a dynamic mechanical analyzer, FT-Rheospectra DVE-V4 (Rheology Co. Ltd.). The strain frequency was fixed to $10Hz$, and the temperature was increased with a rate of $2°C/min$ during the elastic modulus measurements. E' at the rubbery plateau region was used to estimate the molecular weight between crosslinking points.

4. RESUTLS AND DISCUSSION

4.1. κ-carrageenan

κ-carrageenan, a sulfated galactan extracted from red seaweeds, is composed of a disaccharide repeat unit of 1,3-linked β-D-galactopyranose and 1,4-linked α-D-galactopyranose, where most of the former residues are sulfated on the position 4. 1,4-linked α-D-galactopyranose residues are mostly present as 3,6-anhydride, and the occasional inclusion of sulfated 1,4-linked residues is thought to disturb the regular helical structure of κ-carrageenates. κ-carrageenan aqueous solutions form thermo-reversible gel by adding salt, where two types of gels are available as the strong gel and the weak gel conventionally termed from the appearance. The type of gel depends on the added cation species as in general monovalent and bivalent cations induce the strong and weal gel, respectively. Gelation in κ-carrageenanate aqueous solutions may take place by double-helical crosslinking (probably a three-fold helix) and subsequent aggregation of crosslinks, and various models are proposed for its mechanism.

SAXS from monovalent cation κ-carrageenate aqueous solution exhibits a pronounced correlation maximum in its gel state as shown in Figure 1(a), although its maximum becomes less pronounced in the sol state. For example, the correlation maximum was observed for potassium κ-carrageenate aqueous solution at $q = 0.05 \mathring{A}^{-1}$ in the gel state, which shifts to the larger scattering vector ($q = 0.075 \mathring{A}^{-1}$) in the sol state. Bivalent cation κ-carrageenate aqueous solutions exhibit no such a maximum as demonstrated by the scattering envelopes from calcium κ-carrageenate aqueous solution in Figure 1(b). As Figure 2 demonstrates, the addition of salt or the increase of the solute concentration in monovalent cation κ-carrageenate aqueous solutions results the correlation maximum in the scattering envelopes less pronounced, suggesting this apparent maximum is caused by the electrostatic interaction which is screened by increasing the ionic strength.[11] Such an interaction is small and no appreciable change of the scattering envelopes was observed in the bivalent cation κ-carrageenate aqueous solutions by adding salt and/or increasing the solute concentration.

A pronounced maximum/minimum characterizes the distance distribution function of monovalent cation (potassium or cesium) κ-carrageenate aqueous solutions in Figure 3, obtained by the Fourier transformation of the scattered intensities according to eq.(12). The distance distribution function of bivalent cation (calcium or barium) κ-carrageenate aqueous solution in the gel state consists of a single broad maximum around $120 - 130 \mathring{A}$ with a long tail typical of a chain molecule as shown in Figure 4. Since the cross-sectional radius of gyration is evaluated from the cross-sectional Guinier plot eq.(13) as $5 \mathring{A}$ and $9 \mathring{A}$ in the sol and gel state, respectively, for both potassium and calcium carrageenates, monovalent and bivalent cation carrageenates would assume a similar chain structure in solution despite of the apparent difference in the gel state. Here $5 \mathring{A}$ corresponds to the cross-sectional radius of gyration calculated from a double-helical conformation of three disaccharide residues per turn,[12] and $9 \mathring{A}$ to that of a triangular array of three double-helix domains. Thus the distance distribution function of monovalent cation κ-carrageenates is considered to be composed of two functions; (i) a sharp distribution function with a maximum around $20 \mathring{A}$

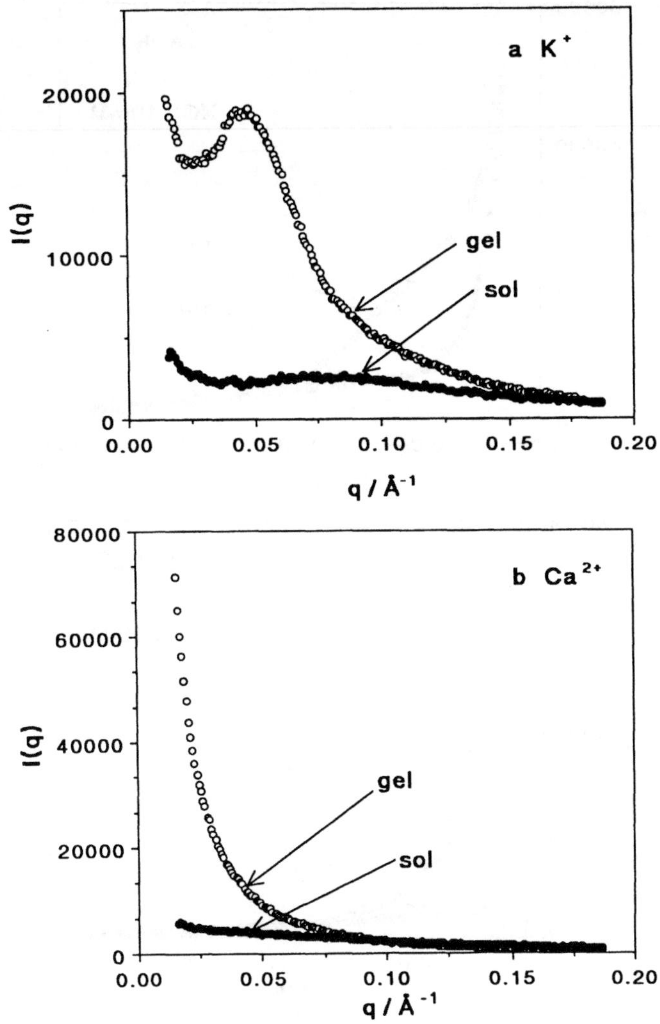

Figure 1 Excess small-angle X-ray scattering from cation κ-carrageenate aqueous solutions. (a) Potassium κ-carrageenate and (b) calcium κ-carrageenate in sol (at $60°C$) and gel (at $5°C$) states, where no salt was added.

and a subsequent correlation hole characteristic of strong repulsive interactions,[13] and (ii) a broad distribution function with a single peak observed in the bivalent cation κ-carrageenate gel. Gel is formed by the association of double-helix domains arranged triangularly in either cation κ-carrageenate aqueous solution. The concentrated charged density due to free sulfate groups on the domain surface may cause a strong repulsive interaction between domains, and be responsible to the correlation hole observed in monovalent cation κ-carrageenate aqueous solutions. The domain association may be resulted as a consequence of the reduction of solvent quality by lowering temperature and/or adding salt. The solvent quality reduction also promotes intra-helix hydrogen bonding as monitored by the positional shift of the correlation hole from $30 - 50\mathring{A}$ (in sol) to $75 - 80\mathring{A}$ in gel.

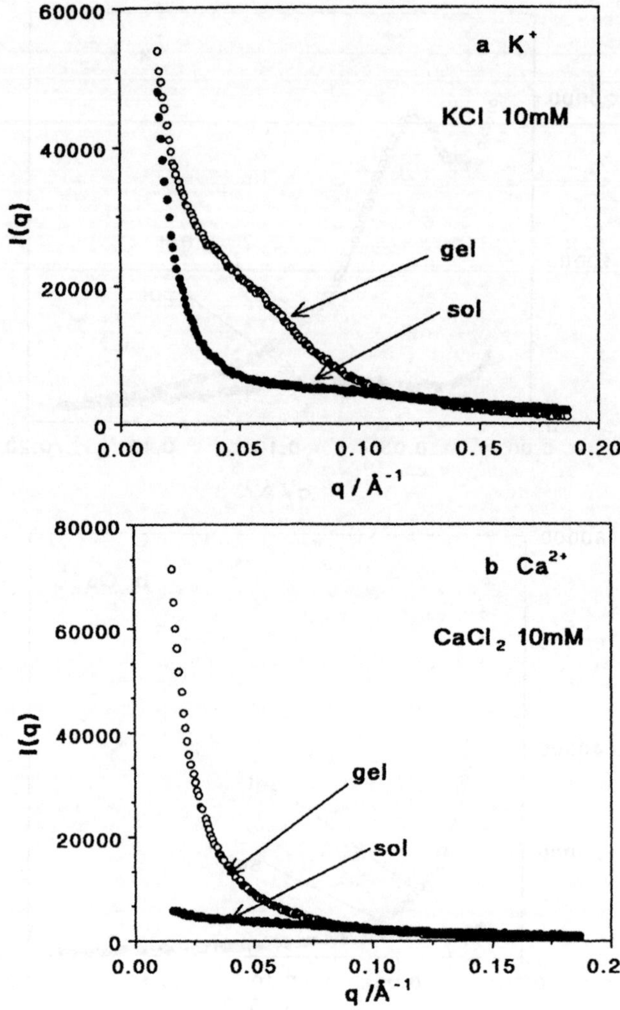

Figure 2 as Figure 1, but salt was added as indicated in the figure.

The scattered intensity scales as $I(q) \sim q^{-1.30}$ and $I(q) \sim q^{-2.55}$ in the sol and gel phases of both monovalent and bivalent cation κ-carrageenate aqueous solutions, respectively. A rather small exponent in the sol phase (1.30) may be due to the stiff nature of carrageenate chain. The observed exponent in the gel phase is larger than that expected from both the fractal and classic theory. This deviation may indicate the different exponent τ (probably smaller) on the mass distribution of clusters (see eq.(3)), resulted from the two-step formation of cross-linking domains by double-helix and its subsequent association.

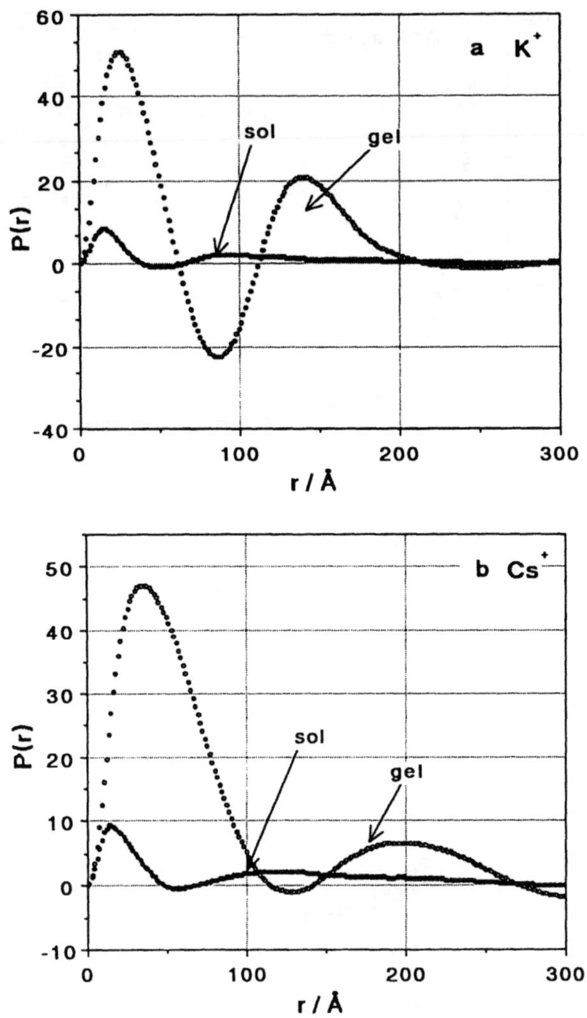

Figure 3 Distance distribution function of monovalent cation κ-carrageenate in sol and gel states. (a) Potassium κ-carrageenate with no added salt, and (b) cesium κ-carrageenate with 10mM CsCl. The screening effect of salt will be seen by comparing (b) with (a).

4.2. Poly(vinyl alcohol)-borate gels

PVA-borate gels were prepared by heating PVA aqueous solutions containing borate ions above the gelling temperature, and then cooling gradually. Two PVA solutions (A12 and A16) of different concentrations ($12.0wt.\%$ and $16.0wt.\%$ respectively) were prepared as described above, where the concentrations of $B(OH)_3$ and NaOH are same for both solutions as 1.67×10^{-2}M and 1.51×10^{-1}N, respectively. Thus the molar ratios of borate to PVA are 0.705 and 0.529 for A12 and A16, respectively. The sol-gel transition temperatures were found as $40°C$ (A12) and $55°C$.

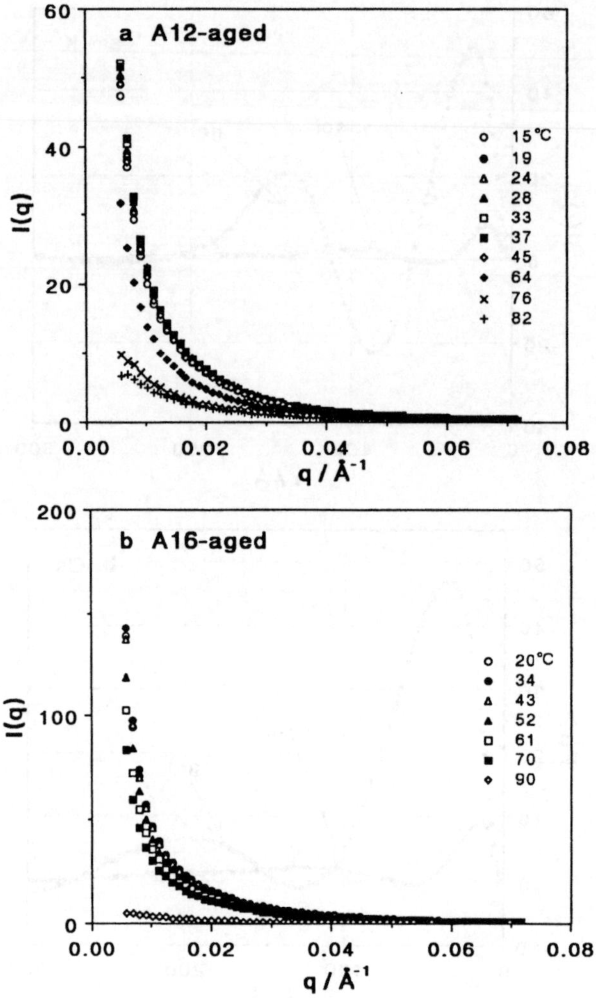

Figure 4 Excess small-angle neutron scattering from PVA-borate solutions at various temperatures indicated in the figure. (a) A12-aged, and (b) A16-aged solutions.

Figure 4 shows the scattered intensity profiles from A12 and A16 solutions at various temperatures. The scattered intensity decreases monotonously with the scattering magnitude q. Although the PVA hydrogels were characterized by the drastic increase of the scattered intensity at a given q by aging,[14] the PVA-borate gels were found to yield almost constant scattered intensities over several months. The scattered intensities scale as $I(q) \sim q^{-2.6}$ and $I(q) \sim q^{-2.8}$ in A12 and A16 gels, respectively, in the range of $q > 0.03\mathring{A}^{-1}$. This scaling exponent μ decreases with increasing temperature in the sol state as shown in Figure 5. The larger exponent than 2.0 observed in the PVA-borate gels may suggest the presence of two types of crosslinking by hydrogen bonding in addition to the borate-aided crosslinks, causing the narrower mass distribution than the expected distribution for fractal gels.

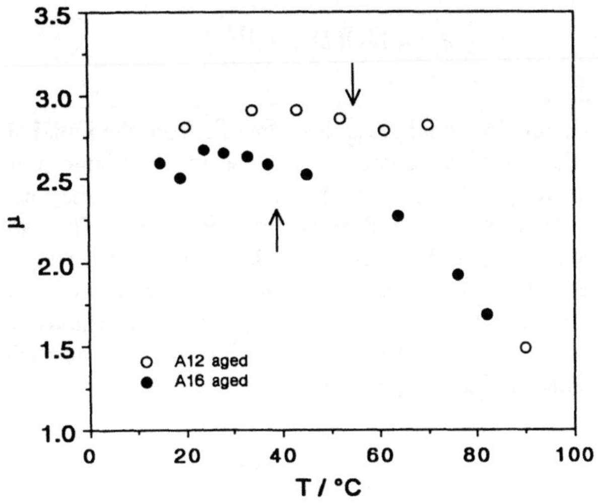

Figure 5 Temperature variation of the scaling exponent μ for A12 and A16 aged solutions where $I(q) \sim q^{-\mu}$. Respective gel points are indicated by arrows in the figure.

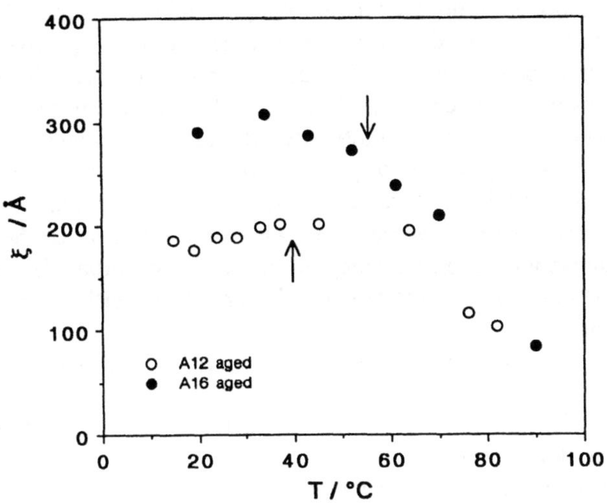

Figure 6 Correlation lengths ξ of A12 and A16 aged solutions as a function of temperature. The arrows indicate the gel-to-sol transition temperature of A12 and A16, as indicated.

Expanding eq.(1) with respect to q^2, the correlation length ξ is given by

$$\xi = [6/D(D+1)]^{1/2}\xi_{app} \tag{14}$$

where the apparent correlation length ξ_{app} is estimated from the initial slope of the $I^{-1}(q)$ vs. q^2 plots. Since the fractal dimension D is close to 2 for large branched polymers in a dilute solution near the gel threshold,[15] the correlation length may be replaced with the apparent correlation length estimated directly from the slope. The estimated correlation lengths in a low q region for A12 and A16 solutions are shown in Figure 6 as a function of temperature, where two arrows indicate the sol-gel transition temperature of A12 and A16, respectively. The unperturbed dimension of present PVA chains was evaluated from the SANS measurement of a 50/50 mixture film of deuterated and hydrogenated PVAs[16] in terms of the z-average radius of gyration as

$$< R_G^2 >_z^{1/2} = 33.5\text{Å} \tag{15}$$

which is reduced to 19.3Å in terms of the correlation length given as $3\xi^2 =< R_G^2 >_z$. The observed correlation length is longer than this value expected for an isolated PVA chain as approximately $180 - 240$Å which suggests multiple chains involved in the inter-crosslink domain.

Eq.(1) implies the scattered intensity profile for gels be superposed to a single master curve with respect to a reduced scattering magnitude, ξq, if the fractal dimension D is invariant. As demonstrated in double logarithmic plots of $I(q)/\xi^3 c$ vs. ξq (Figure 7 where the longitudinal scale is corrected with the contrast factor K), respective intensity profiles fall into a single curve, confirming the D invariance in the gel phase. These plots at high q regions yield the scaling exponents μ (= $D(3-\tau)$) 2.6 and 2.8, respectively, for A12 and A16 gels, as estimated from individual intensity profiles. These values are somewhat larger than the accepted value[15] ($\mu = 1.6$ or $\mu = 2.0$ corresponding respectively to the fractal dimension $D = 2.0$ of very large branched polymers in a dilute solution or $D = 2.5$ for the same branched polymers in the reaction bath where the exponent of the size distribution $\tau = 2.2$ near the threshold). The result may prove the participation of hydrogen bonding in PVA network formation in addition to borate-aided crosslinks, yielding a different exponent on the size distribution.

In conclusion, the PVA-borate gels are composed of fractal clusters specified by a correlation length of $180 - 240$Å. Clusters are probably formed by a number of PVA chains coupled with hydrogen bonding, and the addition of borate ions promotes the gel formation by crosslinking clusters.

Although the direct estimation of the fractal dimension D is not possible, D seems invariant in the gel phase. A large scattering exponent μ may be due to the nature of the crosslinking mechanism where two mechanisms are involved in the gelation of PVA-borate and κ-carrageenate systems. Expected exponents were observed in the system of simple covalent crosslinks such as poly(tetramethoxysilicon)[17] or polyurethane.[18] The mass distribution will scale differently in the system which undergoes two-step crosslinking, and in consequence a larger exponent μ was observed in comparison with the covalent gels produced by random crosslinking. There is no proof whether the fractal dimension also varies by the crosslinking mechanism.

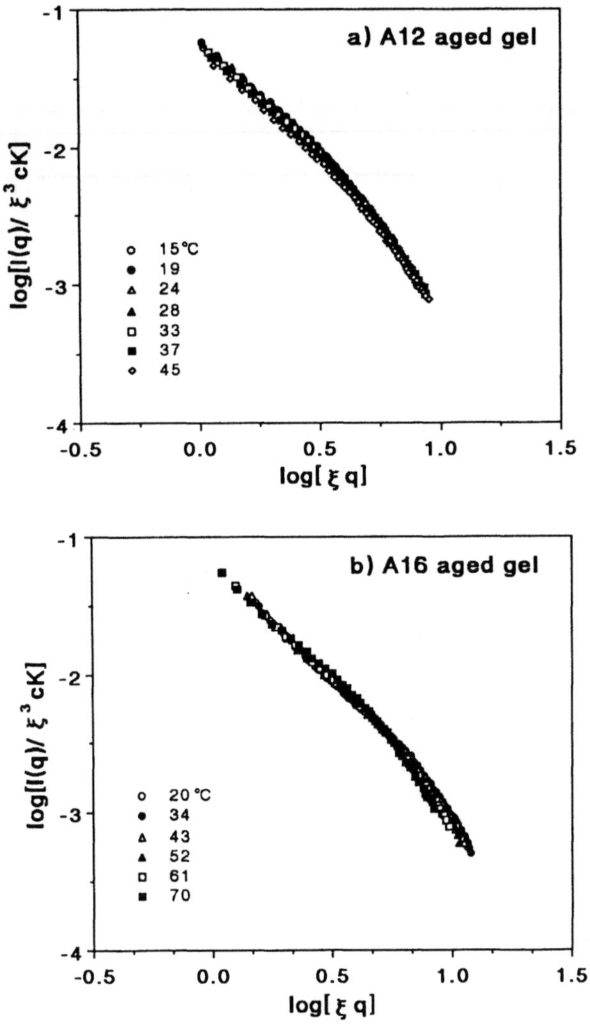

Figure 7 Double logarithmic plots of the reduced scattered intensity ($I(q)/\xi^3 cK$ with K being a contrast factor) against the reduced scattering magnitude (ξq) with respect to the correlation length ξ. (a) A12 gels, and (b) A16 gels.

4.3. Inorganic/organic hybrid gels

Inorganic/organic hybrid gels were prepared as described above, varying the compositional ratio of inorganic (TEOS) and organic (ET-PTMO) components as shown in Table I. Here the inorganic components constitute the crosslinking domains distributed homogeneously in the gel when the molar ratio of organic to inorganic components is large, and the domains are connected by organic components. Assuming the Gaussian network, the molecular weight M_c of the interval chain between crosslinks can be estimated approximately from the elastic modulus of the gel with the aid of the theory of rubber elasticity:[1]

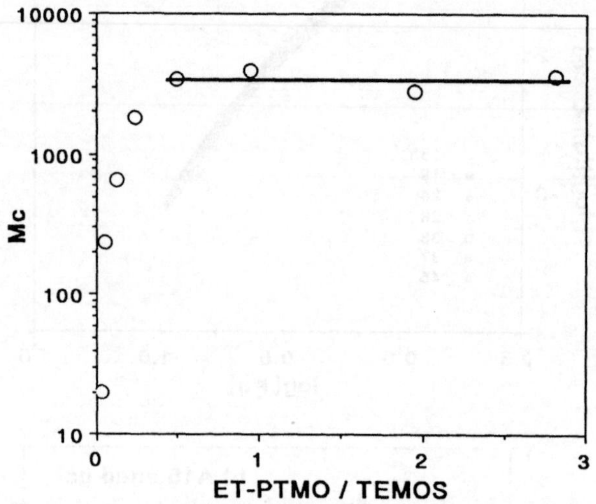

Figure 8 The molar ratio ET-PTMO/TEMOS dependence of the molecular weight of an inter-crosslink chain, calculated from the measured elastic modulus by eq.(16)

Table 1. Chemical composition and elasticity of hybrid gels

p/t	molar ratio of ET-PTMO/TEOS	weight fraction of organic component	log E'	Mc	R/Å
0.02	0.022	0.395			
0.03	0.032	0.494	9.57	20	0.4
0.05	0.056	0.622	8.53	240	13.9
0.1	0.123	0.766	8.05	650	22.9
0.2	0.236	0.844	7.59	1800	38.6
0.5	0.503	0.896	7.34	3400	52.4
1.0	0.948	0.920	7.29	3900	56.2
2.0	1.949	0.935	7.43	2800	47.6
3.0	2.815	0.939	7.30	3600	54.0

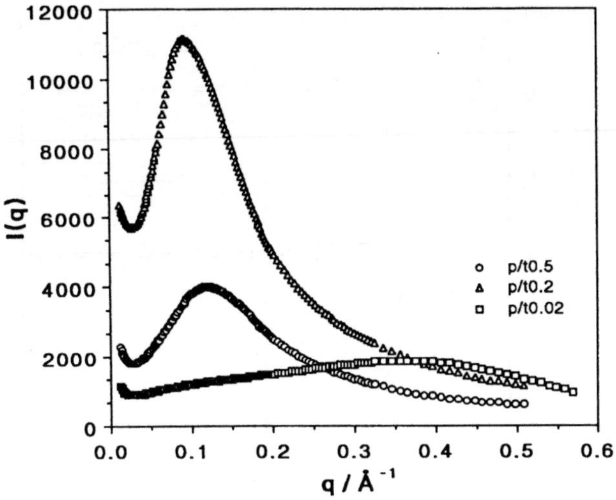

Figure 9 SAXS from three samples of hybrid gel

$$E = 3\rho RT/M_c, \tag{16}$$

where E and ρ denote Young's modulus and the polymer density, respectively. The value of 0.982 was adapted for ρ.[19] Considering the initial molecular weight of an organic prepolymer PTMO ($M_n = 1980$), approximately two prepolymer molecules are coupled to form the inter-crosslink chain while the molar ratio of organic to inorganic components is larger than 0.5, and the inter-crosslink chain becomes shorter with increasing the fraction of inorganic components (see Figure 8) when this molar ratio becomes smaller than this value. As indicated by M_c values smaller than the molecular weight of organic prepolymers, inorganic domains become dominant in the hybrid gels, when the molar ratio of organic to inorganic components is decreased further below 0.2. The last column of Table I lists the interval distance R between crosslink domains which is assumed to be equal to the unperturbed end-to-end distance of poly(tetraoxymethylene) chain of the molecular weight M_c and calculated according to Kurata et al.[20] The SAXS's were observed from three samples of the hybrid gels with varying the molar ratio ET-PTMO/TEOS 0.503 (p/t0.5), 0.236 (p/t0.2) and 0.022 (p/t0.02) as listed in Table I. As seen from Figure 9, the SAXS profiles from hybrid gels are characterized by a correlation peak around $q = 0.1 \mathring{A}^{-1}$ for p/t0.5 and p/t0.2, and by a broad maximum around $q = 0.36 \mathring{A}^{-1}$ for p/t0.02. Here the SAXS's from p/t0.5 and p/t0.2 scale as $I(q) \sim q^{-1.6}$ as expected for random clusters in a dilute solution, implying $\tau = 2.2$ (the exponent of the mass distribution of the sol, eq.(3)) and $D = 2.0$ (the fractal dimension of an individual cluster in a dilute solution).[15,17] Since in these hybrid gels the organic components form a continuous phase in which inorganic clusters are distributed to constitute crosslinking domains, the observed scaling exponent is considered as due to the inorganic crosslinking domains formed by the random branching process of tetra-ethoxysilane. The scattering maximum may reflect approximately the distance between inorganic crosslinking domains. This interfacial distance is evaluated as $61 \mathring{A}$ or $88 \mathring{A}$ for the hybrid gel p/t0.5 or p/t0.2 from the second peak in respective distance distribution function $p(r)$ calculated according to eq.(12) (see Figure 10). The inorganic domains will grow larger with increasing the inorganic fraction as indicated by the shift of

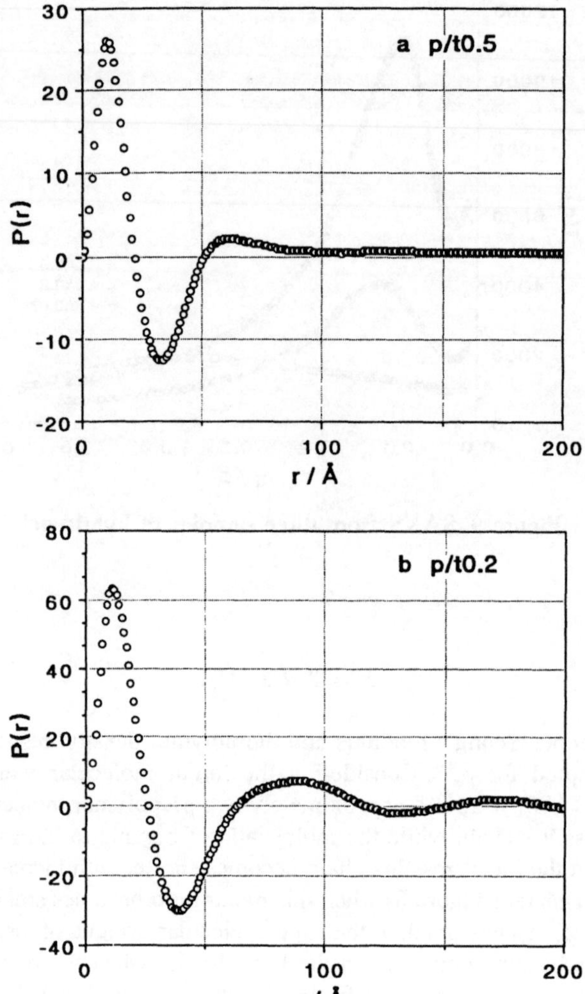

Figure 10 Distance distribution function of hybrid gels; (a) p/t0.5 and (b) p/t0.2

the minimum position in the distance distribution function to a larger distance R, but the interfacial distance also becomes larger to accommodate orgainc components. When the molar ratio of ET-PTMO/TEOS is decreased below 0.25, then the interfacial region becomes less available for organic components to form long network chains and the elastic modulus shows the abrupt increase to the value close to that in the glassy state (or the apparent molecular weight of an interfacial chain decreases to 0 as in Figure 8). The continuous inorganic phase may appear at about ET-PTMO/TEOS = 0.2, corresponding the gel point of inorganic components. The broad maximum in the SAXS from the hybrid gel p/t0.02 corresponds to a short correlation of $24\AA$, which may be caused by thin alternating inorganic and organic layers. The SAXS from the hybrid gel p/t0.02 scales as $I(q) \sim q^{-2.8}$,and this exponent implies the involvement of more than one crosslinking mechanisms as discussed in earlier sections.

5. CONCLUDING REMARKS

Although the fractal domain is not specified explicitly in the gel phase, the scattered intensities scale as $I(q) \sim q^{-\mu}$ where μ depends on the gelling mechanism. The domain is observed if it is distinguished by X-ray or neutron, and its approximate size is evaluated in terms of the correlation length ξ from the fractal argument. The domain is formed probably by the density fluctuation in the gel, resulting the contrast observable by X-ray or neutron. The conventional analysis of the scattered intensities reveals the structure of crosslinking domains in some cases. The domain dimension is consistent between those specified by the correlation length and estimated from the conventional analysis.

REFERENCES

1 See, for example, P. J. Flory, *Principles of Polymer Chemistry,* Cornell U.P., Ithaca, New York, (1953)

2 See, for example, P. G. De Gennes, *Scaling Concepts in Polymer Physics,* Cornell U.P., Ithaca, New York, (1979)

3 M. Muthukumar, private communication

4 J. E. Martin, B. J. Ackerson, *Phys. Rev.,* **A31,** 1180 (1985)

5 D. Stauffer, A. Coniglio, M. Adam, *Adv. Polym. Sci.,* **44,** 105 (1982)

6 K. Kajiwara, M. Gordon, *J. Chem. Phys.,* **59,** 3623 (1973)

7 K. Kajiwara, C. A. M. Ribeiro, *Macromolecules,* **7,** 121 (1974)

8 T. Ueki, Y. Hiragi, Y. Izumi, H. Tagawa, M. Kataoka, M.Muroga, T. Matsushita, Y. Amemiya, *Photon Factory Report,* **1983,** IV 70

9 See, for example, O. Glatter, in *Small-Angle X-ray Scattering* (ed. O. Kratky, O. Glatter), Academic Press, New York, 1982

10 Y. Hiragi, H. Urakawa, K. Tanabe, *J. Appl. Phys.,* **58,** 5 (1985)

11 G. S. Manning, *J. Chem. Phys.,* **51,** 924 (1969)

12 N. S. Anderson, J. W. Campbell, M. M. Harding, D. A. Rees, J. W. B. Samuel, *J. Mol. Biol.,* **45,** 85 (1969)

13 See, for example, A. Guinier, G. Furnet, *Small-Angle Scattering of X-ray,* John Wily, New York, (1955)

14 M. Shibayama, H. Kurokawa, S. Nomura, S. Roy, R. S. Stein, *Polym. Prepr. Jpn.,* **37,** 3052 (1988)

15 E. Bouchaud, M. Delsanti, M. Adam, M. Daud, D. Durand, *J. Physique,* **47,** 1273 (1986)

16 M. Shibayama, H. Kurokawa, S. Nomura, S. Roy, R. S. Stein, W. Wu, *Macromolecules,* in press

17 J. E. Martin, J. P. Wilcoxon, *Phys. Rev.,* **A39,** 252 (1989)

18 M. Adam, M. Delsanti, J. P. Munch, D. Durand, *J. Physique,* **48,** 1809 (1987)

19 I. Bowman, D. S. Brown, R. E. Wetton, *Polymer,* **10,** 715 (1969)

20 M. Kurata, H. Utiyama, K. Kamada, *Makromol. Chem.,* **88,** 281 (1965)

PHYSICAL GELATION OF SYNTHETIC AND BIOLOGICAL MACROMOLECULES

Simon B. Ross-Murphy

Cavendish Laboratory, University of Cambridge
Madingley Rd, Cambridge CB3 OHE, Great Britain

INTRODUCTION - COVALENT AND ENTANGLEMENT NETWORKS

Historically, polymer networks can be divided into two main classes, chemically cross-linked materials (including bulk elastomers), and 'entanglement networks'. The latter are formed by the topological interaction of polymer chains, either in the melt or in solution when the product of concentration and molecular weight becomes greater than some critical molecular weight M_e.[1,2] In this case they behave as 'pseudo gels' at frequencies higher (timescales shorter) than the lifetime of the topological entanglements. This depends for linear chains on M_r^3, where M_r is molecular weight. The covalently cross-linked materials, on the other hand, are formed by a variety of routes including cross-linking high molecular weight linear chains, either chemically or by radiation, by end-linking reactant chains with a branching unit, or by step-addition polymerisation of oligomeric multifunctional precursors. They are true macromolecules, where the molecular weight is nominally infinite, and they therefore possess an infinite relaxation time and an equilibrium modulus.[1]

Rheological discrimination between these two classes of networks can be made by the technique of mechanical spectroscopy. In this, a small deformation oscillatory shear strain of frequency ω is applied to the material, and the real and imaginary parts, G′ and G′′, of the complex shear modulus are measured. Typical spectra are illustrated in Fig. 1; for the entanglement networks, as the frequency is decreased there is a 'cross-over' in G′ and G′′. At very low frequencies, in the 'terminal zone' they flow as high viscosity liquids. As

Polymer Gels, Edited by D. DeRossi *et al.*
Plenum Press, New York, 1991

expected the cross-linked networks show no such effect, and for a rubbery gel (i.e. above its glass transition), G′ and G′′ are parallel, and largely frequency insensitive. A further discrimination can be made by adding excess solvent to the bulk 'gel'. Entanglement network systems will dissolve to form a more dilute polymer solution, whereas the gel will swell, but not dissolve.

The dynamics of entanglement networks have been extensively studied both experimentally and theoretically following the

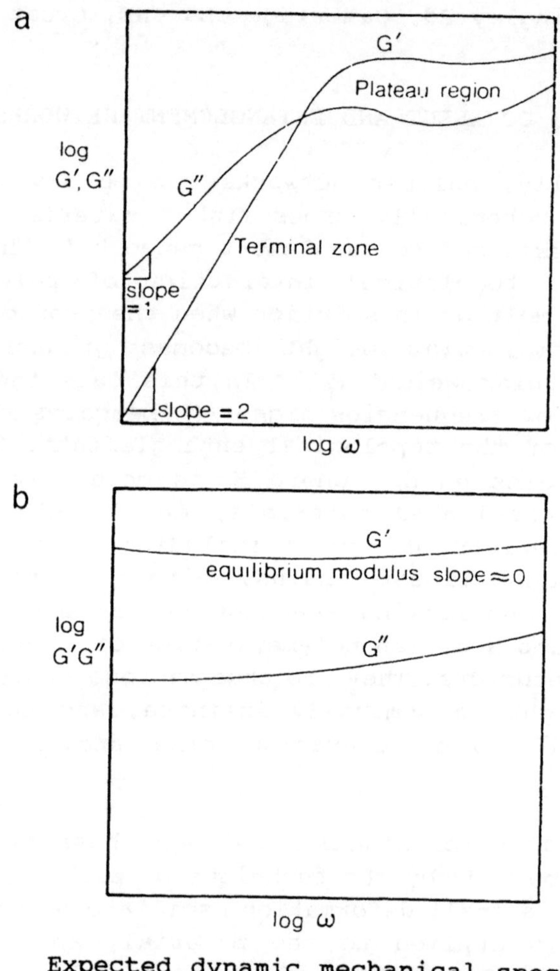

Fig.1. Expected dynamic mechanical spectra of the real (G′) and imaginary (G′′) parts of the shear storage modulus for (a) an entanglement network system (pseudo-gel) (b) a covalently cross-linked network (true gel)

suggestion that their behaviour may be described by 'reptation'. In this approach, pioneered by de Gennes,[3] and by Doi and Edwards,[2] a particular chain wriggles snake-like down a hypothetical 'tube' formed by all the other chains. For solutions and melts of linear polymers the agreement between theory and experiment is usually very good, and efforts are now being made to extend the approach to polymers of more complex architecture, including the effects of introducing star and long chain branching.

Covalently cross-linked gels have also been extensively investigated, and both the small and large deformation mechanical properties have been measured and compared with the expectation of the theory of rubber elasticity. Much of the experimental effort is employed in making networks free from 'defects', for example reducing cycles to give a tree-like network. One of the major controversies in the area of rubber elasticity is still whether or not entanglements contribute to the equilibrium modulus and/or the stress vs. strain behaviour of a cross-linked network. One school, led by the late Paul Flory, argues that they do not, and attribute any deviations from ideal rubber theory to fluctuations in the position of network junctions. Others, perhaps now in the majority, argue that since entanglements clearly contribute to the viscoelastic response of high molecular weight uncross-linked polymers in the melt ('quasi-gels' as these are systems are sometimes called), they must still be present in a cross-linked network formed from these same chains. This argument has still not been completely resolved.

PHYSICAL GELS

Although both classes of networks are still very active areas of interest, the behaviour of both can be said to be understood, at least to zeroth order, and any discrepancies between theory and experiment are largely at the refinable level. However, many systems fall between the above two categories, as they consist of chains which are 'physically' cross-linked into networks, the cross-links themselves being of small but finite energy, and/or of finite lifetime; these are the so-called physical gels, which include both biological and synthetic polymers; for these it is fair to say that only very recently has much effort been employed in trying to elucidate structure-property relations.*

---

* The term 'physical gel' appears to have been introduced by de Gennes[3]. However, the class of materials covered by his description implies thermoreversibility, which is not the case with every system included here. Here we will consider the term to include any non-covalently cross-linked system.

The presence of non-covalent cross-links complicates any physical description of the network properties enormously, because their number and position can, and does, fluctuate with time and temperature. In many cases the nature of the cross-links themselves is not known unambiguously, often involving such disparate forces as, for example, Coulombic, dipole-dipole, van der Waals, charge transfer and hydrophobic and hydrogen bonding interactions.[4] Furthermore the cross-links usually act not at a point on the chain as do covalent cross-links, but rather involve more extended 'junction zones'. A description of these usually becomes even more complex, because the potential number of junction zones per primary chain, the 'functionality', f, and the extent (molecular weight) of the junction along the chain profile, M_J, can be estimated, if at all, only indirectly. In many cases there is a subsequent lateral aggregation of chains, after the initial contact. These factors must influence the actual number of physical cross-links and, consequently, the modulus of the final gel. Since this modulus will usually reflect both entropic (rubber-like elasticity) and enthalpic contributions, an *a priori* description of the modulus and mechanical response of these materials is bound to involve a number of approximations.

Amongst the biopolymer physical gels are, of course, gelatin, the seaweed and plant polysaccharides such as agarose, the carrageenans and pectin, starches and cellulose derivatives. More details of the structure and properties of these is given in our recent review.[5] Physical gels formed from synthetic polymers include isotactic polymers in certain solvents (i-polystyrene in decalin), ionomer systems in solvents of low dielectric constant, and a number of A-B-A type block copolymers - aggregation of heterostructural chains is often a common feature of such materials. In this case block A is compatible with a solvent and block B is incompatible. Under these circumstances, in solution the B-B interaction between units on adjacent chains forms the physical cross-link. As far as the biopolymers, in particular, are concerned, non-covalent cross-links are formed by one or more of the mechanisms listed above, usually combined with more specific and complex mechanisms involving junction zones of known, ordered secondary structure.

COMMON FEATURES OF BIOLOGICAL AND SYNTHETIC PHYSICAL GELS

Real progress in the study of physical gelation of synthetic and biological macromolecules has so far been limited

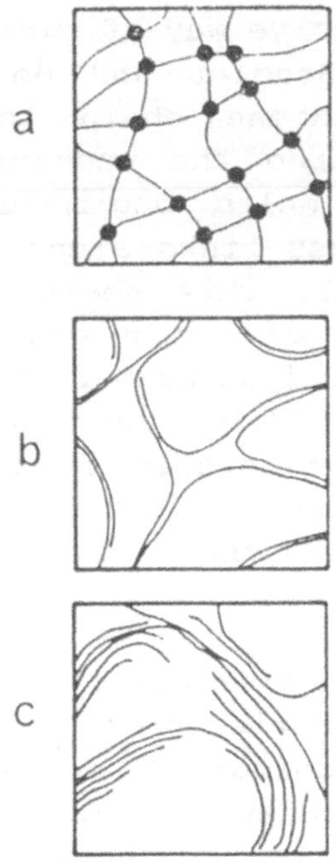

Fig.2. Schematic representation of polymer networks involving (A) point (covalent cross-links) (B) two chain junction zones (C) multiple chain junction zones

by a number of factors, but recently there has been more of an appreciation of some of the common features of synthetic and 'natural' gels.[6,7] From previous remarks it should be clear that any model study of the phenomena of physical gelation would require that the molecular weight M_r (chain length) and polydispersity were known and could be controlled, as could f, the number of sites per chain, and the junction zone length M_j (clearly $fM_j \leq M_r$). However, the majority of measurements in the literature have been obtained for samples of polydispersity which is either unknown, or which is rather broad ($DP_w/DP_n > 2$) compared to accepted standards in polymer research. This applies not only to the biopolymers, but also the synthetic polymer analogues. There is also the uncertainty in the lateral extent of junction zones, and in the distribution of substituent groups along the chain backbone.

At least one attractive way of surmounting these problems has been adopted by Stadler and de Lucca Freitas.[8] They treated model poly(butadiene) chains to give "point" H-bonded cross-links, and measured the mechanical spectra of their modified and unmodified melts. One interesting outcome of this work is that point cross-links alone are not sufficient to produce a physical gel. This seems to add weight to the concept that a degree of chain contour co-operativity is required, ie. that a critical number of adjacent segments need to be involved in an effective junction zone. (One might even speculate that since most covalent bond energies are around 10 times as great as the non-covalent bond energies, this factor could be used to estimate the minimum length of an effective junction zone.)

The same authors have recently described the use of a 'sticky tube' model (modified from the Doi-Edwards theory) to combine the usual topological entanglements with longer lifetime H-bonds.[9] The measured relaxation spectra for blends of different chain lengths are, however, very different from those predicted from the theory. The concentration dependence of zero shear viscosity also has a much higher exponent ($\simeq$5.3) than predicted by the theory (3), or is found for normal polymer solutions ($\simeq$3.4), but is in keeping with early data by Ferry and co-workers on so-called 'hyperentanglement systems'.[10] This term was introduced precisely to describe systems which had a rheological contribution over and above that expected from simple entanglement, due to non-covalent chain-chain interactions.

'STRENGTH' OF NON-COVALENT CROSS-LINKS

It is clear that we need to be able to estimate the 'strength' of physical gel interactions. A first assumption is that a single 'physical' cross-link would have a free energy/lifetime which was of Boltzmann form. However, as discussed above, the actual physical cross-link requires a degree of cooperativity and we would then expect the net free energy to be greater than the sum of component units. If we are to try and relate the modulus of the gel to macromolecular properties we must be able to factorise the contribution from the total number of units cross-linked, ie. the connectivity of the network, from the free energy of any individual junction zone. Further, even though this procedure can be carried out formally, to be confident of the conclusions we require structural detail, at the 'molecular' and 'macromolecular' distance scales (say 0.1-10 nm and 10-10^4 nm respectively). This is because another major difference between covalently and non-covalently cross-linked networks is that the former are effectively homogeneous ('self-similar') over the entire distance scale from say 10nm to 10^7nm, whilst this is certainly not the case for the latter, since as has already ben mentioned, in many cases there is a specific, and often intricate, hierarchy of structures (Fig. 3).

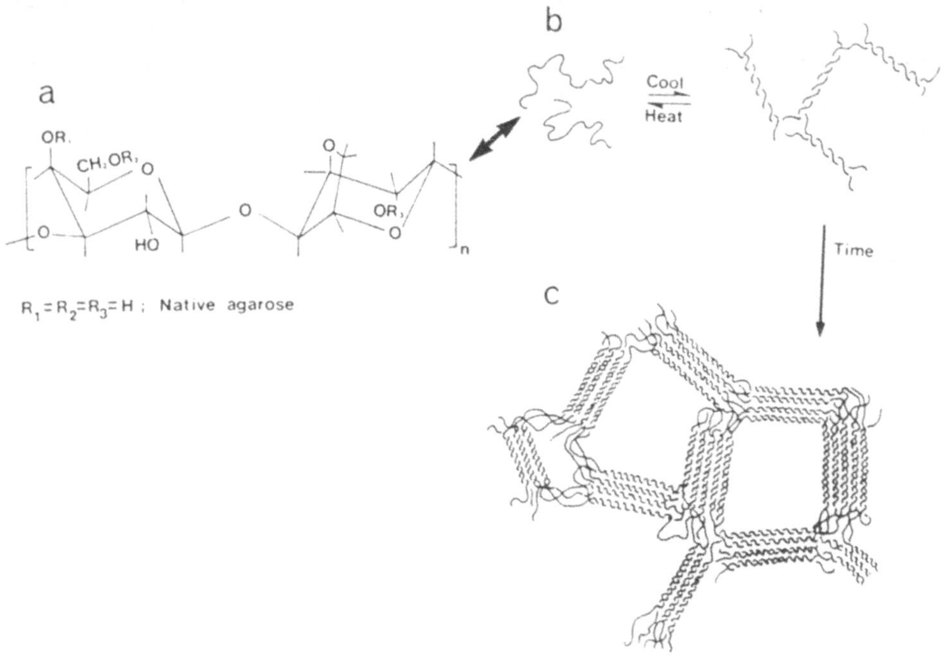

Fig.3. Hierarchy of structure in an agarose gel

Such complexity is often more familiar to molecular biologists than to polymer scientists, although mechanical measurements are now being performed on, for example, the extremely precisely assembled actin and tubulin networks.[11]

PHYSICAL GELATION AND PHASE SEPARATION

One view of physical gelation is that it implicitly involves a degree of heterogeneity. This is undoubtedly true at the level of detail of the chain backbone. One example of this is the heterostructure in the type of block copolymer gel system already mentioned above. Yet another example is in the original Rees model for gels formed by carrageenan double helices.[12] The existence of 'rogue residues' was established chemically. In these, certain saccharide units are in the wrong conformation to allow the helix to propagate. This ensures that, rather than isolated helix pairs being formed, each chain can share a portion of double helix with a number of other chains, and form the gel.

This degree of microscopic heterogeneity or phase incompatibility, has led some workers to suggest that physical gel formation intrinsically involves <u>macroscopic</u> incompatibility so that gelation occurs within the metastable (binodal) region of the phase diagram, when a more concentrated phase exceeds a certain critical concentration, or even that gelation occurs within the unstable region by the process of non-nucleated spinodal decomposition. Whilst there are undoubtedly some cases where one or other of these effects do 'drive' the gelation, and where the phase and gel loci have been carefully mapped, it is certainly not, in the Author's view, true in every case.

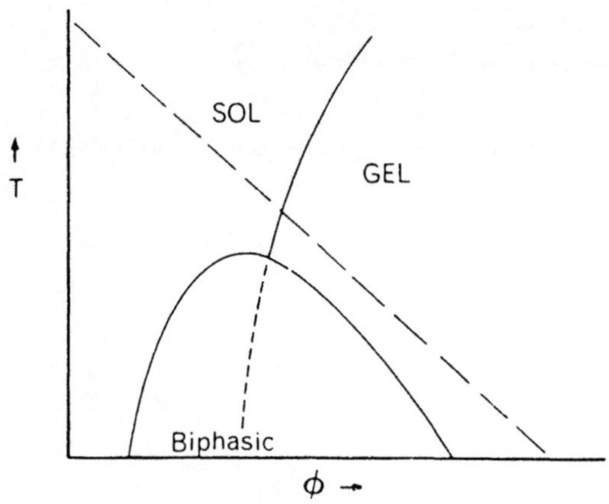

Fig.4. Phase diagram for a sol/gel transition close to a binodal. The dotted line indicates where the gel should become biphasic. The dashed line illustrates the triangular locus obtained for $M_w \rightarrow \infty$.

The interplay between liquid-liquid phase separation and gelation has been treated using a site correlated lattice model by Coniglio, Stanley and Klein.[13] Fig.4. illustrates an idealised phase diagram for a sol/gel transition close to a binodal. It includes one very important aspect - the binodal locus will depend on polymer or aggregate molecular weight. As the gel point is approached and the molecular weight increases, the binodal becomes less and less symmetric, so that for infinite molecular weight it actually becomes triangular.[14] This may explain why there is some controversy between those workers who consider that 'phase separation drives gelation' and those who believe the converse. It may be that when phase separation is seen, it follows the chain-chain aggregation, and in these cases the pre-gel polymer/solvent system is close to, or even below the theta state. Phase separation is then a consequence of the molecular weight increase at gelation.

In some cases, however, the position of the phase boundary has proved crucial. For example atactic polystyrene, which unlike the tactic forms cannot crystallize, forms amorphous thermoreversible gels in both *cis* and *trans* decalin. Arnauts and Berghmans[15] have shown how, in this case, the T_g vs. composition locus crosses the binodal. They suggest that subsequently the more concentrated of the two concomitant phases becomes 'frozen' below its T_g, giving a gel with glassy cross-links. Miller and co-workers have suggested that poly(γ,benzyl-L-glutamate) -PBLG - forms gels by the spinodal decomposition route, but other workers have challenged this view. In this case the exact mechanism is complicated by PBLG's tendency to form both liquid and solid crystals.[17]

THE CRITICAL GEL AND OVERLAP CONCENTRATIONS

There is much discussion in the Literature as to the relationship between the critical gelling concentration C_0 and the overlap concentration C^* (defined by $C^* = S/[\eta]$, with S a constant of order unity). Some workers have argued that the two concentrations are the same (or very nearly so) and confusion is spread by the tendency of some workers to use the notation C^* for the former, even when there is no evidence or implication that the two are the same. However, a simple argument shows that the implied equality cannot always be true. C^* refers to chain overlap of a time averaged volume of rotation - in order for two mutually 'sticky' chains to form a cross-link then they do not have to be in each others sphere of influence except for that instant when the cross-link is formed, provided that it subsequently remains intact. In other words their mutual translation up to this time argues for a much larger sphere of influence than given by C^* at a given concentration. For finite energy cross-links we affirm that $C^0 < C^*$ - this assertion seems to be borne out by the majority of experiments in which the hypothesis has been carefully tested.

THERMOREVERSIBILITY AND THE CONCENTRATION DEPENDENCE OF GEL MELTING

Many, but by no means all, physical gels are thermoreversible, and the Eldridge-Ferry theory, originally proposed some 40 years ago,[18] to describe the melting of micro-crystallite junction zones, can still prove invaluable. Interestingly the method was originally devised for and is still being applied to gelatin, although there is now good evidence that gelatin gels do not involve extensive lateral aggregation and subsequent crystallisation of junction zones.[19] Nevertheless, even though the method has been dismissed as being too *ad hoc*, it can still be applied in a wide variety of systems. One expectation of the theory is that the gel time will increase as the melting temperature is approached, but there are some, although rather specialised, exceptions to this rule, including the interesting poly(carbodiimide) gels studied by Rogovina,[20] which form <u>more</u> rapidly as the melting temperature is approached. More generally, some systems appear to 'soften' significantly before melting, others melt very sharply. Although this observation could be related to the original cooperativity of junction zone formation, no detailed studies seems to have been undertaken in this area.

Where the small-deformation moduli and the gel time of biopolymer gels is of interest, the freedom to choose a range of polymer concentrations prior to the introduction of cross-links, has a number of important consequences. If temperature alone is the driving force for gelation [cf. the poly(carbodiimide) gels] both gel time and ultimate gel rigidity are crucially related to the concentration of polymer present, the gel time diverging logarithmically to infinity at C_0 and the equilibrium modulus tending to zero. Nonetheless for systems allowed to gel under a well-defined thermal regime, and allowed to achieve a limiting degree of cross-linking, a characteristic modulus-concentration and reciprocal gel time- concentration dependence is found. For example, for the modulus G, near the critical concentration a large and variable power law dependence of G on C is observed, whilst at much higher concentrations, a constant, and approximately C^2, limiting law emerges. Such behaviour has been noted historically. In more recent literature, some data appear to fall closer to the asymptotic scaling prediction[3] $G \propto C^{2.25}$ (although discrimination between these values is a testing task!) In the next two sections we will describe the application of two simple models to describe the modulus-concentration and gel-time concentration profiles from C_0 to the high concentration asymptote.

THE CONCENTRATION DEPENDENCE OF GEL MODULUS

Whereas the elasticity of ideal rubbers is essentially entropic in origin (modulus is proportional to absolute temperature), in most physical gels (and to a lesser extent in real covalent systems) enthalpic contributions are very important. Nevertheless there is a strong argument to be made that the number of active network chains is independent of the contribution per chain to the gel modulus, so that combinatoric network theories can still be employed. A number of approaches have made in this area, most recently by Clark and the present author,[5,21,22] and also by Oakenfull.[23] Here we will describe only the former method, the latter does differ in a number of ways, and a detailed comparison has recently been published.[24]

In our work, we have used the cascade theory[25] to count elastically active chains (EANCs) and formulated the modulus-concentration relationship in a 'universal' way. To do this we abandoned the conventional (ideal entropy) assumption of a contribution RT to the modulus per mole of an EANC, and introduce a non-ideality parameter a such that aRT units would be the actual contribution. (It should be noted from the outset that a itself will generally be temperature dependent since the product aRT is a free energy and will in principle involve enthalpic as well as entropic contributions.) The proposed factorisation into terms a and RT merely reflects a particular choice of energy units and (unlike the case of ideal rubbers) is <u>not</u> a statement about the expected temperature dependence of the gel modulus.

Over the frequency range usually investigated, say 10^{-3} to 10^{1} Hz, the mechanical spectrum of physical gels usually follows that for permanently cross-linked gels (Fig.1b), with $G' > 10G''$, and no real frequency dependence. If we make the assumption that the making and breaking of cross-links occurs within the time window of our rheological experiment, we can follow the strategy of Eldridge and Ferry[18] and of Hermans,[26] and define an equilibrium constant for the reversible conversion of a cross-link into two free sites in the form of the Ostwald dilution law.

In more detail, if α is the proportion of cross-links formed from two free sites coming together (at this stage we attribute no molecular detail to either cross-links or sites) we can say

$$K' = [(1 - \alpha)^2/\alpha]N_0 f \qquad (1)$$

where K' is an equilibrium constant, f is the functionality introduced above, and N_0 is the number of chains per unit volume. For the case of 'weak binding', $K' >> 1$ and $\alpha << 1$, this immediately gives α as

$$\alpha = N_0 f/K' = fKC/M_r \qquad (2)$$

Here K is an association constant (= $1/K'$), and C is the concentration. From this the behaviour at the gel point is given by

$$\alpha/\alpha_c = C/C_o \qquad (3)$$

where α_c is the critical proportion of cross-links for gelation, and C_o is the critical gel concentration already mentioned above. This solution implies that f is very large, an observation which cannot always be true in practice. The more general case when K′ is not >> 1 is only a little more complex, and results in a solution valid for any f.[21]

Subsequent steps in the theoretical formulation of the modulus-concentration relationship are now very much those which we have described previously. The equilibrium shear modulus G is written in the form,

$$G = N_e(\alpha, f)\,{}_aRT \qquad (4)$$

where N_e, the number of EANCs per unit volume, is a function of the degree of cross-linking in the system. The parameter $_a$ is the non-ideality parameter discussed earlier and R and T have their usual meanings.

The expression used for N_e may be derived from the following cascade expression for the number of EANCs per mole of polymer chain

$$N_e = f\alpha(1 - v)^2(1 - \beta)/2 \qquad (5)$$

where β is given by,

$$\beta = (f - 1)\alpha v/(1 - \alpha + \alpha v) \qquad (6)$$

and v is the so-called extinction probability obtained as the lowest positive root of the equation,

$$v = (1 - \alpha + \alpha v)^{f-1} \qquad (7)$$

The final result may be written in the 'universal' form

$$\frac{GK(f - 1)(f - 2)}{{}_aRT} = \frac{(f - 1)^2\alpha(1 - v)^2(1 - \beta)}{2(f - 2)} \times \frac{C}{C_o} \qquad (8)$$

where α, v, and β are functions of f and C/C_o only.

We have found in a number of different systems that the small deformation results do follow the concentration dependence of modulus predicted from this network theory above the critical gel concentration, and also that the enthalpic "front factor" $_a$ is related to chain stiffness in a predictable way.

Fig.5. illustrates the application of the method to a range of amylose gels,[27] which were synthesised to have a narrow molecular weight distribution[28] (thus satisfying at least one of our requirements for a model system with which to test theory).

In amylose it has been established experimentally that the chain length at which precipitation replaces gelation is equal to 110 sugar residues. This is therefore an estimate of the distance between the start of consecutive junction zones and so the functionality, f, was assumed to be equal to $DP_n/110$. Under these conditions the fitted values of K and a (~ 10) are essentially independent of chain length. For other polysaccharide and some protein gels it was concluded that a could be a factor of ten or more, although for gelatin it was found to be quite close to the classical value of unity.

The apparent success of this method requires us to examine the assumptions and conclusions very carefully. Since 2 (or 3

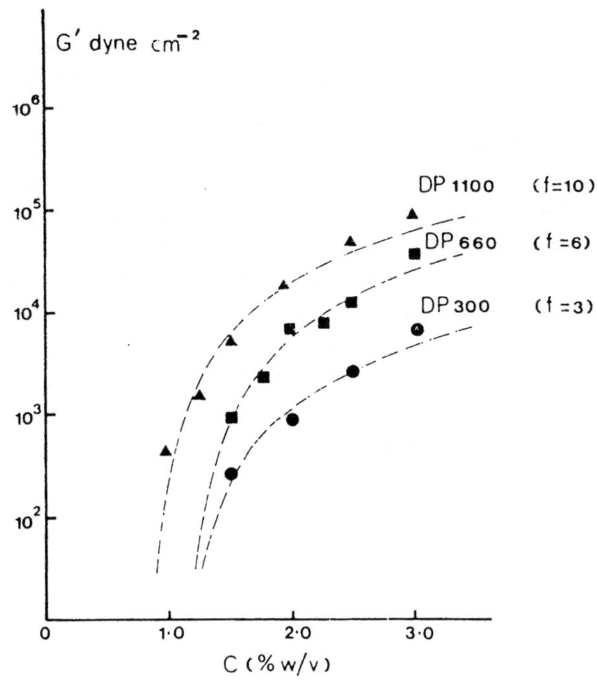

Fig.5. Experimental G′ vs. concentration (% w/v) data for amylose gels (•) DP = 300, (▪) DP = 660, (▲) DP = 1100. Theoretical fits obtained using the theory outlined above with f = 3,6, and 10 respectively. Reproduced with permission from Macromolecules 22:346, copyright American Chemical Society 1989.

if f is unconstrained) parameters are employed, it is not surprising that a reasonable fit can be obtained, especially when relatively few data points are available (typically 5 - 10). The first assumption which may be challenged is that of a cross-linking equilibrium, especially since recent work on the de-swelling of agarose gels has suggested that cross-links in this system are effectively permanent,[29] and that the degree of cross-linking in any system is largely under kinetic control. However, calculations in which cross-linking progresses by a pairwise irreversible association mechanism have been performed,[22] and analyses of simulated modulus-concentration data obtained in this way suggest, that provided the gel curing process proceeds for long times, and for the same time for each system concentration (i.e. isochronal curing is applied), the same equations and fitting procedures can be adopted as in the equilibrium situation. In this case the constant K no longer has the significance of an equilibrium constant, but is a function of the forward rate constant and time.

The implications of introducing the parameter a must also be examined, especially as an alternative Oakenfull theory does not relax the requirement that a= 1.[23] Whilst no independent evidence for the its particular significance can be deduced, recent theoretical progress has been made on Nishinari's model of unzipped chains or "reel-chains",[30,31] and this establishes that, purely from entropy considerations, a parameter essentially equal to a, does indeed increase in approximately the expected manner as the chain becomes stiffer. A similar conclusion was reached from a quite different approach, viz. a computational investigation of the contribution of very short chains to network elasticity.[32]

THE CONCENTRATION DEPENDENCE OF GEL TIME

Another essentially simple model can be deduced for the concentration dependence of the gel time, t_c.[33] In this we assume that t_c corresponds to a non-zero gel modulus, which is assumed to occur at a small ratio $[J]/[J_c]$ ($= J_r$) of the molar concentration of junction zones [J] to the critical ratio. In the 'critical' regime, when $J_r < 1$, then the modulus must vary as

$$G \simeq (([J]/[J_c])-1)^p \tag{9}$$

Here the 'critical exponent' p is 3 in the classical theory of percolation on a treelike or Bethe lattice [25] and around 1.8 for percolation on a cubic lattice.[34] These exponents are only appropriate extremely close to the gel point, at higher degrees of cross-linking there is a 'wastage' effect, whereby not all of the cross-links lead to the formation of elastically effective chains. In other words if we are to apply such a formula for $[J]/[J_c] > 1$, the measured exponent will be less than this 'critical value', and as a first estimate we assume p to be ~ 2 if we use the classical exponent of 3 at the gel point.

If we again apply a second order kinetic model for the concentration of cross-linking loci L, we have

$$-d[L]/dt = d[J]/dt = kC^2 \qquad (10)$$

Here there is no specific wastage mechanism to give a critical concentration, this will be incorporated by choosing an arbitrary $[J_c]$. [Much more sophisticated, and realistic solutions to this problem are available, including that by Peniche-Covas et al. specific to the junction zone model,[35] but are not used here.]

Now the gel point will be defined in terms of $[J_c]$, and so t_c will be given by the time when $[J] = [J_c]$. (Note that for a system under kinetic control, at concentrations below the critical concentration C_o, the ratio J_r must always be less than 1, however long the time. In the absence of another process in the kinetic scheme, either a back reaction leading to an equilibrium state, or an alternative 'wastage' mechanism, such as intramolecular cross-linking, this is necessarily a somewhat artificial constraint.)

However, the simplest way of calculating J_r actually does not require us to solve any kinetic equations at all. Instead when we assume that J_r is very small the ratio $[J]/[J_c]$ is well approximated by the differential ratio $d[J]/d[J_c]$. Since the concentration corresponding to $[J_c]$ is the critical concentration C_o, we can use Eqn. 10 and say

$$J_r \approx d[J]/d[J_c] \approx (d[J]/dt)/(d[J_c]/dt) = kC^2/kC_o^2 \qquad (11)$$

If we substitute Eqn.11 into Eqn.9,

$$G \approx [((C/C_o)^2-1)^p] \qquad (12)$$

and finally since as $C \to C_o$, then $G \to 0$ and $t_c \to \infty$, we now approximate G by saying it is $\propto 1/t_c$, so that

$$t_c \approx K/[((C/C_o)^2-1)^p] \qquad (13)$$

where K is a proportionality constant. In practice, the inverse relationship between G and t_c is quite well obeyed.[36]

To test the applicability of Eqn.13, we reproduce some of our own previously published data for heat set protein gels (BSA).[36] In this dataset there are a total of 13 points collected using two different techniques. It is clear that this data does behave in the manner predicted, and Fig.6. illustrates the fit obtained firstly constraining p to be 2 and secondly by allowing the three parameters K, p and C_o to float independently. The best fit is actually obtained for p ≈ 1.70.

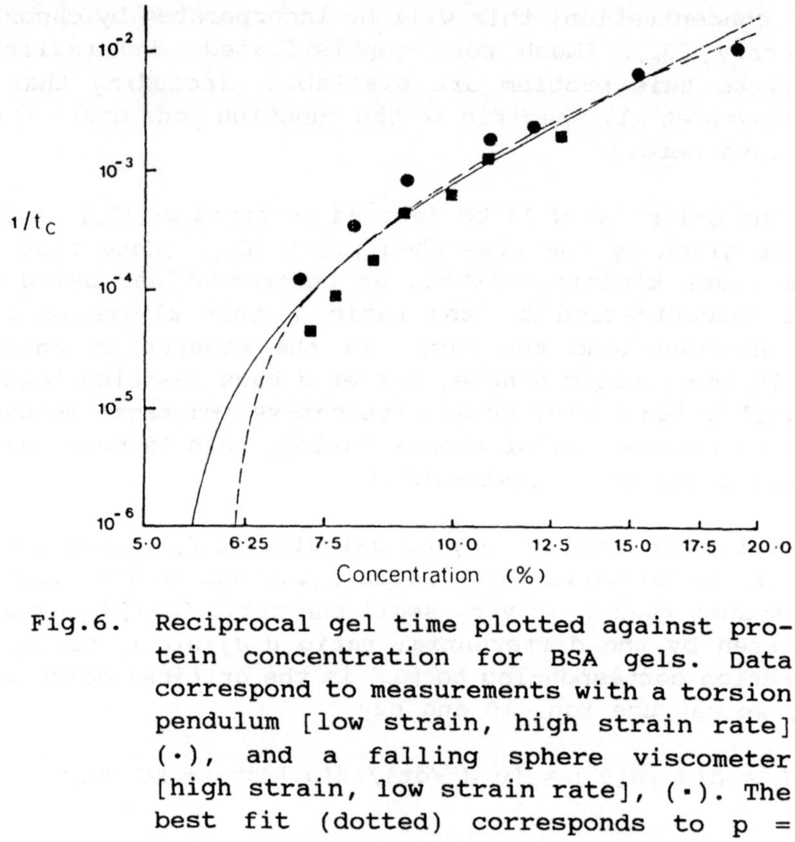

Fig.6. Reciprocal gel time plotted against protein concentration for BSA gels. Data correspond to measurements with a torsion pendulum [low strain, high strain rate] (•), and a falling sphere viscometer [high strain, low strain rate], (▪). The best fit (dotted) corresponds to p = 1.70, C_o = 6.04; fit constrained to p = 2 (solid line), C_o = 5.36.

The best fit p is very close to that expected from the percolation model, but in view of the actual extent of the concentration axis (to $C/C_o > 3$), this must be regarded as coincidental. If we accept that a single averaged exponent can describe the behaviour over such a wide range, this would tend to support a considerably larger exponent very close to C_0.

CONCLUDING REMARKS

The relationship between gelation concentration and overlap in single phase systems has already been mentioned above. Accepting that the two are distinct, what this implies is that for physical gels there is some discontinuity in behaviour above and below C_0, just as there is for covalently cross-linked systems. However, if gelation is really equilibrium controlled, any discontinuity in behaviour would be much less pronounced. The nature of this critical concentration in the case of agarose, a system believed to be under kinetic control has already been mentioned. It is also important to point out that, as in covalently cross-linked systems, the existence of such a critical concentration may merely reflect the competition between intra- and intermolecular cross-linking, the balance of which will itself depend upon dilution. Interestingly, this would not greatly effect the model for the concentration dependence of modulus described above, because the shift in the gel point would, to first order, be scaled out in exactly the correct manner.

On another topic, for both gelatin and agarose solutions there is, intriguingly, evidence of pre-gel clustering at very low concentrations. In other words even well below the gel concentration there is an apparent low stress elastic response, which cannot be attributed to artifacts.[37] Although more data are required to examine the generality of this phenomenon, for gelatin in particular, the effect seems well established.[38] Whether or not this can be related to the clustering reported from dynamic light scattering of semi-dilute solutions,[39] can only be the subject of speculation. It may also be that this is an illustration of a tendency for systems to become bi-phasic in very dilute solution (cf. the previously mentioned triangular phase diagram), so that a very tenuous network becomes dispersed in a second more dilute phase.

What should be clear, however from this, and the preceding sections, is that the physics and the chemistry of non-covalently cross-linked systems is still not greatly understood, and by this criterion must remain an active area for future research. It may be that, in view of the very many different systems involved, there really are no generalities of behaviour, and thus to look for these is fruitless - in this case the present author will not be alone in his disappointment.

ACKNOWLEDGEMENTS

The Author thanks Dr A. Yamauchi, Prof Y. Osada, and the Research Group on Polymer Gels, The Society of Polymer Science, Japan for their hospitality and the opportunity to participate in this Symposium.

REFERENCES

1. J.D. Ferry, "Viscoelastic Properties of Polymers," 3rd Edn. John Wiley, New York (1980)
2. M. Doi and S.F. Edwards, "The Theory of Polymer Dynamics," Clarendon, Oxford (1986)
3. P.G. de Gennes, "Scaling Concepts in Polymer Physics", Cornell U.P., Ithaca, N.Y. (1979)
4. E. Tsuchida and K. Abe, Adv. Polym. Sci. 45:1 (1982)
5. A.H. Clark and S.B. Ross-Murphy, Adv. Polym. Sci. 83:60 (1987)
6. W. Burchard, R. Stadler, L.L. Freitas, M. Möller, J. Omeis and E. Mühleisen, in: "Biological and Synthetic Polymer Networks," O. Kramer ed. Elsevier Applied Science, London (1988) p.3
7. W. Burchard and S.B. Ross-Murphy, eds. "Physical Networks - Polymers and Gels," Elsevier Applied Science, London (1990)
8. R. Stadler and L. de Lucca Freitas, Colloid Polym. Sci. 264:773 (1986)
9. R. Stadler and L. de Lucca Freitas, in: "Physical Networks - Polymers and Gels," W. Burchard and S.B. Ross-Murphy, eds., Elsevier Applied Science, London (1990) p.91
10. T.E. Newlin, S.E. Lovell, P.R. Saunders and J.D. Ferry, J. Colloid Sci. 17:10 (1962)
11. T.P. Stossel, P.A. Janmey and K.S. Zaner, in: "Cytomechanics," J. Bereiter-Hahn, O.R. Anderson and W.E. Reif, eds., Springer, Berlin (1987)
12. D.A. Rees, Adv. Carbohydr. Chem. Biochem. 24:267 (1969)
13. A. Coniglio, H.E. Stanley and W. Klein, Phys. Rev. B 25:6805 (1982)
14. M. Kurata, "Thermodynamics of Polymer Solutions," Harwood Academic, New York, N.Y. (1982)
15. J. Arnauts and H. Berghmans, Polymer Commun. 28:66 (1987)
16. W.G. Miller, L. Kou, K. Tokyama and V. Voltaggio, J. Polym. Sci.(Symp) C65:91 (1978)
17. A. Hill and A.M. Donald, Polymer 29:1426 (1987)
18. J.D. Ferry and J.E. Eldridge, J. Phys. Chem. 53:184 (1949)

19. I. Pezron, T. Herning, M. Djabourov and J. Leblond, in: "Physical Networks - Polymers and Gels," W. Burchard and S.B. Ross-Murphy, eds., Elsevier Applied Science, London (1990) p.231
20. L.Z. Rogovina, in: "Physical Networks - Polymers and Gels," W. Burchard and S.B. Ross-Murphy, eds., Elsevier Applied Science, London (1990) p.133
21. A.H. Clark and S.B. Ross-Murphy, Brit. Polymer J. 17:164 (1985)
22. A.H. Clark, in: "Food Structure and Behaviour," J.M.V. Blanshard and P.J. Lillford eds., Academic Press, London (1987) p.13
23. D.G. Oakenfull, J. Food Sci. 49:1103 (1984)
24. A.H. Clark, S.B. Ross-Murphy, K. Nishinari and M. Watase, in: "Physical Networks - Polymers and Gels," W. Burchard and S.B. Ross-Murphy, eds., Elsevier Applied Science, London (1990) p.209
25. M. Gordon and S.B. Ross-Murphy, Pure Applied Chem. 43:1 (1975)
26. J.R. Hermans, J. Polym. Sci. A3:1859 (1965)
27. A.H. Clark, M.J. Gidley, R.K. Richardson and S.B. Ross-Murphy, Macromolecules 22:346 (1989)
28. B. Pfannemüller and W. Burchard, Makromol. Chem. 17:1636 (1984)
29. A. Lips, P.M. Hart and A.H. Clark, Food Hydrocolloids, 2:141 (1988)
30. K. Nishinari, S. Koide and K. Ogino, J. Phys. (Paris) 49:763 (1985)
31. P.G. Higgs and R.C. Ball, Macromolecules 22:2432 (1989)
32. J.E. Mark and J.G. Curro, J. Chem. Phys. 81:6408 (1984)
33. S.B. Ross-Murphy, Carbohydr. Polym. 13:000 (1990)
34. D. Stauffer, A. Coniglio and M. Adam, Adv. Polym. Sci. 44:103 (1985)
35. C.A.L. Peniche-Covas, S.B. Dev, M. Gordon, M. Judd, and K. Kajiwara, Farad. Discuss. Chem. Soc. 57:165 (1974)
36. R.K. Richardson and S.B. Ross-Murphy, Int. J. Biol. Macromol. 3:315 (1981)
37. J.-P. Busnel and S.B. Ross-Murphy, Int. J. Biol. Macromol. 10:121 (1988)
38. P.L. Privalov, I.N. Serdyuk, and E.I. Tiktopulo, Biopolymers 10:1777 (1971)
39. M. Eisele and W. Burchard, Macromolecules 17:1636 (1984)

EFFECTS OF SHEAR MODULUS OF POLYMER GELS

Yong Li and Toyoichi Tanaka

Department of Physics and Center for Material Science and Engineering
Massachusetts Institute of Technology
Cambridge, Massachusetts 02139

I. INTRODUCTION

Gels exhibit mechanical properties characteristic of both solid and liquid-gas systems. From the compressibility (bulk modulus) point of view, a gel network system behaves like liquid-gas system. The value of the bulk modulus of a gel is small in the dilute (corresponding to the gas) phase and is large in the dense (corresponding to the liquid) phase. However, the shear modulus of a liquid-gas system is zero, but is finite for a gel (although much smaller than solids). Thus, the gel network behaves like a solid from the rigidity (shear modulus) point of view. These special properties make the gel system a unique material for theoretical study and practical applications. Compared with the liquid-gas, the free energy of a gel is

$$dF = -SdT - PdV - YdX \,. \tag{1}$$

Where Y and X are the shear stress and the shear deformation. The bulk (K) and shear (μ) moduli are related with the second derivatives of F with respect to V and X, respectively [1]. The shear modulus given by Flory's theory is $\mu = \nu_e \mathrm{RT}$, with ν_e the effective number of chains per unit volume at Θ temperature [2]. In most of the experiments, the gel is free of macroscopic shear constraints and hence the last term in Eq.(1) can be neglected when the system is far from the critical point. For this reason the analogy between the phase transition of the gel system and liquid-gas system is often used in discussion of the gel phase transitions. Near the critical point, this term will suppress the density fluctuations of the system.

The effect of the shear modulus had been overlooked until the recent publication of a series of papers by A. Onuki [3,4,5] and our study of the gel swelling kinetics [6]. The longitudinal and the transverse moduli of density fluctuations of the network are $(K+4\mu/3)$ and μ, respectively [1,7]. The conventional definition of the critical point is the point at which $K = 0$. At this point, however, both of these moduli are non-zero hence the density fluctuations are finite. This is in contradiction with the conventional picture of the critical point, at which the density fluctuation diverges. It is interesting to point out that the divergence of the scattered light intensity from a gel occurs when $(K + 4\mu/3) = 0$, which is an unstable state since $K < 0$ (assume $\mu > 0$) [8]. This unusual behavior makes the critical point of the network system special and interesting. Because of the suppression of the fluctuations, some the the critical exponents (like δ) were predicted to be the same as the mean-field values [3].

The effect of the shear modulus is to keep the system "in shape", i.e., to have isotropic swelling or shrinking. This contradicts the Tanaka-Fillmore (TF) theory [9], which treats the swelling process as a diffusion like process. In the recent gel kinetic theory, it was found that

Polymer Gels, Edited by D. DeRossi *et al.*
Plenum Press, New York, 1991

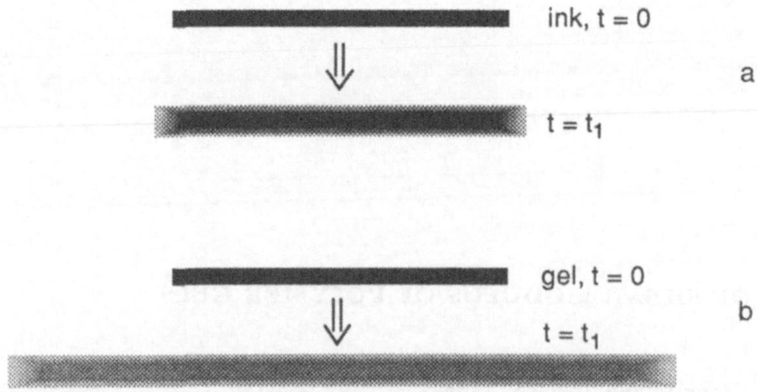

FIG.1 A schematic description of [a] a diffusion process of ink molecules in water and [b] a swelling process of the gel. For the former, the relative change of the length is negligible compared with that of the diameter. In the swelling of a long cylindrical gel, the relative changes of the length and the diameter are the same.

the shear modulus reduces the effective collective diffusion constant in the volume relaxation process ([6], also Sec.II of this article). The reduction is shape dependent.

In this article, we will discuss the effect of the shear mudulus on the swelling and shrinking kinetics of gels. This has direct application in the current effort in making artificial muscles and actuators using gel fibers. We will also show the experimentally determined values of some critical exponents and compare them with the theoretically predicted values. At the end, we will point out some other unsolved questions regarding the effect of gel shear modulus.

II. KINETICS OF GELS

An equilibrium state of gel can be represented by a state point in the phase diagram. In equilibrium, the external osmotic pressure is balanced by the internal osmotic pressure of the system, i.e., $\pi_{ext} = \pi_{int}$. When an external parameter (e.g. temperature, solvent, etc) is changed at $t = 0$, the equilibrium state is redefined by a different state point in the phase diagram. This new equilibrium state will not be reached until $t/\tau \rightarrow \infty$, with τ the relaxation time. This process is usually a relaxation process of volume and the internal stress (π_{int}). The kinetics of this process was first studied by Tanaka and Fillmore (TF) [9] and later by Peters and Candau [10]. It was shown by those authors that the speed of the process depends on the elasticity strength (K, μ) of the network and the friction coefficient (f) between the network and the solvent. In those calculations, the swelling kinetics were treated as diffusion-like process with the collective diffusion constant $D_0 = (K + 4\mu/3)/f$, which can be directly measured by dynamic light scattering [7]. We will show later that as a consequence of the existence of the shear modulus, the speed of the process also depends on the shape of the sample.

A. Isotropic Swelling of Gel Surface

The theory presented by Tanaka and Fillmore describes the kinetics of the gel swelling by [9]

$$\frac{\partial \mathbf{u}}{\partial t} = D_0 \nabla^2 \mathbf{u} + (D_0 - \mu/f) \nabla \times (\nabla \times \mathbf{u}) . \tag{2}$$

Here $\mathbf{u}(\mathbf{r}, t)$ is the displacement vector measured from the final equilibrium state. The last term is zero for gels with special shapes such as sphere, long cylinder and discs, in which cases the independent displacement vector components are decoupled. Without this term, the above equation coincides with an ordinary diffusion equation, and is called the collective diffusion equation.

However, in Eq.(2), the most important effect of the shear modulus has been neglected by neglecting the motion of the solvent [4,6]. Because of the existence of the shear modulus, the swelling kinetics is not a pure diffusion-like process. This can be demonstrated intuitively by comparing the diffusion of molecules, say ink, in water and the swelling of a gel. As shown in Fig.1a, imagine a long string of ink is placed in water at time zero. A pure diffusion process will occur and the ink string will become thicker and longer, but the relative change of the length will be negligible compared to that of the diameter. In the case of gel, however, experiments show that the relative change of the diameter and the length are the same (Fig.1b). This indicates that the TF theory is insufficient to describe the swelling process.

As has been mentioned earlier, the shear modulus acts to minimize the overall shear deformation. In general, the total energy of a gel can be separated into a bulk energy and a shear energy [1]. The bulk energy of the system is related to the volume change, which is controlled by diffusion. The shear energy, on the other hand, can be minimized by readjusting the overall shape of the gel without any volume change. Since there is no relative motion between the network and the solvent, this process is instantaneous. The shear energy of a gel of arbitrary shape is [1]

$$F_{sh} = \mu \int_V \left[\left(u_{xx} - \frac{T}{3} \right)^2 + \left(u_{yy} - \frac{T}{3} \right)^2 + \left(u_{zz} - \frac{T}{3} \right)^2 \right] dV \,, \tag{3}$$

where $T = (u_{xx} + u_{yy} + u_{zz})$ is the trace of the strain tensor u_{ik}, $u_{ik} = (\partial u_i/\partial k + \partial u_k/\partial i)/2$. Without losing any generality, we have chosen the coordinate system to be the one in which u_{ik} is diagonalized. At any instance, this energy is minimized. As long as the shear modulus μ is not zero, the change of the total shear energy in response to any small change in shape that maintains constant volume for every volume element within the gel should be zero,

$$\delta F_{sh} = 0 \,. \tag{4}$$

This equation plus the TF diffusion equation (2) should determine the kinetics of the gel swelling and shrinking processes. Every tiny diffusion process determined by TF equation is coupled to a small shear relaxation process governed by Eq.(4). Physically, Eq.(4) takes care of the fact that in general the solvent inside the gel moves in the swelling/shrinking process [6].

In the case of a sphere, Eq. (4) is a trivial relation because the shape of the sphere is maintained automatically by the diffusion process and the shear energy is already at its minimum with respect to its shape. Because of this, the calculations by TF and Peters and Candau in the case of spherical gels are valid [10].

For an infinitely long cylindrical gel, u_{zz} is position independent. For any given $u_{rr}(r)$, u_{zz} is determined by minimizing the total shear energy with respect to u_{zz}, i.e., $\partial F_{sh}/\partial u_{zz} = 0$. Since u_{zz} is independent of r for an infinitely long gel, from Eq. (4) we have

$$u_{zz} = \frac{1}{A} \int \left[\frac{u_{rr} + u_{\phi\phi}}{2} \right] dA. \tag{5}$$

Therefore, the element u_{zz} is the average of the other diagonal elements. Substituting $u_{rr} = (\partial u_r/\partial r)$ and $u_{\phi\phi} = u_r/r$ into Eq. (5) and integrating by parts, we get

$$\frac{u_z(z,t)}{z} = \frac{u_r(a,t)}{a} \,, \tag{6}$$

where a is the radius of the gel.

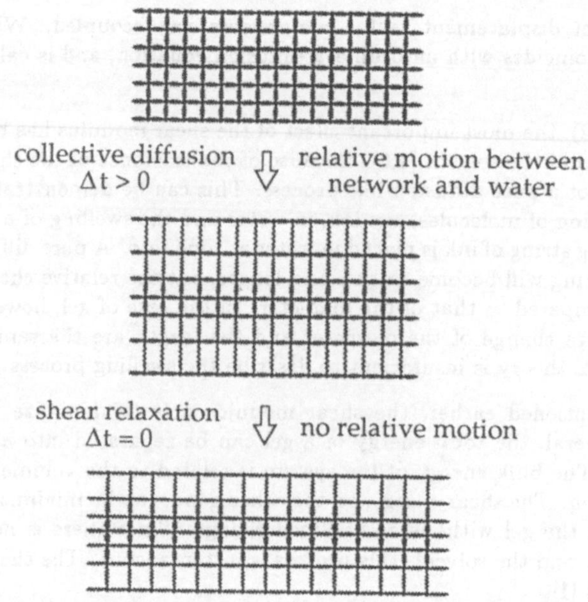

FIG.2 A infinitesimally small swelling process of a cylindrical gel network can be considered as a two-step process. The diffusion process makes the gel thicker but builds up shear energy (shear stress). Then the shear relaxation process will take place to minimize the shear energy. The second process is instantaneous since there is no relative motion between network and solvent.

Similarly, for any given $u_z(z,t)$ of a disc with thickness equal to $2a$, $u_r = u_{rr}r$ is determined by

$$\frac{u_r(r,t)}{r} = \frac{u_z(a,t)}{a}, \tag{7}$$

with $u_{rr} = u_{\phi\phi}$. Eq. (6) and (7) clearly indicate that the relative change of these gels at the boundary is isotropic, i.e., the swelling rate in the axial and radial directions at the surface of a cylinder (or a disc) are the same. This is exactly what we have observed experimentally.

B. Effective Diffusion Constant

In this section, we intend to present the kinetic theory by intuitive understanding of the problem. The mathematical details can be found in reference [6].

As seen in the previous section, because of the shear modulus, the swelling or shrinking process of gels is described by two coupled equations,

$$\frac{\partial \mathbf{u}}{\partial t} = D_0 \nabla^2 \mathbf{u}, \tag{8}$$

$$\delta F_{sh} = 0. \tag{9}$$

The first equation (differential) is related with local motion, the second equation (integration) is related with the global behavior. If we define operator $\mathcal{D}$ to represent a small diffusion process and $\mathcal{S}$ to represent a small shear relaxation process, the above two equations are equivalent to

$$\mathbf{u}(\mathbf{r},t) = \left[\prod_i \mathcal{S}(\mathbf{r},i)\mathcal{D}(\mathbf{r},i)\right] \mathbf{u}(\mathbf{r},0), \tag{10}$$

where the index i represents time, $t = i\Delta t$ ($i = 1, 2, \cdots, t/\Delta t; \Delta t \rightarrow 0$). Notice that the two operators $\mathcal{S}$ and $\mathcal{D}$ do not commute, which can be seen from Fig.2.

We will use the swelling of cylindrical gel to further demonstrate the physical meaning of this equation. The kinetics of the swelling of a long cylindrical gel can be understood in the following manner. Let us consider an infinitesimally small swelling process as a combination of two consecutive small processes, as shown in Fig.2. The first process is a collective diffusion process described by Eq.(8), in which the diameter swells to a certain degree and the change of the length is negligible (zero for an infinitely long cylinder). During this process, the shear energy builds up. The second process is the shear relaxation process described by Eq. (9). In this process, the length of the gel will expand and the diameter will shrink to minimize the shear energy, with the volume kept constant. The consequence of the constant volume restriction is that there is no relative motion between the network and the solvent, hence no friction in the shear process. The shear process is therefore instantaneous. Notice that in this approach, it is in the shear relaxation process that the solvent moves with the network.

Assume the change of the displacement vector $\mathbf{u}(\mathbf{r})$ of the cylindrical gel in the radial and axial directions after the first process (diffusion) are ($\delta\bar{u}_r$, 0). The second process will reduce the change in the radial direction and increase the change in the axial direction. Assume after the second process (shear relaxation), the changes are (δu,δu_z). Since the second process does not take any time, ($\delta\bar{u}_r$, 0) and (δu,δu_z) occurred in the same time interval δt. The collective diffusion constant reflects the rapidness of the process,

$$\frac{D}{D_0} = \frac{\delta u_r}{\delta t} / \frac{\delta \bar{u}_r}{\delta t}.$$

Where D is the effective diffusion constant. So we find

$$D = D_0 \frac{\delta u_r}{\delta \bar{u}_r}. \tag{11}$$

The two quantities δu_r and $\delta\bar{u}_r$ are related. In the second process, the volume of each individual volume element is unchanged. So the change of the volume from the initial state is a constant

$$\begin{aligned} \delta ln(V) &= 2\frac{\delta\bar{u}_r}{a} = 2\frac{\delta u_r}{a} + \frac{\delta u_z}{z} \\ &= 2\frac{\delta u_r}{a}(1 + \frac{a}{2z}\frac{\delta u_z}{\delta u_r}). \end{aligned}$$

From this equation,

$$D = \frac{D_0}{1+\lambda} \tag{12}$$

with

$$\lambda = \frac{a}{2z}\frac{\delta u_z}{\delta u_r}. \tag{13}$$

Using Eq.(6), we find that $\lambda = 1/2$ at the surface. Thus the effective diffusion constant of a cylindrical gel at the surface is $2D_0/3$. Following the similar argument, we find that for a disc gel, D is $D_0/3$.

The above results can also be derived qualitatively by a simple dimension counting argument. The diffusion, in which a network undergoes a relative motion with respect to a fluid, occurs in all three dimensions for spheres, two dimensions (radial) for cylinders, and only one dimension for discs (axial). Because of the existence of the shear modulus, the volume change caused by diffusion occurring in these directions is shared by the remaining dimensions through the shear relaxation process. Effectively, this reduces the rate of the diffusion in these directions. The reduction factor is directly related to the ratio of the diffusion dimensions to

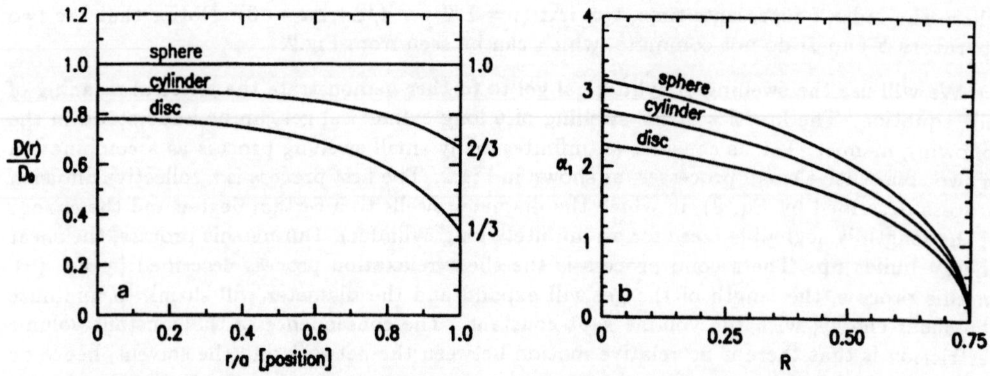

FIG.3 [a] The position dependence of the normalized effective collective diffusion constant. At the boundary, the values for sphere, cylinder, and disc are 1, 2/3, and 1/3 respectively. [b] The first eigenvalues the kinetic processes of gels with different shapes. At $K = 0$ ($R = 3/4$), they are all equal to zero.

the total dimensions, which is 3. The reduction factors are, therefore, 3/3, 2/3, and 1/3, for sphere, cylinder and disc, respectively.

The above argument is only valid for $r = a$. In general, $D = D(r,t)$ depends on both r and t,

$$D(r,t) = \frac{1}{t}\int_0^t \frac{D_0}{1+\lambda(r,t)}dt\,. \tag{14}$$

The parameter λ is

$$\begin{aligned}\lambda(r,t) &\equiv \frac{1}{2}\left(\frac{1}{z+u_z}\frac{\partial u_z}{\partial t}\right) / \left(\frac{1}{r+u_r}\frac{\partial u_r}{\partial t}\right) \\ &\approx \frac{r}{2z}\left(\frac{\partial u_z}{\partial t}\right) / \left(\frac{\partial u_r}{\partial t}\right).\end{aligned} \tag{15}$$

Fig.3a shows the position dependence of $D(r,t)$. The solution of this problem is

$$u_z(z,t) = \frac{\Delta}{a}z\sum_n A_n J_1(\alpha_n)e^{-D_a q_n^2 t}\,, \tag{16}$$

$$u_r(r,t) = \Delta\sum_n A_n J_1(q_n r)e^{-D q_n^2 t}\,, \tag{17}$$

Where $D_a = D(a,t)$ is a constant (independent of t), $\Delta = u_r(a,0)$ is the total change of the radius of the gel. The eigenvalues α_n are determined by the boundary condition $\sigma_{rr}(a) = 0$, which is equivalent to

$$R = \frac{1}{4}\left(1 + \frac{\alpha_n J_0(\alpha_n)}{J_1(\alpha_n)}\right) \tag{18}$$

with $R = \mu/(K + 4\mu/3)$. The first eigenvalues of sphere, cylinder, and disc gels are plotted against R in Fig.3b.

In the case of discs,

$$D = \frac{1}{t}\int_0^t \frac{D_0}{1+\lambda^{-1}}dt\,, \tag{19}$$

and

$$u_z(z,t) = \Delta\sum_n A_n \sin(q_n z)e^{-D q_n^2 t}\,, \tag{20}$$

$$u_r(r,t) = \frac{\Delta}{a}r\sum_n A_n \sin(q_n a)e^{-D_a q_n^2 t}\,. \tag{21}$$

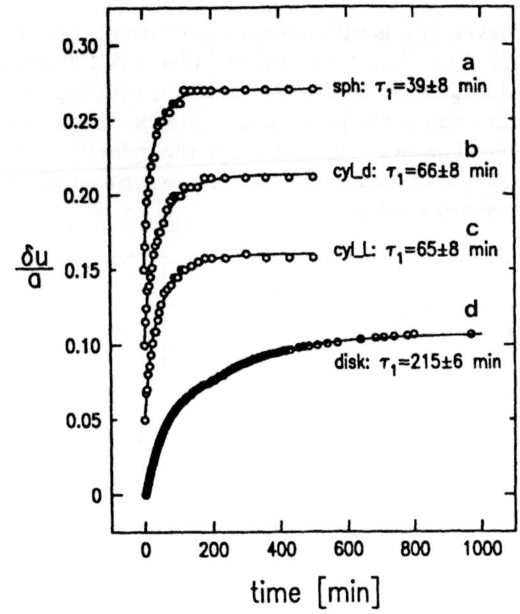

FIG.4 Experimental measurements of the swelling process of acrylamide gels. [a] Diameter of a short cylindrical gel with equal length and diameter; [b] diameter of a long cylinder with length roughly thirty times the diameter; [c] the length of the long cylinder of [b]; [d] the thickness of a disc with the side twenty times the value of the thickness. Each of the top three curves has been shifted from the one below it by 0.05 for clearness.

C. Discussion and Experimental Results

In the theory presented above, other than the shape of the system, only two parameters appeared. One is the network collective diffusion constant $D_0 = (K + 4\mu/3)/f$. The other is $R = \mu/(K+4\mu/3)$,the ratio of shear (transverse) modulus to longitudinal modulus. Since R is the ratio of two moduli, it should mainly depend on the nature of the polymer chains and the topology of the network. The parameter R is equal 1/2 for Gaussian (linear spring) network [11], which can be verified readily for the simple-cubic network structure. Geissler and Hecht experimentally confirmed this and suggested that R is around 1/3 for a self-avoiding network [11].

It is interesting to notice from Fig.2 that at the beginning of the swelling the inner portion of the volume elements are compressed in the radial direction and stretched in the axial direction. This is a direct consequence of the shear relaxation process. This behavior is reflected in the time dependence of the effective diffusion constant at r. This can also been seen from $\partial u_r/\partial t$

$$\begin{aligned}\frac{\partial u_r}{\partial t} &= -\frac{r}{z}\frac{\partial u_z}{\partial t} - D_0\Delta\sum_n A_n q_n^2 J_1(q_n r)e^{-Dq_n^2 t}, \\ &= \frac{2r}{3a}D_0\Delta\sum_n B_n q_n^2 J_1(q_n r)e^{-D_a q_n^2 t}\left[1 - \frac{3a}{2r}\frac{J_1(\alpha_n r/a)}{J_1(\alpha_n)}e^{-(D-D_a)q_n^2 t}\right].\end{aligned}$$

We have used Eq.(14)-(17) in deriving this result. For small r, the leading order terms of the expression in the square bracket is $[1 - 1.5exp\{\cdots\}]$, which is always negative for small enough t. This gives $\partial u_r/\partial t > 0$, meaning the gel shrinks in the radial direction near the center at the beginning.

Table.I Kinetic behavior of gels with different geometries. The last two columns are experimental results. The diameter of the cylinder was 1.35mm with aspect ratio equal to 30. The thickness of the disc was 1.33mm with aspect ratio equal to 20. The two numbers in each of the theoretical columns correspond to μ/K=0 (left) and μ/K=∞ (right). The two values of τ_1 for the cylinder in column 5 represent the diameter and the length measurements. The experimental results satisfactorily agree with the theoretical results.

	Theory			Experiment	
	α_1	D_a/D_0	$\tau_1/\tau_{1,sph}$	$\tau_{1,exp}(min)$	$\tau_1/\tau_{1,sph}$
sphere	π, 0	1	1	39±8	1
cylinder	2.74, 0	2/3	2.0, 1.9	(66, 65)±8	1.7±0.3
disc	2.29, 0	1/3	5.7, 5.0	215±6	5.5±0.9

The relaxation time τ is defined by

$$\tau = \frac{a^2}{D_a \alpha_1^2} . \tag{22}$$

Where the first eigenvalue $\alpha_1 = \alpha_1(\mu/K)$ is a function of μ/K only. When $K = 0$ ($R = 3/4$), α_1 is zero (Fig.3b). So at the critical point, kinetics slowing down will be observed (this is not the conventional critical slowing down). This was also discussed by A. Onuki [3,4].

It is clear from the derivation of this theory that the driving force of the swelling or shrinking is the diffusive relaxation of the elastic stress. So the overall speed of the process is determined by the diffusion process. The diffusion process, however, depends on the square of the smallest effective size of the sample. This effective size is the diameter for a long cylinder, thickness for a disk. For a dumb-bell shaped gel, for example, the effective size would somewhat larger than the diameter of the handle but smaller than the balls.

Swelling kinetics of gels of various shapes were measured to examine the above theory. Acrylamide gels of standard composition (5g of acrylamide and 0.133g of bisacrylamide in 100ml of water) prepared by free radical polymerization were used in our experiments. The relaxation times obtained for the long cylinder were 66±8 and 65±8 minutes for the diameter and the length, respectively. For the spherical gel, the relaxation time was 39±8 minutes. The relaxation time of the disc was 215±6 minutes. The diameters of the spherical gel, cylindrical gel, and the thickness of the disc gel are all equal to 1.35mm. The data are presented in Fig.4.

Experiments clearly show that for a long cylindrical gel, the swelling ratio is the same for both the length and the diameter, regardless of time. The relaxation times of a long cylinder and large disc are 1.7 ± 0.3 and 5.5 ± 0.9 times longer than that of a short cylinder with the same diameter. These values are in satisfactory agreement with the theoretical predictions (Table I).

III. CRITICAL BEHAVIOR OF NIPA GELS

Associated with the volume phase transition of gel network system, there is a critical point ($K = 0$) at which the discontinuity of the volume change disappears. The study of the critical point is of great theoretical and practical interest. It was known decades ago that the critical exponents of many systems can not be explained by the classical mean-field theories [12]. The discrepancy was not understood until the introduction of renormalization group (RG) theory [13]. According to renormalization group theory, the critical exponents are related with only

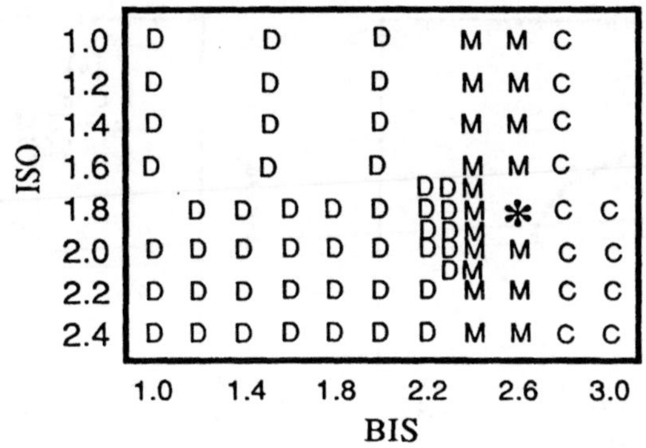

FIG.5. Isobar behavior of NIPA gels. In this figure, 'D' and 'C' stand for discontinuous and continuous volume transition, respectively. 'M' stands for marginal transition (hard to tell). In both axes unity means the standard concentrations for ISO (78mg/cc) and BIS (1.33mg/cc).

a few parameters of the system and are independent of the details. Because of this, totally different systems can have identical exponents when they are in the same universality class. By measuring the critical exponents, one can test the existing theories and obtain information about the possible universality class to which the system may belong.

The critical behavior of gel system was first observed in the scattered light intensity by Tanaka et al. [8] and Hochberg et al. [14] from acrylamide gel. Recently we obtained several critical exponents of N-isopropylacrylamide (NIPA) gel [15]. The observed critical phenomena are still not well understood. The most controversial aspect of the understanding comes from the shear modulus. As has been pointed out by Onuki, near the critical point (where $K = 0$), the fluctuations should be suppressed and hence no short wavelength fluctuations should be observed [3,4]. Onuki assumed that the scattered light intensity diverges at $K+4\mu/3 = 0$. We will see later that this assumption does not agree with experimental results. Onuki suggested that the mean-field theory should be applicable in describing the gel critical point because of the suppression of the critical fluctuation due to the shear modulus [3,4]. This suggestion was also made by Golubović and Lubensky [16]. However, this suggestion does not agree with our experimental results. We will consider our experimental results here and compare them with the recent theoretical predictions.

In most of the gel experiments the gels are immersed in a large volume of solvent, therefore the external osmotic pressure is kept at zero. Depending upon the property of the network, the zero external pressure can be higher or lower than the critical osmotic pressure of the network. If it is higher , the isobar curve is continuous (if lower, discontinuous). Fig.5 is a two dimensional map of the isobar behavior of N-isopropylacrylamide (NIPA) gels with different monomer (ISO) and crosslinker (BIS) concentrations. We find from this figure that for BIS=2.6, the critical osmotic pressure is around zero. The asterisk $*$ in Fig.5 is the composition of the gel we used as the critical sample in the following experiments on critical behaviors.

The critical exponents α, γ, etc, are conventionally defined along the critical isochore. In gel experiments, however, the most convenient path is the critical isobar. Since the isochore path is the same as the co-existence path, along the isochore we have $\delta\rho \sim (\delta T)_\rho^\beta$. Along the critical isobar, we have $\delta\rho \sim (\delta T)_\pi^{1/\delta}$. For the same $\delta\rho$, we have $(\delta T)_\rho \sim (\delta T)_\pi^{1/\beta\delta}$. Accordingly, the exponents related with δT should be re-scaled by the factor $1/\beta\delta$. This result can also be obtained from the scaling argument of the free energy [15]. Thus the exponent α_π measured along the isobar is related to the conventional exponent α through $\alpha_\pi = \alpha/\beta\delta$. Using the

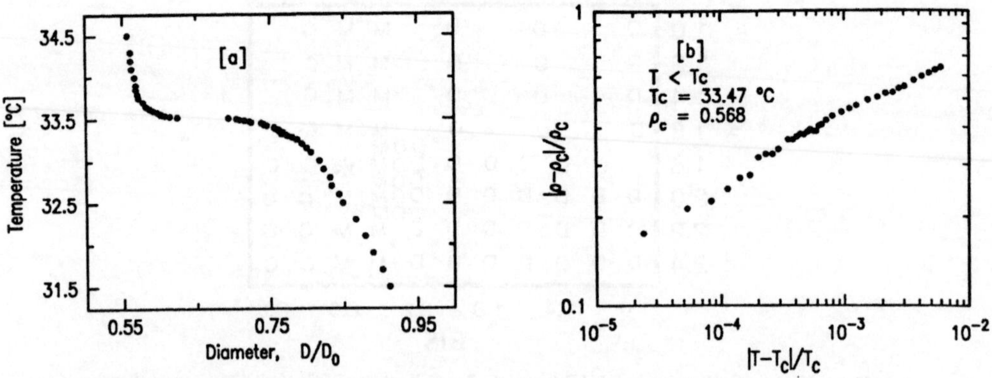

FIG.6. [a]. Isobar of NIPA gel with BIS concentration 3.45mg/cc and ISO concentration 156mg/cc. [b]. Density temperature relationship of the critical sample. This figure is generated from Fig.6a. The critical temperature and density are 33.47 and 0.568mg/cc (see the comment in Ref.12). The slope of this figure is $1/\delta$.

scaling laws [17], if one knows α_π and δ, then the other exponents can be obtained through relations such as $\alpha = 2\alpha_\pi\delta/(\delta + 1 + \alpha_\pi\delta)$.

A. Exponent δ

From Fig.5, we find that the non-ionic NIPA gel with BIS concentration equal to 3.45mg/cc passes through the critical point [15]. Whether an isobar is continuous or discontinuous is relatively insensitive to the concentration of ISO. In all the experiments on the gel critical behavior, the samples we used were taken from a large gel with BIS and ISO concentration 3.45mg/cc and 140mg/cc, respectively.

Fig.6a is the swelling ratio of a sample which passes through the critical point. Fig.6b is a replot of Fig.6a ($T < T_c$ only) in log-log scale. The T_c and ρ_c were chosen so that the line for $T > T_c$ is parallel with $T < T_c$. The curve in Fig.6a defines the critical exponent δ through $\delta\rho \sim \delta T^{1/\delta}$. This can be seen from the fact that the derivative of the curve is the thermal expansion coefficient, which diverges as $\delta T^{-\gamma_\pi}$. This gives $\delta\rho \sim \delta T^{-(\gamma_\pi+1)}$. Using the scaling laws, one can readily show that $(\gamma_\pi + 1) = 1/\delta$.

Mean field theory predicts that $\delta = 3$. From Fig.6b we find that the exponent δ is 4.2 ± 0.5. This is much larger than the mean field value but very close to the value obtained from Ising systems. The T_c and ρ_c determined for this sample were 33.47 °C and 0.568mg/cc.

B. Specific Heat C

The critical behavior of the specific heat is one of the most intensively studied response functions both theoretically and experimentally. Often more conclusive results can be drawn from the study of the specific heat than from others partly because the exponent is quite different for different universality classes. Not only the specific heat exponent, but also the ratio of the amplitudes and the quantities related with the correction-to-scaling term, are very informative [18].

In this experiment, the heat capacity $C = \delta Q/\delta T$ of the sample (holder + solvent + network) were measured along the critical isobar path. The gel sample consisted of hundreds of small pieces of crushed gels of linear size $\sim$ 0.7mm. The sample was brought slowly to equilibrium at temperature T. Then a small amount of heat, δQ, was deposited into the system and the temperature response, $\delta T(t)$, was measured. The temperature jump, δT, was

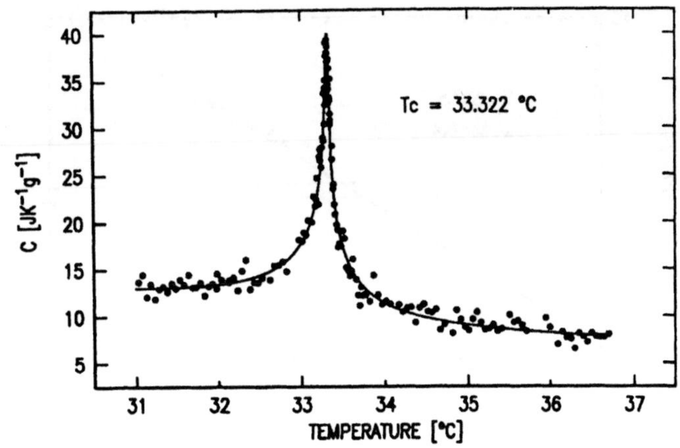

FIG.7. Specific heat of the critical gel along the critical isobar path. The solid curve is the fit to the data. The contribution from the sample holder and the solvent has been subtracted.

obtained by fitting $\delta T(t)$ to an single exponential function and extrapolating the fit to $t = 0$ point. The heat δQ was fixed at 6.7mJ, and the typical δT was 2~4mK. The specific heat of the network was obtained by knowing the mass fraction of the network and the heat capacity of the blank sample (for details, see [15]).

The specific heat result is shown in Fig.7. The sample is the same as the one used to obtain Fig.6. The data were fitted to the standard equation [19]

$$C = A|t|^{-\alpha_\pi}[1 + D|t|^{\Delta}] + B + c(T - Tc)\,. \tag{23}$$

The correction-to-scaling exponent, Δ, was fixed at 0.5 [18]. From the fitting parameters we noticed that other than the slow background, the data are symmetric about the critical temperature. The amplitude ratios A_-/A_+ and D_-/D_+ were found to be 0.99 and 0.90, respectively ($A_- = 257$, $D_- = -1.16$), where the subscripts + and − indicate the temperature range above and bellow T_c. The fact that the data are symmetric about T_c is a direct result of the path we took. Along the isobar path, the system never enters the two phase co-existence region and hence the amplitudes above and below T_c are the same. For Ising system along isochore, $A_-/A_+ = 1/2$.

The exponent obtained from this data is $\alpha_\pi = -0.05 \pm 0.13$. The data can also be fitted very well to a logarithmic function. The mean-field theory, in general, predicts a step discontinuity in C ($\alpha = 0$), rather than a singular peak. The T_c determined from this measurement is 33.32°C.

Sanchez, Meichle, and Garland [19] pointed out that heat capacity data of nonequilibrium systems exhibit hysteresis between warming and cooling measurements. In our experiment, no hysteresis was observed. We also compared the data taken 3 hours and 4.5 hours after the temperature had been set. We did not see any appreciable difference (less than a few percent). All these indicate that the data were taken under equilibrium conditions; that is, at $t = 0^-$, the system was at equilibrium. This experiment was repeated on other samples with very similar chemical composition. All the experimental results reported here were fully reproducible.

The interpretation of this data is quite difficult because of the condition under which C was measured. The average size of the sample pieces was about 0.7mm. Using the result in table 1, the estimated volume relaxation time of these pieces should be about 10 minutes. This value is comparable with the temperature relaxation time of δT [20]. This means the

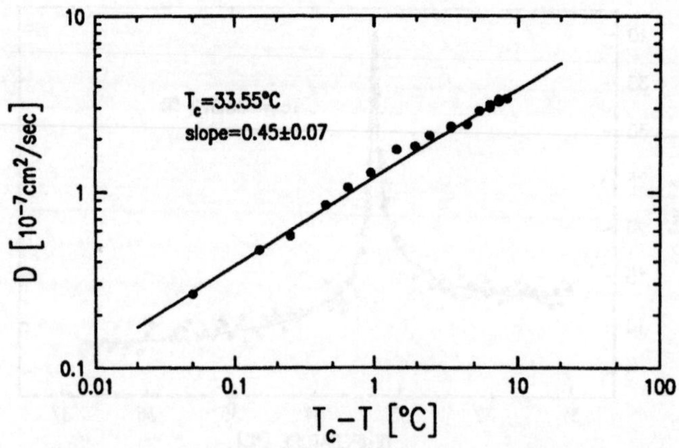

FIG.8. Temperature dependence of the collective diffusion constant of NIPA gel.

heat capacity we have measured is a mixture of C_v and C_π. Near the critical point, since the volume relaxation time diverges (see Fig.9b), the measured C is basically C_v. We believe that the basic behavior of C_v near the critical point should be similar to the one given in Fig.7.

C. Light Scattering Spectroscopy

Laser light scattering spectroscopy has been proven to be a powerful tool in studying gel systems. Fig.8 shows the collective diffusion constant as a function of $(T - T_c)$ of the sample used in obtaining Fig.6 and Fig.7. As the critical temperature is approached, D goes to zero.

Tanaka, Hocker and Benedek (THB) [7] had shown that the scattered intensity (with background subtracted) and the collective diffusion constant D are related with the shear modulus μ and bulk modulus K of the gel through

$$I \propto \frac{kT}{K + 4\mu/3} \quad (24)$$

$$D = \frac{K + 4\mu/3}{f}. \quad (25)$$

If we neglect the shear modulus μ, (or if μ goes to zero by the same power as or faster than K near the critical point), then I and D diverge with power γ_π ν_π, respectively [21]. From Fig.8, we find that $\nu_\pi = 0.45 \pm 0.07$. Again this value disagrees with the mean field result, which is 1/3.

D. Discussion

All the measured exponents are summarized in table II. The measured isobar exponents have been converted to isochore exponents.

In the above, we have assumed that the order parameter of the gel system is the network density. At equilibrium, the gel is always uniform and we can use the density alone to describe the system. However, in the case of swelling kinetics, we find that the density alone is not enough to describe the process. Instead, the displacement vector **u** should be used. Although the choice of the order parameter is often obvious, there is no precise definition or standard way of identifying it [22]. One of the differences between the network system and liquid-gas system is that the network has shear modulus. The effect of the shear modulus on the nature

Table.II Comparison between the measured gel exponents and the mean field results. The typical 3-d Ising results are also listed. The fourth row are the gel results after Fisher exponents renormalization. The last row are critical temperatures determined from these experiments.

system	α	δ	ν
mean-field	disc.	3	1/2
3-d Ising	0.11	4.2	~0.6
NIPA Gel	-0.08±0.19	4.2±0.5	0.7
	0.09	4.2	.65
T_c [°C]	33.32	33.47	33.55

of the order parameter is not clear to us at present. Onuki believes that the universality class of the gel should not be the Ising class [3,4]. He suggested that gel is in the same universality class as the solid solutions of $Sm_{1-x}Y_xS$ or $Ce_{1-x}Th_x$ [23], or the phase transition of hydrogen in Niobium [24].

IV. FURTHER DISCUSSION ON THE SHEAR MODULUS OF GELS

A. Onuki's Interpretation

The basic assumption of Onuki's theory is that near the critical point, the bulk modulus K is zero but the shear modulus is finite and small. However, the result by Hirotsu [25] shows that near the critical point the value of μ is comparable with the value far from the critical point.

The conventional critical point corresponds to $K = 0$. According to Onuki [3,4], at $K = 0$, the fluctuation of the gel network density is suppressed by the finite shear modulus μ and the observed scattered intensity should diverge at $K + 4\mu/3 = 0$. From reference [8], we find that the change of the intensity is about 100 times. Also from Fig.8, D decreases by a factor of 50. Experiments [6,25,10,11] show that μ and K of the sample used to obtain Fig.8 are of the same order. Assuming μ is a constant in the experimental temperature range [25], we need $K \sim -4\mu/3$ to explain the large change of the scattered intensity I and the collective diffusion constant D. It is hard to believe that near T_c, K can have such a large negative value. Therefore what we should observe following Onuki's assumption is that I and D change by a much more moderate factor, rather than 100 times.

The critical temperatures listed in table II are very close to each other. The second one corresponds to $K = 0$, the third corresponds to $K + 4\mu/3 = 0$. The difference here is too small, however, to be used as convincing evidence that the observed singularities occurred at different temperatures due to the shear modulus.

B. Relaxation time τ and thermal expansion coefficient α

We measured the thermal expansion coefficient $\alpha = (1/V)(\partial V/\partial T)_\pi$ and the swelling relaxation time τ of NIPA gel near the critical point. Fig.9a is the temperature dependence of the swelling ratio. The chemical composition of the sample is the same as the sample used in Sec.III. The thermal expansion coefficient α is obtained by taking the derivative of Fig.9. The relaxation time τ was measured by measuring evolution of the length of the gel after a small temperature jump was given to the sample. The two curves in Fig.9b were obtained at the same time, therefore no aging effect, if any, need to be considered between these two curves.

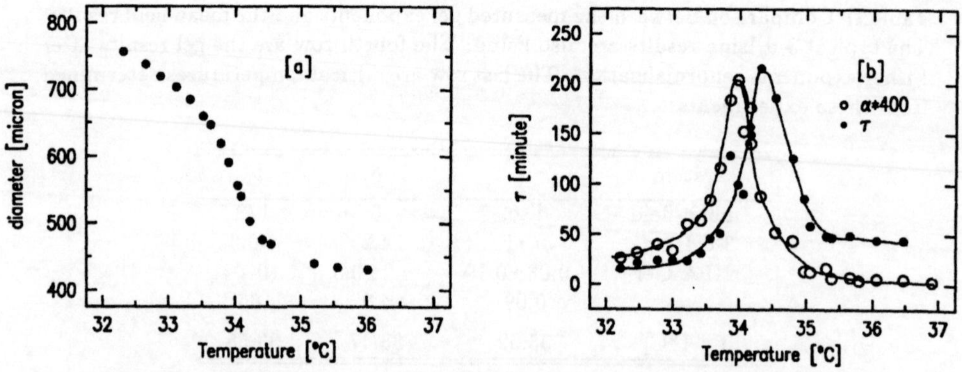

FIG.9. [a]. The swelling ratio of NIPA gel used in obtaining the Fig.9b. [b]. Thermal expansion coefficient α and the swelling relaxation time constant τ. They peak at different temperatures.

Clearly, the two curves peak at different temperatures. This same result was reproduced with a number of other samples.

Since $\alpha \sim (\partial V/\partial T)_\pi \sim K^{-1}(\partial \pi/\partial T)_v$ is proportional to the inverse of the bulk modulus K, the peak in α corresponds to the critical point. As we have mentioned in Sec.II, τ should also diverge at $K = 0$. Following the current understanding of the swelling kinetics and the critical point of gels, the two peaks in Fig.9b should be located at the same position, in contradiction to this experiment. The peak τ occurs at higher T (by 0.4 °C). This experiment may indicate that there is something missing in our current understanding of the nature of the gel system. (In this experiment, since Fig.9a describes a continuous transition, K is always positive.)

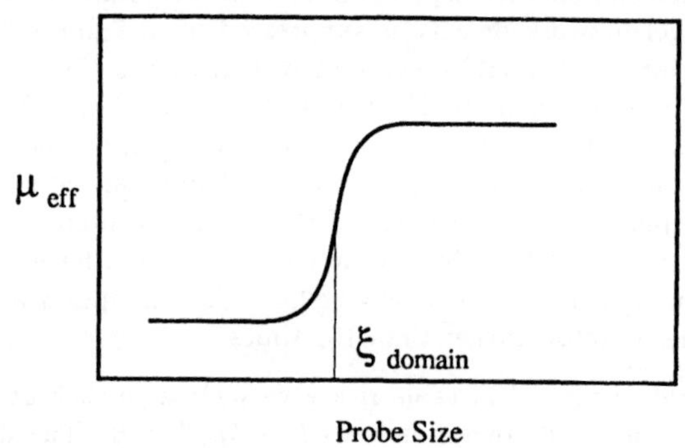

FIG.10. The effective shear modulus as a function of the probing size of the dynamic experiment.

C. Effective Shear Modulus

It had been shown by Onuki [5] that the effect of the shear modulus on the density fluctuations of the constrained gels vanish at the critical point. It is known [26] that there are inhomogeneities with characteristic length scale $\sim 30\mu$m in NIPA and acrylamide gels. In general, each of these domains may not be fully relaxed even in the equilibrium state due to the fact that the mechanical properties of each domain are different. Hence each domain can be treated as a constrained sub-systems (with temperature dependent boundary condition). A gel element with volume smaller than the domain size behaves like a constrained gel; elements with size larger than the domain size behave the same as free gels. In other words, there may be no shear modulus effect in the light scattering experiment. The critical behavior of each of these sub-systems is not affected by the shear modulus. As far as the critical fluctuation is concerned, we can treat the shear modulus as zero, i.e., $\mu(q < \xi^{-1}) \sim 0$. Fig.10 illustrate this hypothesis. If we take this argument, the difficulty in Onuki's assumption discussed above may be resolved.

Onuki [5] and Hirotsu and Onuki [27] have shown that the transition temperature is stress dependent. We believe, however, that the stress caused by the inhomogeneity in the network is too small to create any appreciable dispersion in the transition temperature.

We would like to thank Prof. A. Onuki and Dr. S. Gorti for valuable discussions. The suggestions by F. Ilman and K. Wasserman regarding this manuscript are greatly appreciated. This work was supported by the National Science Foundation through grant number NSF DMR87-19217.

References

[1] L. D. Landau and E. M. Lifshitz, *Theory of Elasticity*, 1986 Pergamon Press, Oxford.

[2] P. J. Flory, *Principles of Polymer Chemistry* (Cornell Univ. Press., Ithaca and London, 1953).

[3] A. Onuki, Phys. Rev. A. **38**(4), 2192(1988).

[4] A. Onuki, *Formation, Dynamics, and Statistics of Patterns*, (edited by K. Kawasaki et al., World Science, 1989).

[5] A. Onuki, J. Phys. Soc. Jpn. **57**(3), 699(1988) and **57**(6), 1868(1988)

[6] Y. Li and T. Tanaka, J. Chem. Phys. **92**(2), 1365(1990); Y. Li and T. Tanaka, *Dynamics and Patterns in Complex Fluids*, (edited by A. Onuki, Springer, in print).

[7] T. Tanaka, L. O. Hocker, and G. B. Benekek, J. Chem. Phys. **59**(9), 5151(1973).

[8] T. Tanaka, S. Ishiwata, and Coe, Ishimoto, Phys. Rev. Lett. **38**(14), 771(1977).

[9] T. Tanaka and D. Fillmore, J. Chem. Phys. **70**(3), 1214(1979).

[10] A. Peters and S. J. Candau, Macromolecules **19**(7), 1952(1986); A. Peters and S. J. Candau, Macromolecules **21**(7), 2278(1988). We believe the calculation in the second reference is not correct. In the first reference, due to a mistake made in the initial condition, the coefficients A_n given by Eq. (6) are not correct. The correct answer should be

$$A_n = 4\frac{(\alpha_n^2 - 3)\sin\alpha_n + 3\alpha_n\cos\alpha_n}{2\alpha_n^2 + \alpha_n\sin(2\alpha_n) - 4\sin^2\alpha_n}\,.$$

The boundary condition of spherical gel, $R = \alpha_n^2/(4 - 4\alpha_n\cot\alpha_n)$.

[11] E. Geissler and A. M. Hecht, Macromolecules **13**, 1276(1980) and **14**, 185(1981).

[12] P. Heller, Rept. Prog. Phys. **30**, 731(1967).

[13] See, for instance, S-K. Ma, *Modern Theory of Critical Phenomena*, (Benjamin/Cummings Publishing Compand, Inc., 1976).

[14] A. Hochberg, T. Tanaka, and D. Nicoli, Phys. Rev. Lett. **43**(3), 217(1979).

[15] Y. Li and T. Tanaka, J. Chem. Phys. **90**(9), 5161(1989). The critical density given in this paper is not correct. The correct value should be 0.568mg/cc.

[16] L. Golubović and T. C. Lubensky, Phys . Rev. Lett. **63**(10), 1082(1989).

[17] see H. E. Stanley, *Introduction to Phase Transition and Critical Phenomena*, (Oxford University, Oxford and New York, 1971).

[18] C. Bervillier, Phys. Rev. B **34**, 8141(1986), and the references cited therein.

[19] G. Sanchez, M. Meichle, and C. W. Garland, Phys. Rev. A. **28**, 1647(1983).

[20] In the analysis presented in reference [11], we used the relaxation time of acrylamide reported by Tanaka and Fillmore in reference [6].

[21] T. Tanaka, *Dynamic Light Scattering* (ed. Pecorra, Plenum Press, New York, 1985).

[22] P. W. Anderson, *Symmetries and Broken Symmetries*, (ed. N. Bocarro, Idset, Paris, 1981).
[23] A. Onuki, Phys. Rev. B. **39**(16), 12308(1989).
[24] H. Zabel and H. Peisl, Phys. Rev. Lett. **42**(8), 511(1979).
[25] S. Hirotsu, preprint.
[26] S-T. Sun, Y. Li, E. Sato-Mutsuo, and T. Tanaka, to be published. Y. Li and T. Tanaka, to be published.
[27] S. Hirotsu and A. Onuki, J. Phys. Soc. Jpn. **58**, 1508(1989)..

INTERACTIONS BETWEEN SOLVENT MOLECULES AND NETWORKS IN BIOGELS

-VISCOELASTICITY, STRENGTH-

Kunio Nakamura

Department of Polymer Science
Hokkaido University
Sapporo 060, Japan

INTRODUCTION

It has long been known that the rheological properties of gels are affected by the interactions between solvent molecules and networks. The nature and properties of these interactions have been studied extensively during the last 40 years, and in many types of gel systems the frameworks of gel can now be predicted reasonably[1]. The objective of this report was to investigate the relationship between dynamic viscoelastic properties and gelation.

DYNAMIC MECHANICAL PROPERTIES OF BIOGELS

Hen's egg white consists of the thick and thin white. The thick white has gel structure which is held by the fibres composing of ovomuchin. On the other hand, the thin white does not have gel structure. Brooks et al.[2] suggested that the gel structure of thick white was kept by the network composed of ovomuchin-lysozyme complex. The thick white of Hen's egg changes from gel to liquid during the storage. It is called as the liquefaction phenomenon[3].

In order to investigate the disintegration process of gel structure, we studied the changes of viscoelastic behavior in the stored shell egg[4] by use of sphere rheometer. The complex compliance of the thick white of Hen's egg for various storage time at 25°C was measured over the frequency range from 20 to 0.01 Hz. A nickel particle of 0.15 mm diameter was inserted in a transparent layer of the thick white. Frequency-storage time superposition of the complex compliance J^* data can be achieved in the high frequency region when a shift factor f (M_c correspondingly changing 1/f) was used on the vertical axis(Fig. 1).

It has already been pointed out by Ferry that maxima in J" are characteristic of network structure. With increasing the storage time, the maxima in J" becomes broader in shape. The magnitude of the loss tangent at low frequencies increases rapidly with increasing the storage time. The rate of this egg thinning depends on the temperature at which the egg is kept. We found that the values of the loss tangent at $\omega = 1$ could be used as a measure of liquefaction of egg white(Fig. 2).

We have shown that the dynamic viscoelastic properties of the thick white of a newly laid egg in the main peak region could be formulated with analytical form of the complex compliance $J^*(\omega)$:

Polymer Gels, Edited by D. DeRossi *et al.*
Plenum Press, New York, 1991

$$J^*(\omega) - J_\infty = (J_o - J_\infty)/[1 + (i\omega\tau')^{1-\alpha}], \quad 0 < \alpha < 1 \qquad 1)$$

where J_o and J_∞ are the values of $J'(\omega)$ at $\omega = 0$ and $\omega = \infty$, respectively. τ' is a characteristic retardation time. J_o, τ' and α obtained from J' and J" data of master curves for a newly laid egg are 0.018 cm^2/dyn, 50 ms and 0.07, respectively.

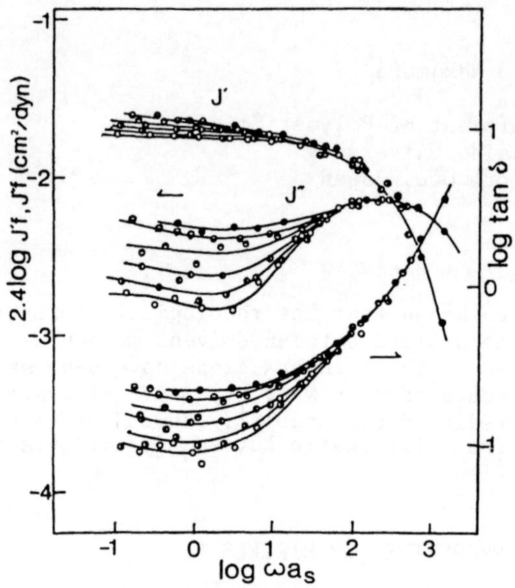

Fig. 1. Reduced plots of the loss tangent, and the complex compliance with the vertical shift factor f included for the thick white of various storage time S at 25°C. The egg was kept at 25°C; the reference state was taken as S= 0 day(the newly laid egg). (○): 0 day, (◑): 1, (◐): 3, (◒): 6, (◓): 9, (●): 17.

The gelation implies the process which transforms a liquid solution of polymers into an elastic solid. The mechanism of gelation increases in the number of bonds between molecules or particles, until a continuously connected network forms. If the bonds is of physical origin, the gels are reversible by changes in temperature, solvent power, etc.

Analysing the G' and loss modulus G" data of gelation of gelatin, Djabourov et al.[5] proposed a method of determining the gel point. They consider the 3-elements mechanical model, that is, Zener model for such soft gels. We define that the material at the gel pointis at its visco-elastic liquid-solid transition. At low frequencies, we have an expression for gel state as the inequality G'>G" or loss tangent tan δ <1, and for liquid state as the inequality G'<G" or tan δ >1. It requires that close to the gel point: tan δ=G"/G'=1. Gelation process of gelatin were interpreted within the framework proposed by de Gennes of the percolation theory. Dynes et al.[6] in 1982 have shown that gel time can be determined as the equality G'=G" in a study of thermosetting of epoxy resins. Therefore, we could qualitatively relate the values of the loss tangent to the extent of liquefaction of egg white.

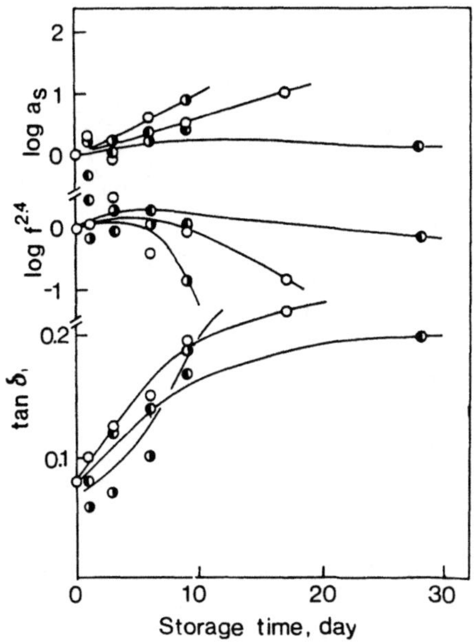

Fig. 2. 2.4log f, log a_s and loss tangent at ω=1 rad/sec plotted against the storage time. (○): egg kept at 25°C, (◐): egg kept at 5°C, (◑): egg kept at 35°C.

The eggs of most animal organisms are surrounded by various investments or integuments during early embryonic development. The integument for frogs takes a form of jelly layers. At some time during development the embryo must escape or hatch from investiments to continue normal development. Bles[7] postulated in his study on the life history of Xenopus laevis that hatching involved an enzyme secreted by the frontal gland. Carrol et al.[8] have shown that hatching can be affected by chemically(enzymatic) and physical rupture of integments.

We are recently interested in studying this hatching process viscoelastically. Sexually mature Rana chensinensis(Ezoakagaeru) were obtained and we suddenly determined the complex compliance by using a sphere rheometer. A preliminary result is shown in Fig. 3. A steel ball of 0.6 mm diameter was held near the embryo in the jelly. The decrease in storage compliance is obvious with increasing tadpole size. In the high frequency region, J" decreases approximately with ω for the newly laid jelly. This means that the dynamic response of the newly laid jelly is expressed by the single relaxation model of eq. 1 with $\alpha = 0$.

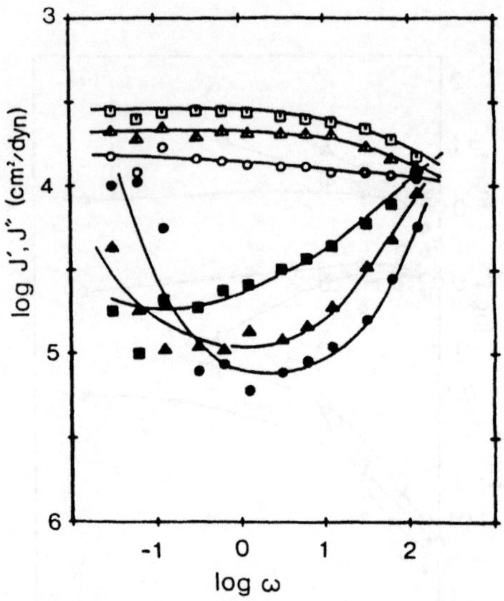

Fig. 3. The storage and loss compliance measured by the sphere rheometer at 20°C for the jelly of Rena chensinensis (Ezoakagaeru) plotted against the angular frequency. A steel ball of 0.6 mm diameter was placed near the embryo; the newly laid jelly having the embryo of 1.2 mm diameter (○), the jelly was kept for 1 day (a tadpole of 2 mm length)(△), and 5 days (that of 6 mm length)(□)in the cell of the rheometer at 20°C. Open symbols denote J' and filled symbols J".

DYNAMIC MECHANICAL PROPERTIES OF AQUEOUS XANTHAN SOLUTIONS

Xanthan is an ionic polysaccharide produced by the bacterium Xanthomonus campestris. The rheology of xanthan solutions is of great practical interest because of its widespread use in food and agriculture industries. Its most important behaviors are high viscosity at low shear rate and shear thinning. In dilute regime, xanthan undergoes a thermal transition[9], which is dependent on the concentrations of salts. This transition has been identified as a cooperative change on cooling from disordered to ordered chain geometry. The backbone conformation of xanthan in ordered state has been the subjects of much discussion in the literature[8,10,11].

The present work is concerned with the concentrated regime(1% wt.) in which stress nonlinearities and liquid crystal type behavior were observed[12]. It is known that in this regime the solution exhibits an yield stress, where the shear stress appears to approach a finite value as the shear rate goes to zero[13]. The existence of a yield stress is an indication of structure or gel-like behavior.

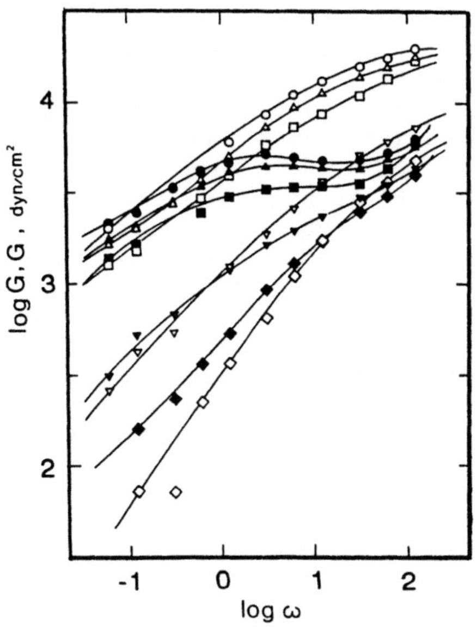

Fig. 4. Log G' and G" plotted against log ω for a 1% Xanthan (Kelzan) solution in water at various temperatures: (◇) 80°C, (▽) 70, (□) 60, (△) 50, (○) 40. Measurements carried out on cooing. Open and filled symbols denote G' and G", respectively.

Experimental

Xanthan used was Kelzan(Kelco Company). This polymer was purified by a modification of the procedure of Holzwarth[14]. Polymer was dissolved to 0.3g/l in water and stirred overnight. NaCl(0.34 M), EDTA(0.0025 M) and 300 ml/l of water were added and stirred for 2 hr. The solution was centrifuged at 13000 rpm for 3 hr to remove cells and undispersed polymer. Ethanol(440 ml/l of water) was added to precipitate the polymer as soft gels. The precipitate thus obtained was separated by centrifugation at 7500 rpm for 20 min.

This procedure repeated one more time. The soft gel obtained again was dissolved in water. The solution was passed through an ionexchange column (Amberlite IR-120 Na form) and dialyzed against water at room temperature. The polymer was obtained by freeze-drying.

The intrinsic viscosity at 25°C of xanthan sodium salt in 0.01 M aqueous NaCl solution was 18.4 dl/g. The viscosity average molecular weight was 1500000.

The shear dynamic moduli were measured by the method of annular pumping between coaxial cylinders in the frequency range from 20 to 0.01 Hz and the temperature range 30 to 80°C. A small amount of silicone oil was placed on the sample solution in the gap of glass cylinders to prevent vaporization of water.

Representation in the Complex Space

The storage modulus G' and loss modulus G" for a 1% solution of original Kelzan in water at various temperatures are shown in Fig. 4. For concentrated xanthan solutions, the viscoelastic properties are affected not only by the course of the solution preparation, but also by the time and temperature course of rheological measurements.

The data shown in Fig. 4 were obtained as follows; Kelzan was dissolved in water overnight at room temperature. At first, the sample solution to be measured was carefully poured in the cell of the rheometer, allowed to relax for several hours at room temperature, then was kept at 80°C for 1 hr. G' and G" were determined on cooling in time intervals of one hour by step 5°C, including the time of measurements(about 15 min).

We consider the mechanical model of viscoelastic liquid in the terminal region for the concentrated polymer solutions. The interpretation of dynamic rheological properties will be based on the representation of the complex viscosity η^* in the complex plane. This representation can be written as the following equation:

$$\eta^*(\omega) = \eta/[1+ (i\omega\tau)^{1-\alpha}], \quad 0 < \alpha < 1 \qquad 2)$$

which is typical of a model comprising two elements arranged in series: namely a spring, representative of the pure elastic component G in the rubbery plateau region, and a dash-pot, representative of the viscous component η in the terminal region. η can be related to G by the following equation:

$$\eta = G\tau(i\omega\tau)^{-\alpha} \qquad 3)$$

where τ is a characteristic relaxation time, and α a parameter which characterizes the width of the relaxation time distribution,

If α is independent of temperature for a given solution, considering temperature and frequency as variables, the behavior of the system fulfills the usual reduced variables. Therefore, by choosing an appropriate viscosity η_p, the dynamic data at various temperatures can be normalized in the complex space. The reduced Cole-Cole plots thus obtained are shown in Fig. 5 for a 1 % aqueous Kelzan solution at various temperatures. The Cole-Cole representation forecasts the variation of the dynamic mechanical properties with a change of temperature. We can see two different regions. One is a circular arc region in which the reduced variables method is successful.

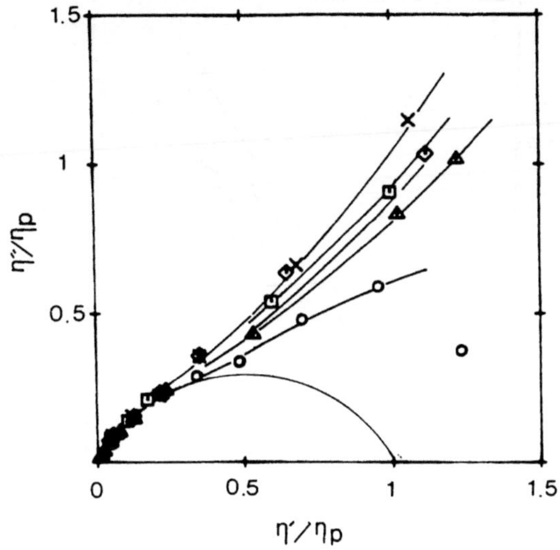

Fig. 5. Reduced Cole-Cole plot of a 1 % xanthan (Kelzan) in water at various temperatures. (O) 80°C, η_p=600(poise); (△) 70, η_p=4000; (□) 60, η_p=22000; (◇) 50, η_p=25000; (×) 40, η_p=30000. The circular arc characterized by α=0.3.

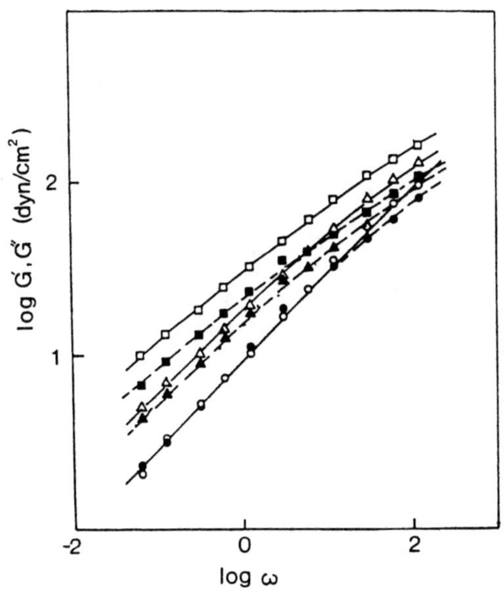

Fig. 6. Log G' and G" plotted against log ω for a 1% Na xanthan solution in water at various temperatures: (O) 80°C, (△) 60, (□) 40. Measurements carried out on cooing. Open and filled symbols denote G' and G", respectively.

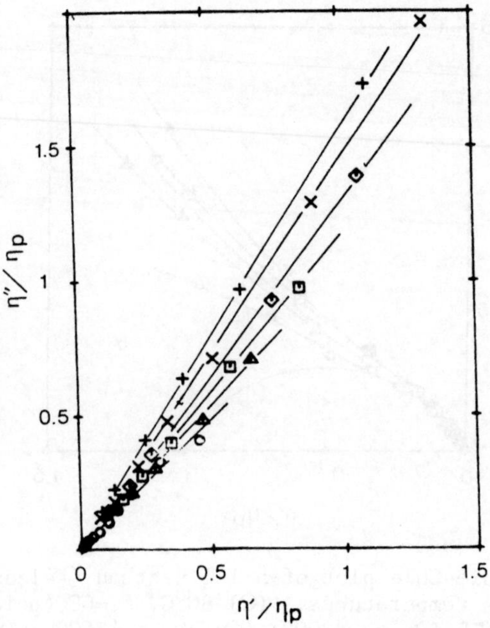

Fig. 7. Cole-Cole plot of a 1 % Na xanthan in water at various temperatures. (O) 80°C, (△) 70, (□) 60, (◇) 50, (×) 40, (+) 30. All data of η' and η" were divided by η_p= 80 (poise).

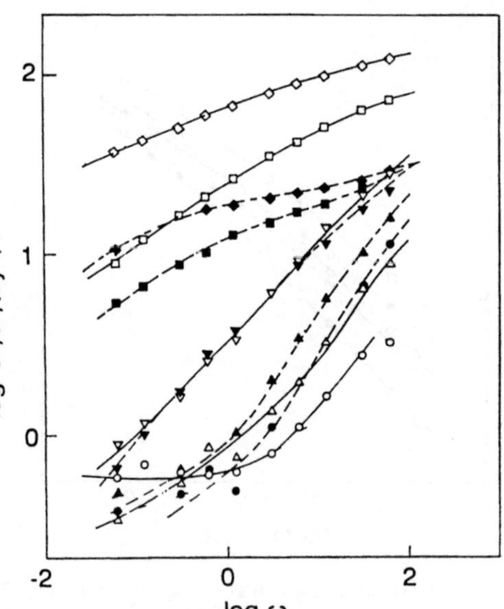

Fig. 8. Log G' and G" plotted against log ω for a 1% Na xanthan solution in water at various temperatures: (O) 80°C, (△) 70, (▽) 60, (□) 50, (◇) 40. Measurements carried out on heating. Open and filled symbols denote G' and G", respectively.

The other is a region in which the loss viscosity deviates upward from a circular arc. This behavior does not predict the steady state flow in the frequency range measured, but reflects the elastic contribution in the system, because loss viscosity which is equal to G'/ω does not fall to zero as the frequency decreases. Formally, data at low frequencies could not be combined by the reduced variables method. The structural change in temperature is obvious for aqueous xanthan solution. Accordingly, in the high frequency, that is, in rubbery plateau region, the effect of the structural change in temperature does not much affect to the rheological properties.

Next, G' and G" of purified Kelzan, xanthan sodium salt(Na-xanthan) are shown in Fig. 6. It is seen that G' and G" values of Na-xanthan solution at the same temperature are much smaller than those of Kelzan solution. It may be considered that the gelling effect of the possible existence of divalent[15] cations(calcium and magnesium ions)in the original Kelzan decreases in Na-xanthan. The most remarkable difference in G' and G" data between Kelzan and Na-xanthan is the change in frequency dependency. For example, G' is equal to G" at 80°C below $\log \omega = 1.5$, and its slope is close to 0.5 in double-logarithmic scale.

Cole-Cole plots of these data are shown in Fig 7. Data expressed by a circular arc of eq. 2 were so small that the reduced variables method described before is practically impossible. For data at low frequencies, instead of a circular arc, a linear relation between η' and η'' is obtained. It follows from the result that the frequency dependence of G' and G" can be written as the following power law equations:

$$G' = A' \omega^t \quad 4)$$

$$G'' = A'' \omega^t \quad 5)$$

where A' and A" is a constant.

In Fig. 7, a slope of η' vs. η'' plot at 80°C is close to 1. This means that A' is equal to A". In this case, the exponent t at 80°C(Fig. 6) is 0.5. Accordingly, G' and G" can be expressed by the equation:

$$G' = G'' = A'(=A'') \omega^{0.5} \quad 6)$$

The slope, which is equal to the reciprocal of the loss tangent(G"/G')(Fig. 7), increases with decreasing temperature. In parallel to this, the exponent t (Fig. 6) decreases with decreasing temperature. Therefore, the G' and G" behavior below 80°C for Na-xanthan solutions can be described by the following power law equations:

$$G' = A'\omega^t, \; G'' = A''\omega^t, \; A' > A'', \; t < 0.5 \quad 7)$$

The power law behavior expressed by eq. 7 has already reported in a study of the dynamic viscoelasticity of bovine vitreous body[16].

Fig. 8 illustrates the frequency dependence of G' and G" on heating. The measurements were carried out as follows. Na-xanthan was dissolved in water and stirred over night at room temperature. The sample solution was held in the cell, and then was kept for 2 hours at 30°C. G' and G" were measured on heating in the time interval of 1 hr by step 5°C,including the measurement time.

The frequency dependence of G' and G" was appreciably changed in comparison with the data of the cooling measurement(Fig. 6). The power law behavior characterized by eq. 6 is observed at 60°C. G' and G" at

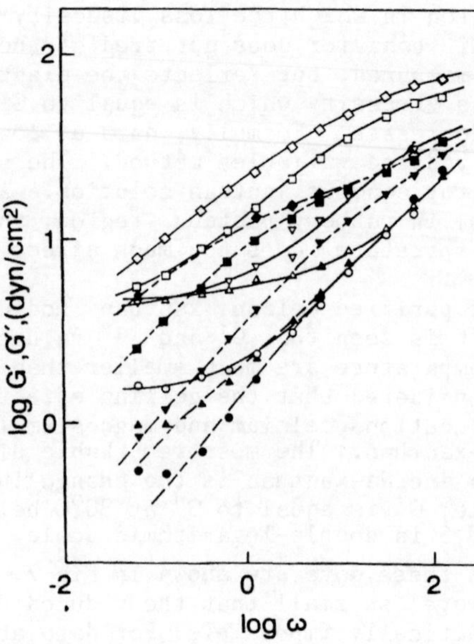

Fig. 9 Log G' and G" plotted against log ω for a Na xanthan(1 %wt.) in 20 mM aqueous NaCl at various temperatures: (○) 80°C, (△) 70, (▽) 60, (□) 50, (◇) 40. Measurements carried out on cooling. Open and filled symbols denote G' and G", respectively.

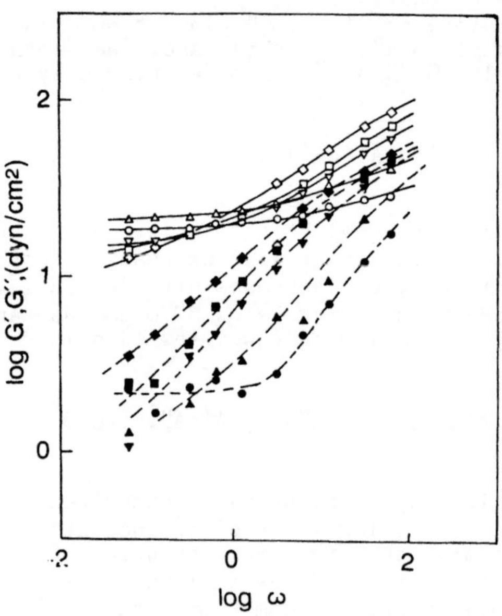

Fig. 10 Log G' and G" plotted against log ω for a Na xanthan(1 %wt.) in 20 mM aqueous NaBr at various temperatures: (○) 80°C, (△) 70, (▽) 60, (□) 50, (◇) 40. Measurements carried out on cooling. Open and filled symbols denote G' and G", respectively.

70^oC at the frequencies(1-100 rad/s) shows the power law behavior expressed by eq. 5. The exponent t is obviously higher than 0.5 and G' is lower than G". This behavior can be written as the following equations:

$$G' = A'\omega^{t},\ G'' = A''\omega^{t},\ A'<A'',\ t>0.5 \quad 8)$$

Possibly these power law behaviors(eqs. 5-7) are a characteristic of gel-like response. Moreover, it is clearly seen that G' at 80^oC at the frequencies(0.06-1 rad/s) is independent of ω. This means that aqueous Na-xanthan solution at 80^oC behaves as a gel rather than gel-like on heating.

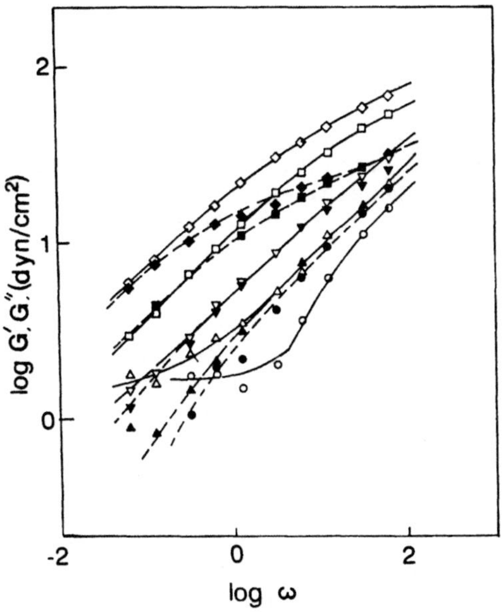

Fig. 11. Log G' and G" plotted against log ω for a Na xanthan(1 %wt.) in 20 mM aqueous NaI at various temperatures: (○) 80^oC, (△) 70, (▽) 60, (□) 50, (◇) 40. Measurements carried out on cooling. Open and filled symbols denote G' and G", respectively.

A aqueous 1% xanthan solution behaves as follows: The viscoelastic properties in high frequency and low temperature region, that is, in rubbery plateau region in which G' is characterized by the molecular weight between the tie point in the solution can be expressed by the relaxation of viscoelastic liquid model(eq. 2).

The rheological behavior at low frequencies and high temperatures corresponds to that of solid. It was probably thought that modulus of this very week gel must be interpreted in terms of the percolation theory. The properties in intermediate frequency and temperature region in which the relaxation was dominated by the power law reflect those in gel-like state.

The Effect of the Addition of Salts

As Na-xanthan is a polyelectrolyte, the solution rheology are greatly affected by the ionic strength. When salt is added to the solution in the dilute regime, xanthan molecules were collapsed down due to charged screening effects, and consequently the ordered conformation would be expected to stabilize. The molecule with which the electrostatic repulsions are screened aligns easily and associate strongly due to the hydrogen bonding and ionic interactions to form a gel structure.

Figs. 9-11 illustrate the frequency dependence of G' and G" on cooling for aqueous Na-xanthan solutions in the presence of 20 mM NaCl, NaBr, and NaI. In these cases it is seen that gel plateau is observed at 80°C, in contrast with Na-xanthan in water (Fig. 6). This clearly indicates that the effect of these ions is to enhance the gelation of xanthan solutions, perhaps by the charge screening effect.

Moreover, the effect of enhancement of gelation depends on the anion species. As far as the rheological measurements are concerned the structuring, as seen in Figs. 9-11, is in the order, water $< I^- < Cl^- < Br^-$, which led us to the tentative suggestion that ionic interaction was involved as well as hydrogen bonding. The order of these ions does not fit into place as part of a general anion lyotropic (Hofmeister) series. Recently, attention has been focused on the effect of anions on the disorder-order transition such as κ-carrageenan[17].

Power Law Gel

The simple power law relaxation(t=0.5 and A'=A") over wide frequency range has already shown in a study of dynamic behavior during chemical gelation and vulcanization of polymer by Chambon and Winter[18]. Recently, this critical behavior was interpreted in terms of the theory of polymeric fractals[19].

In the case of aqueous xanthan solutions, the various power law behaviors expressed by eq. 6-8 was observed in the wide temperature range. However, the simple power law relaxation was observed in the limited temperature and frequency region, that is, obtained for the aqueous Na-xanthan solution and the aqueous Na-xanthan in 20 mM aqueous NaCl at 80°C. For a solution of Na-xanthan in 20 mM aqueous NaCl at 80°C one finds a crossover from the solid behavior(G'=G= constant) to a power law behavior at ω=1 rad/s, as was suggested by Hess et al[19].

At least, Na-xanthan in the solution at this temperature will take a disordered conformation. Liu et al.[20] have suggested in a study of thermally induced conformation change of xanthan in 10 mM aqueous NaCl in dilute regime that the disordered conformation at 80°C was well represented by the model in which four wormlike chains were joined together by a short double-helical segment. The dimer does not dissociate to single chains.

We postulate that the transition from networks formed by the domain type structure (a coaxial double-helix or nested single helices) to networks formed by the hydrogen-bonding type structure of four wormlike chains occurs in concentrated regime on heating. It would be useful to study the relationship between the critical behavior and xanthan conformation in context of polymeric fractals.

DYNAMIC MECHANICAL PROPERTIES OF AQUEOUS PMA SOLUTIONS

Much study has been devoted to the gelation of biopolymer solutions such as gelatin[21]. The study of synthetic water-soluble polymers, in particular of polyelectrolyte, may be useful to get a clear relationship between viscoelastic properties and chemical constitution.

Kawanishi et al.[22] have extensively studied the sol-gel transition in poly(vinyl alcohol). They have discussed the thermoreversible sol-gel transition in polymer solutions in terms of liquid-liquid phase separation by spinodal decomposition, crystallite formation of polymer chains, and a hydrogen-bonding type association of the polymer chains.

Eliassaf and Silberberg[23] have reported that concentrated solutions of poly(methacrylic acid)(PMA) in water form a thermoreversible gel, and coagulate like albumin and after sufficient time phase separation occurs on heating. They have suggested that hydrogen bridges between carboxyl groups are responsible for forming the gel, because PMA is a weak acid. the phase separation diagram for PMA(the degree of polymerization DP=6900) obtained by them was given in Fig. 12. It consists of three regions:

A region S in which the solution is a sol, a region G in which it is a gel, and an unstable region U in which it separates into polymer-rich and polymer-poor regions. PMA solutions had a LCST(lower critical solution temperature) type phase diagram. Moreover, PMA solutions display negative thixotropy.

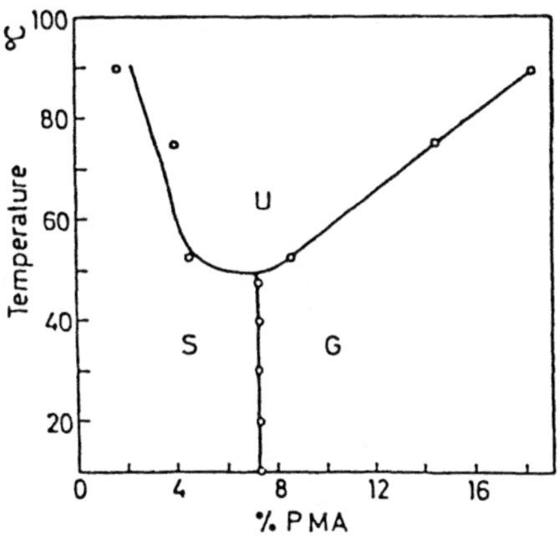

Fig. 12. Phase diagram of PMA solutions (DP=6900)[23]. In region S the solution is a so, in region G it is a gel and in region U a gel and a sol coexist.

Experimental

PMA was synthesized by the procedure of Silberberg et al. Its viscosity average DP was 3900. Dynamic shear moduli of aqueous PMA solutions were determined over the frequency range from 0.159 to 0.0029Hz and the temperature range from 10 to 80°C by the method of torsion between coaxial cylinders. The detailed procedure of measurements will described elsewhere[24].

The temperature dependence of G' and η' at ω=0.618 rad/s for various concentrations are shown in Fig. 13. In each temperature dispersion curve, the maximum in G' is observed around 70°C. The peak temperature does not depend on the polymer concentration. It has been known that the dynamic measurements for "strong gel" for which the Zener model is to be valid, will show the characteristic inflection in G' and maximum in loss tangent. With successive raising of temperature, G' increases only from 400 to 800 dyn/cm^2 for the 8.33% solution. The reversible gel of PMA will be belong to the week one. The decrease in G' at 80°C corresponds to the phase separation. η' varies with temperature in parallel with change in G'. In this temperature region, probably gelation as well as the related phase separation occurs.

In order to classify the domain of applicability of the superposition principle, a representation of the experimental results in the complex space was done. Cole-Cole diagrams are constructed considering temperature and time as variables. Reduced plots of the PMA solutions at different temperatures and concentrations are shown in Fig. 14. Here, η' and η'' are divided by a constant η_p so that each curve of the Cole-Cole plot of a solution at different temperatures has a maximum at the abscissa, η'/η_p=0.5.

For 7.00% solution, data at 10, 20, 30, 40°C constructs a circular arc expressed by eq. 2, having α=0.26, indicating that η_p is equal to steady state viscosity. Data at low temperatures and low concentrations constructs a circular arc, as seen in Fig. 14. It follows from the result that the solution at these temperatures and polymer concentrations behaves as sol, and is miscible and homogeneous.

Data at 50, 60, 70, 80°C for 7.00 % solution constructs an asymmetric circular arc. This change of the relaxation from a circular arc to an

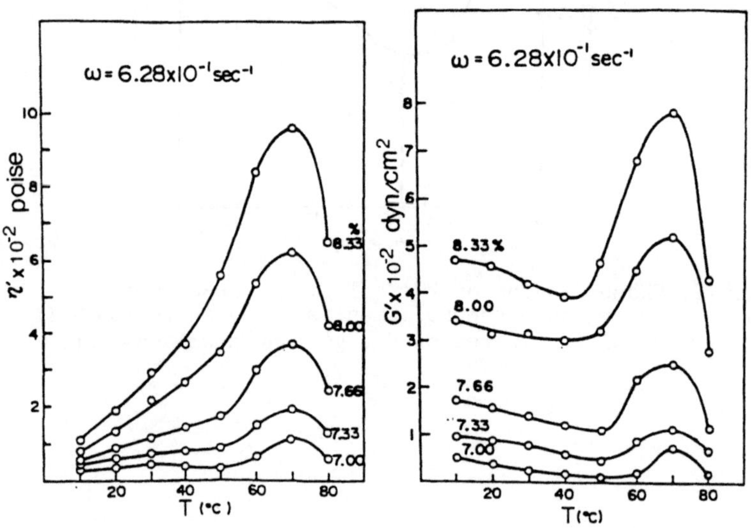

(a) (b)

Fig. 13. The temperature dependence of the dynamic viscosity (a) and the storage modulus (b) for aqueous PMA solutions at a constant angular frequency. The number in figure denotes the polymer concentration.

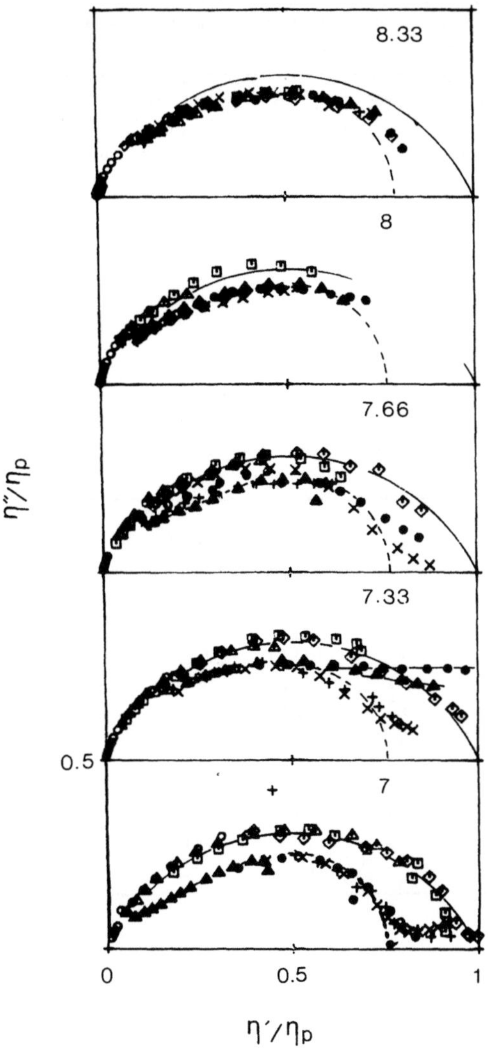

Fig. 14. Reduced Cole-Cole plot of aqueous PMA solutions at various temperatures: (○) 80°C, (△) 70, (+) 60, (×) 50, (◇) 40, (□) 30, (△) 20, (○) 10. The number in this figure is polymer concentration in percent weight The arcs expressed by solid and broken line have the value of α =0.26 and 0 in eq. 2, respectively.

asymmetric circular arc caused by the increase in temperature must be related to the gelation and phase separation. This relaxation behavior is again characterized by a linear relation between η" and η' at high frequencies(left-hand side of the arc). And a circular arc with α =0 is observed at high frequencies(right-hand side).

The linear relation between η" and η' in the high frequency region gradually become less clear with increasing the polymer concentration. As described in the previous section the former was a characteristic of the power law behavior expressed by eq. 4 and the latter was those of the viscoelastic liquid. Therefore, the asymmetric circular arc observed in the higher temperature region can be explained as the crossover from a power law behavior to a typical liquid, from the standpoint of the principle of time-temperature superposition. η_p values obtained by superposition are given in Table 1.

Table 1. η_p Values(poise) Obtained by Reduced Cole-Cole Plot of Aqueous PMA Solutions in Fig. 14

Temperature (^{o}C)	polymer conc.(wt%) 7.00	7.33	7.66	8.00	8.33
10	1900	8000	22000	80000	120000
20	450	2400	5000	20000	28000
30	165	880	2000	6000	8500
40	80	380	1000	3300	4000
50	70	330	700	4700	5100
60	130	620	2500	6500	8000
70	1200	850	3150	7000	8600
80	135	430	1100	2900	3700

Hess et al[19]. have defined as the pre-gel, in contrast to the post-gel observed in the case of aqueous Na-xanthan solutions. However, Ajji et al.[25] have shown in a study of the rheological properties of PS/PVME blends an asymmetric circular arc with immiscible blends similar to those with the 8.33% PMA solution. they discussed in terms of the multiplicity of the relaxation.

It is noticed that the data of η' and η" for the 7.33 % solution at 70 and 80°C can not be combined by the method of reduced variables. Especially, the steady state viscosity for the 7.33 % solution at 80°C could not be determined, because η" at low frequencies was parallel to the abscissa. The frequency dependence of the complex viscosity for the 7.33 % in the high temperature region is shown in Fig. 15. At 60°C, typical liquid behavior was observed at low frequencies, indicated by $\eta'=\eta=$ constant and $\eta'' \sim \omega$. on the other hand, the frequency dependence of the

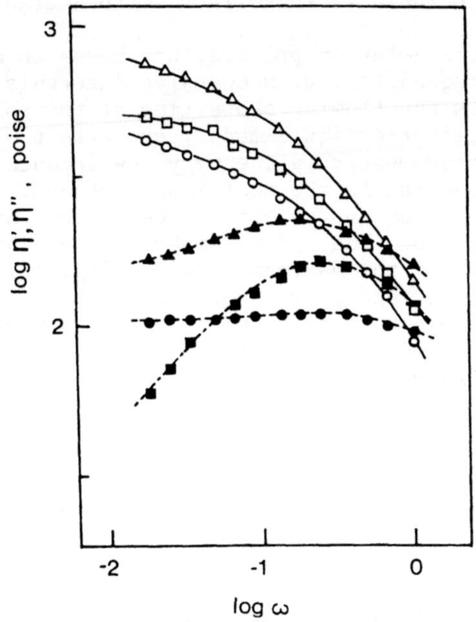

Fig. 15. The frequency dependence of the complex viscosity for 7.33 % aqueous PMA solutions at various temperature: (□) 60°C, (△) 70, (○) 80. Open and filled symbols denote dynamic and loss viscosty, respectively.

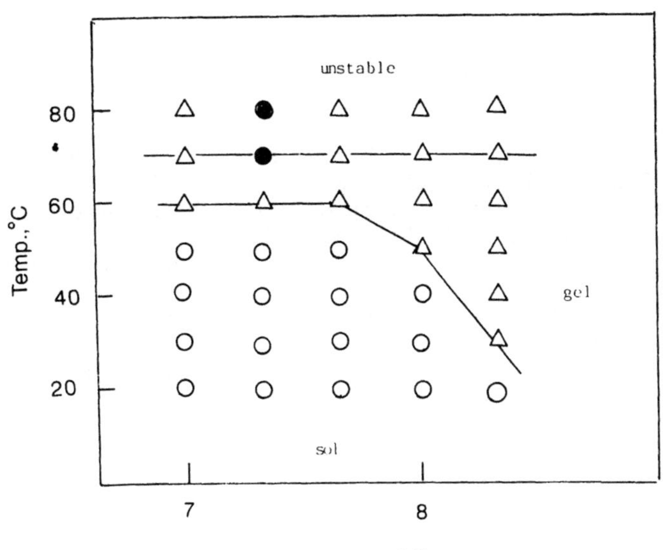

Fig. 16. Phase diagram of aqueous PMA solutions(DP=3900). Open circles denote the circular arc relaxation, triangles, the asymmetric circular arc, and filled circles, non-circular arc.

dynamic viscosity at 80°C in the low frequency region is characterized by η''= constant and the appreciable increase in η' with decreasing the frequency.

In order to analyse the gelation process, one needs to measure G' and G" on the wide range of frequencies, at successive intervals. As gelation of the PMA solution affects the flowing properties of the solution, which was described by the viscoelastic liquid model, the effect of gelation on G' and G" are expect to occur mostly in the very low frequency range. The investigation at low frequencies is the most important one.

Phase diagram obtained due to the difference in the rheological behavior is shown in Fig. 16. The onset temperature of the phase separation is taken as that of the maximum in G' in Fig. 13. It is noticed that gelation curve terminate into phase separation curve. Eliassaf et al[23]. have suggested that in their phase diagram(Fig. 12) a multiple critical phenomenon, namely gelation and phase separation occurs for aqueous PMA solution. The rheological behavior around 80°C for ca. 7.33 % (DP=3900) may lead us to the tentative suggestion that phase separation was involved as well as gelation.

Finally, the viscoelastic properties of biogels such as the thick white of hen's egg and frog's jelly were characterized by the Zener model. The rheological behavior of the newly laid gel was close to the single relaxation, indicated by J'=1/G=constant, $J'' \sim \omega$. In liquefaction process of the thick white for hen's egg, the values of loss tangent can be used as a measure of egg thinning.

In the studies of the dynamic mechanical properties of aqueous Na-xanthan solutions in the disordered-ordered transition region, and aqueous PMA solutions in a reversible sol-gel transition region, the rheological behavior can be interpreted as indicating of the power law relaxation during gelation. The former behaves as a post-gel and the latter remains as pre-gel. The Cole-Cole plot appears to be a sensitive to distinguish between sol state and gel state. For the solution in sol state, this representation gives a unique circular arc, and for the solution in pre-gel state, an asymmetric arc.

REFERENCES

1. P. H. Hermans, Gels, in: "Colloid Science", H. R. Kryst, ed., Elsevier New York (1949).
2. J. Brooks and H. P. Hale, The Mechanical Properties of the Thick White of the Hen's Egg, Biochim. Biophys. Acta, 37:237 (1959).
3. W. F. Holst and H. J. Alquist, Measurements of Deterioraction in the Stored Hen's Egg, Higardia, 6:49 (1931).
4. K. Nakamura, T. Ishii, and H. Uedaira, The Dynamic Mechanical Properties of the Thick White of Hen's Egg by Magnetic Microrheometer, J. Jpn. Soc. Biorheol., 2:207 (1988).
5. M. Djabourov, J. Leblond, and P. Papan, Gelation of Aqueous Gelatin Solutions. II. Rheology of the Sol-Gel Transition, J. Phys.,49:333(1988).
6. C-Y M. Tung and P. J. Dynes, Relationship between Viscoelastic Properties and Gelation in Thermosetting Systems, J. Appl. Polym. Sci., 27:569 (1982).
7. E. J. Bles, The Life History of Xanopus laevis, Trans. Roy. Soc. Edin., 41:789 (1905).
8. E. J. Carrol,Jr. and J. L. Hedrick, Hatching in the Toad Xenopus laevis: Morphological Events and Evidence for a Hatching Enzyme, Developmental Biology, 38:1 (1974).
9. I. T. Norton, D. M. Goodall, S. A. Frangou, E. R. Morris, and D. A. Rees, Mechanism and Dynamics of Conformational Ordering in Xanthan Polysaccharide, J. Mol. Biol. 175:371 (1984).
10. M. Rinaudo and M. Milas, Polyelectrolyte Behavior of a Bacterial Polysaccharide from Xanthmonas campestris, Biopolymers, 17:2663 (1978).

11. T. Sato, T. Norisue, and H. Fujita, Double stranded helix of Xanthan, Macromolecules, 17:2696 (1984).
12. T. Lim, J. T. Uhl, and R. K. Prud'homme, Rheology of Self-Associating Concentrated Xanthan Solutions, J. Rheology, 28:367 (1984).
13. P. J. Whitcomb and C. W. Macosco, Rheology of Xanthan Gum, J. Rheology 22:493 (1978).
14. G. Holtzwarth, conformation of the extracellular Polysaccharide of Xanthmonas campestris, Biochem., 15:4333 (1976).
15. R. K. Richardson and S. B. Ross-Murphy, Non-linear Viscoelasticity of Polysaccharide Solutions. 2: Xanthan Polysaccharide Solutions, Int. J. Biol. Macromol.,9:257 (1987).
16. M. Tokita, Y. Fujiya, and K. Hikichi, Dynamic Viscoelasticity of Bovine Vitreous Body, Biorheol. 21:751 (1984).
17. K. R. J. Austen, D. M. Goodall, Anion Effects on Equilibria and Kinetics of the Order-Disorder Transition of κ-Carrageenan, Biopolym., 27:139 (1988).
18. F. Chambon and H. Winter, Stopping of Crosslinking Reaction in PDMS Polymer at the Gel Point, Polym. Bull.,13:499 (1985).
19. W. Hess, T. A. Vilgis, and H. H. Winter, Dynamical Critical Behavior during Chemical Gelation and Vulcanization, Macromolecules,21:2536(1988).
20. W. Liu, T. Norisue, and H. Fujita, Thermally Induced Conformational Change of Xanthan in 0.01 M Aqueous Sodium Chloride, Carbohydr. Research, 160:267 (1987).
21. T. Tanaka, G. Swislow, and I. Ohmine, Phase Separation and Gelatin Gels, Phys. Rev. Lett., 42:1556 (1979).
22. K. Kawanishi, M. Komatsu, and T. Inoue, Thermodynamic Consideration of the Sol-Gel Transition in Polymer Solutions, Polym.,28:980 (1987).
23 J. Eliassaf and A. Silberberg, The Gelation of Aqueous Solutions of Polymethacrylic Acid, Polymer,3:555 (1962).
24. T. Itoh, K. Nakamura, and T. Nakagawa, Manuscript in Preparation.
25. A. Ajji, L. Chopplin, and R. E. Prud'Homme, Rheology and Phase Separation in Polystyrene/Poly(Vinyl Methyl Ether)Blends, J. Polym. Sci., Polym. Phys., 26:2279(1988).

Synthesis and Properties of Inorganic/Organic Hybrid Gels by the Sol-Gel Process

Shinzo Kohjiya, Kenichiro Ochiai, and Shinzo Yamashita

Department of Material Science
Kyoto Institute of Technology
Matsugasaki, Kyoto 606, Japan

ABSTRACT

The preparation of inorganic/organic hybrid gels was carried out using tetraethoxysilane and organic polymers via the sol-gel process. Hydroxyl terminated siloxane polymers were reactive enough to be incorporated into the silica networks. However, hydroxyl terminated organic polyethers, i.e., poly(oxyethylene), poly(oxypropylene), and poly(oxytetramethylene), failed to react, thus most of them were extracted by organic solvents. The introduction of triethoxysilyl groups in these polymers was found very effective to prepare truly hybrid gels by the sol-gel process. The resulted gels were optically transparent and mechanically more or less elastic. The structures and mechanical properties of these flexible materials were investigated. The organic polymer or silica glass formed the matrix depending on the composition of the gel, and the other component dispersed in the matrix. Dynamic mechanical properties of the flexible gels in which organic polymers were predominant components, were explained in accord with the structural considerations. Noise and vibration attenuating materials are suggested as one of the possible applications of these gels from their mechanical loss behaviors.

1. INTRODUCTION

Recently, the sol-gel method has been paid much attension as a novel process for the preparation of ceramic precursors and inorganic glasses[1-10]. The process is divided into two steps; hydrolysis of metal alkoxides to produce metal hydroxides, followed by polycondensation of hydroxyl groups as shown in Figure 1. The process is usually conducted at relatively low temperature, compared with those at which inorganic glasses have been prepared. Because of the low temperature, it seems reasonable to try to copolymerize metal alkoxides with hydroxyl terminated polymers to prepare hybrid gels, i.e., inorganic glass networks containing organic polymer as their network chains.

Polymer Gels, Edited by D. DeRossi *et al.*
Plenum Press, New York, 1991

Some attempts were already made along this line to synthesize new inorganic/organic hybrid gels[11-19]. In these studies, preparations and some physical properties of the gels were reported. However, there have been no discussions on the unreacted polymers. It is necessary to show experimentally whether or not all the organic polymers in the feed are chemically incorporated into inorganic networks. If one assumes that the composition of the product, i.e., the resultant inorganic/organic network, is identical with that of the feed, all the components should be reacted. In other words, the reaction has to be complete, especially in regard to organic polymers in the feed material. Even if the reaction is not complete, we can recover the product the composition of which is identical with the feed. In that case, most organic polymers can be removed by solvent extraction, because the polymers are just physically trapped in the silica network.

On the other hand, vulcanization of rubber with alkoxysilane has been known to rubber chemists and technologists under the name of moisture cure[20,21]. Several years ago we reported a successful moisture cure of halogenated rubbers by 3-amonopropyltriethoxysilane[22]. From a view point of rubber chemistry, especially reinforcing fillers for rubber, in situ formation of silica particles in rubber matrix is also an interesting application of the sol-gel process[23,24].

In this study, triethoxysilyl as well as hydroxyl terminated polymers were used in order to make the reaction with tetraethoxysilane proceed as quantitatively as possible to obtain new hybrid gels consisting of inorganic and organic network chains. Thermal and mechanical properties were investigated to elucidate the structure and to develope the gels for some applications. The small angle X-ray scattering study of the gels is reported in a separate chapter in this book[25].

2. EXPERIMENTAL

2.1. Materials

Tetraethoxysilane (TEOS) was used as the inorganic component of the hybrid gels throughout this study. TEOS was purchased and used as received. Three hydroxyl terminated poly(dimethyl siloxane)s, (HT-PDMS), were from Chisso Co.: PS-340, molecular weight (MW) = 1700; PS-340.5, MW = 3200; PS-341, MW = 4200. Three kinds of hydroxyl terminated polyethers (HT-polyether) were examined in the sol-gel reactions: Poly(oxytetramethylene) glycols (PTMG, from Sanyo Kasei Kogyo Ltd.) whose MWs were 1000 and 2000 were used after drying at 50°C. Poly(oxypropylene) glycol (PPG) of MW = 2000 was obtained from Asahi-Olin Ltd. Two poly(oxyethylene) glycols (PEG) were used also after drying at 50°C: One was from Nakarai Tesque Co. (MW = 600), the other from Wako Pure Chemicals Co.(MW = 2000).

Some analytical results on these hydroxyl terminated polymers were summarized in Table 1. As you can see in the table, the observed MWs were a little different from those listed above which are taken from the catalogues received from the suppliers. However, the nominal MWs described in these catalogues will be conveniently used for the designation of the materials in this paper. The functionality shown in the last column is a number of functional groups per

Table 1. Properties of hydroxyl terminated polymers

Code	Mn[a)]	Mw/Mn[b)]	$[OH]^{c)} \times 10^4$ (mol/g)	Functionality
HT-PDMS	1700	-	-	2.4
	3200	-	-	2.0
	4200	-	-	2.1
PTMG	700	1.34	-	-
	980	1.37	18.6	1.8
	1850	1.47	9.89	1.8
	2840	1.58	7.00	2.0
PPG	1960	1.05	10.26	2.0
PEG	600	1.05	-	-
	1790	1.07	-	-

a) MW by vapor pressure osmometer.
b) Measured on GPC.
c) Titration in non-aqueous media.

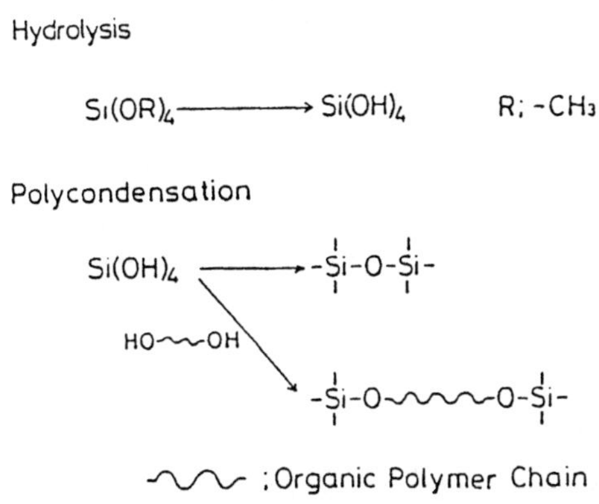

Fig. 1. The sol-gel process.

$$HO(CH_2)_4O(CH_2CH_2CH_2CH_2O)_n(CH_2)_4OH \xrightarrow{K} K^{+-}O(CH_2)_4O(CH_2CH_2CH_2CH_2O)_n(CH_2)_4O^-K^+$$

PTMG

$$\xrightarrow{CH_2=CHCH_2Br} CH_2=CHCH_2O(CH_2CH_2CH_2CH_2O)_{n+2}CH_2CH=CH_2$$

AT-PTMO

Fig. 2. Synthetic route of ET-PTMO from PTMG. ET-PPO and ET-PEO were synthesized by the same route from PPG and PEG, respectively.

molecule. In the present samples, the value very close to 2.0 means telechelic, i.e., a functional group is present respectively at each polymeric chain end.

2.2. Synthesis of Triethoxysilyl Terminated Polyethers

Triethoxysilyl terminated poly(oxytetramethylene) (ET-PTMO), poly(oxypropylene) (ET-PPO) and poly(oxyethylene) (ET-PEO) were synthesized to use as prepolymers for the sol-gel reaction. ET-PTMO, ET-PPO and ET-PEO were synthesized from PTMG, PPG and PEG, respectively. The synthetic route is shown in Figure 2, where description is for ET-PTMO, but ET-PPO and ET-PEO were also synthesized via allyl-group terminated polyethers (AT-PPO and AT-PEO) using the reactions shown in Figure 2.

2.3. Sol-Gel Reaction

Sol-gel reactions were carried out in the following manner: TEOS and a prepolymer were mixed in an ethanol/THF mixture in a glass flask, and hydrogen chloride was added to the same solvent mixture, followed by mixing and refluxing for one hour. The reaction mixture was poured into a Teflon mould, and it was kept at 50°C or at room temperature in open air. After gelling, the sample was dried first at room temperature, and then at 50°C under vacuum to constant weight. The sol-gel reaction product was then subjected to a Soxhlet extraction using tetrahydrofuran (THF) or a solvent extraction at 50°C for 5 days with four renewals of the solvent varying from 2 to 20 in dielectric constant.

The reaction conditions were listed in Tables 2-4. Table 2 is those for HT-PDMS, Table 3 for HT-polyethers (PTMG, PPG and PEG) and Table 4 for ET-polyethers prepared from PTMG, PPG or PEG. The reaction temperature was either 50°C or room temperature.

2.4. Thermal Properties

Thermal properties, i.e., the differential scanning calorimetry (DSC) and thermogravimetry (TG), were measured on Seiko DSC 20 and Rigaku TG instruments, respectively, in order to elucidate the phase transitions, if any, and the thermal stability of polymeric chains in the glass matrix. Sample amount was 10 mg, and the temperature increment was 10 ^{o}C/min.

2.5. Spectroscopic Measurements

Infrared spectra were obtained on Shimazu IR-408, and proton NMR and carbon-13 NMR spectra on Varian XL-200. They were used to characterize all the polymers except ET-polyethers. The alkoxysilyl groups were found to be easily hydrolyzed by atmospheric moisture. Therefore, they were not always subjected to analytical measurements, and used as soon as possible for the sol-gel preparations.

2.6. Dynamic Mechanical Measurements

The dynamic moduli, E' and E" (tensile) or G' and G" (shear) and tan δ (dynamic loss) were obtained on a dynamic mechanical analyzer, FT-Rheospectra DVE-V4 (Rheology Co.

Table 2. Reaction conditions of the sol-gel process of TEOS/HT-PDMS

No.	TEOS (g)	HT-PDMS MW	HT-PDMS (g)	THF/EtOH (ml/ml)	$TEOS/HT\text{-}PDMS/Sol/H_2O/HCl$ (molar ratio)
1	10.41	-	-	16.2/11.6	1 / - / 8 / 2 / 0.01
2	2.09	1700	4.25	4.1/ 2.9	1 / 0.25 / 10 / 4 / 0.01
3	1.00	1700	4.00	1.5/ 1.1	1 / 0.49 / 8 / 2 / 0.01
4	2.09	3200	8.00	6.5/ 4.7	1 / 0.25 / 16 / 2 / 0.01
5	2.08	4200	5.26	11.4/ 8.1	1 / 0.13 / 24 / 2 / 0.01
6	2.08	4200	10.51	11.4/ 8.1	1 / 0.25 / 24 / 2 / 0.01

Solv = solvent.

Table 3. Reaction conditions for the sol-gel process of TEOS/HT-Polyethers

Polym	TEOS (g)	HT-Polym (g)	THF/EtOH (ml/ml)	$TEOS/Polym/Sol/H_2O/HCl$ (molar ratio)
PTMG1	1.00	4.8	1.6 / 1.1	1/ 0.55/ 8/ 2/ 0.03
2	1.04	20.0	6.5 / 4.7	1/ 2.2 /32/ 4/ 0.03
PPG 1	1.00	4.81	1.6 / 1.1	1/ 0.52/ 8/ 2/ 0.03
2	1.04	20.0	6.5 / 4.7	1/ 2.1 /32/ 4/ 0.03
PEG 1	1.00	4.80	1.6 / 1.1	1/ 0.57/ 8/ 2/ 0.03
2	1.04	20.0	6.5 / 4.7	1/ 2.3/ 32/ 4/ 0.03

Polym = polymer, Solv = solvent.
Molecular weight of HT-Polym = 2000.

Table 4. Reaction conditions for the sol-gel process of TEOS/ET-Polymers

Polym	TEOS (g)	ET-Polym MW	ET-Polym (g)	THF/EtOH (ml/ml)	$TEOS/Polym/Sol/H_2O/HCl$ (molar ratio)
PTMO 1	1.85	1000	5.52	2.9/2.1	1/ 0.52/ 8/ 7/ 0.03
2	3.47	2000	1.04	0.7/0.5	1/ 0.03/ 1/ 4/ 0.02
3	1.81	2000	2.05	1.4/1.0	1/ 0.13/ 4/ 5/ 0.02
4	1.01	2000	4.51	1.6/1.1	1/ 0.50/ 8/ 7/ 0.03
5	0.40	2000	3.50	2.4/1.8	1/ 1.0 / 32/ 10/ 0.05
6	0.15	2000	4.02	2.8/2.0	1/ 3.0 /100/ 22/ 0.11
PPO	1.07	2000	6.03	1.6/1.2	1/ 0.64/ 8/ 7/ 0.03
PEO	0.67	2000	2.99	2.1/1.5	1/ 0.05/ 16/ 7/ 0.04

Polym = polymer, Sol = solvent.

Ltd.), at the frequency of 10 Hz unless otherwise stated. Temperature increment was $2^{o}C/min$ when the temperature dispersion was investigated.

3. RESULTS AND DISCUSSION

3.1. Syntheses of Triethoxysilyl Terminated Polyethers from Hydroxyl Terminated Ones

As will be described later in Section 3.3 of this paper, hydroxyl terminated organic polyethers, i.e., PTMG, PPG, and PEG failed to react in the sol-gel process with TEOS contrary to the results previously reported[12,26]. We, therefore, prepared organic polyethers in which the both terminals of their polymeric chains were chemically modified from hydroxyl to triethoxysilyl groups. The synthetic route was already shown in Figure 2. Essentially the same route was followed for three kinds of organic polyethers. ET-polyethers carry alkoxysilyl groups, and are expected to react with TEOS.

Figure 3 shows carbon 13 NMR spectra of PTMG (MW = 2000) and AT-PTMO therefrom. Two peaks due to allylic carbons were clearly observed. Allylic protons were also detected by proton NMR, and assayed by this NMR. Due to the relatively small abundance of allyl groups that are present only at the polymeric chain ends, the accumulation over several hundreds times was necessary. Analytical results on AT-polyethers are shown in Table 5. It is noted that almost quantitative conversions were achieved from hydroxyl to allyl groups.

The resulted AT-polymers were subjected to hydrosilylation by triethoxyhydrogensilane using hexachloroplatinic acid (the Speier's catalyst, $H_2PtCl_6(IV)\cdot H_2O$). As a catalyst, Pt-112 (trans$[PtCl_2(Et_2S)_2]$, from Johnson Matthey Chemicals, Ltd.) was also employed, but the former gave more satisfactory results for all AT-polyethers. Proton NMR spectrum of ET-PTMO (MW = 2000) is shown in Figure 4. Allylic protons disappeared, and the protons on ethoxy group appeared at 1.3 ppm (triplet, methyl) and 3.85 ppm (quartet, methylene). Infrared spectra were all in accord with the NMR results.

In case of PEG, the yield of ET-PEO was lower than the other two, although the products from TEOS/ET-PEO were also obtaind and some of their properties were already studied. Therefore, the details of the synthetic reactions and the sol-gel preparation of inorganic glass/poly(oxyethylene) hybrid gels will be reported separately. The physical properties of the gels will also be published. The reported results[12,26] so far were not of the hyblid gels but of physical mixtures, we assume.

3.2. Sol-Gel Reactions

Reaction conditions for the preparation of gels from TEOS /HT-PDMS or one of the polyethers were shown in Tables 2-4. Almost all the reactions were conducted as described in Section 2.3. Typical results of the sol-gel reactions at $50^{o}C$ are shown in Table 6. High pH conditions using catalysis by ammonia were also conducted, but gellings were not observed under the reaction conditions described in these tables. Hence, hydrochloric acid was used throughout this work. The

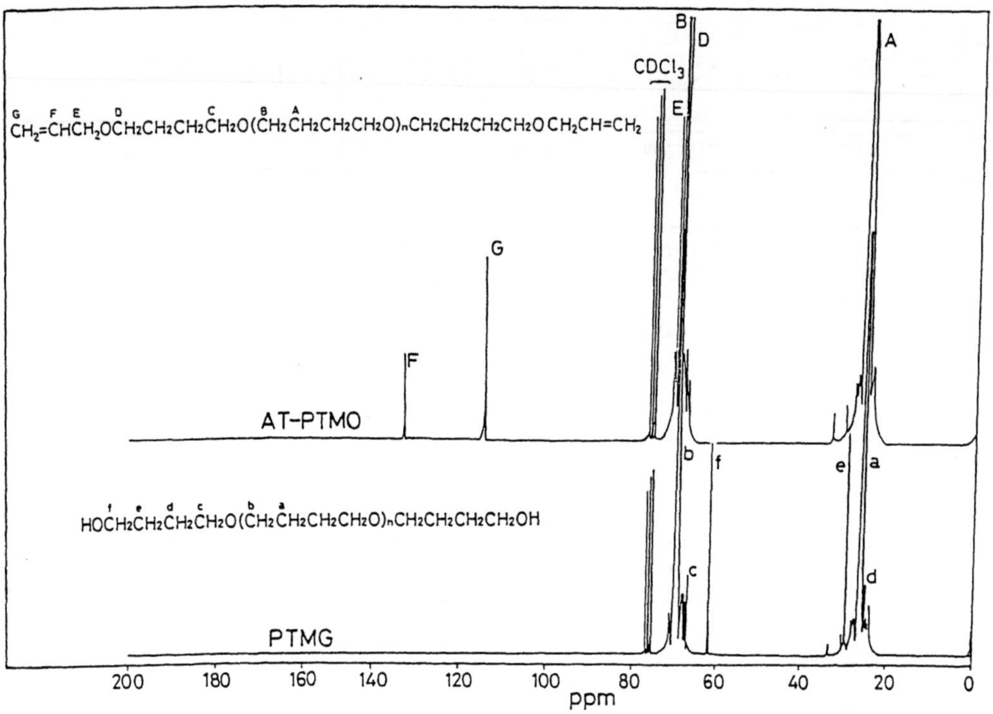

Fig. 3. Carbon-13 NMR spectrum of AT-PTMO
(See no. 2 in Table 5).

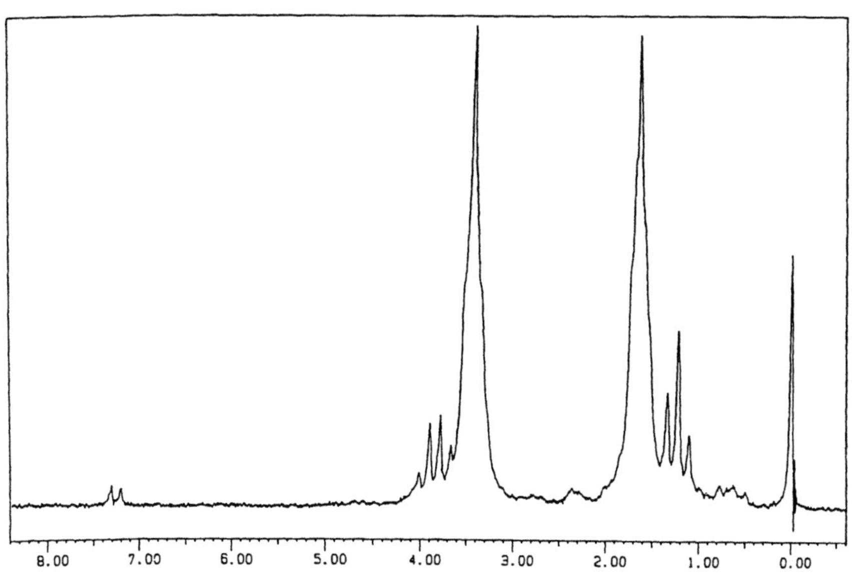

Fig. 4. Proton NMR spectrum of ET-PTMO prepared from AT-PTMO
(See no. 1 in Table 5).

Table 5. Properties of AT-Polyethers

Code	Starting polym	Polym MW	Yield (%)	AT-Polym Mn[a)]	Mw/Mn[b)]	Functionality
1	PTMG	1000	86.5	990	1.37	2.0
2		2000	96.6	1870	1.45	2.2
3	PPG	2000	89.8	1660	1.06	2.0
4	PEG	600	61.4	1490	1.08	1.6
5		2000	50.5	2590	1.08	1.6

a), b) See annotations in Table 1.
Polym = polymer.

Table 6. Results of sol-gel reactions of TEOS/Polymer at 50°C

Code	Polym (nominal MW)	TEOS/Polym/Sol/H_2O/HCl (molar ratio)	Gelation time (h)	Appearance of gel
1	ET-PTMO (1000)	1/ 0.52/ 8/ 7/ 0.03	0.2	Transparent
2	(2000)	1/ 0.50/ 8/ 7/ 0.03	0.4	Transparent
3	ET-PPO (2000)	1/ 0.64/ 8/ 7/ 0.03	0.4	Transparent
4	ET-PEO (2000)	1/ 0.50/ 8/ 7/ 0.03	24	Opaque

Polym = polymer, Sol = solvent.

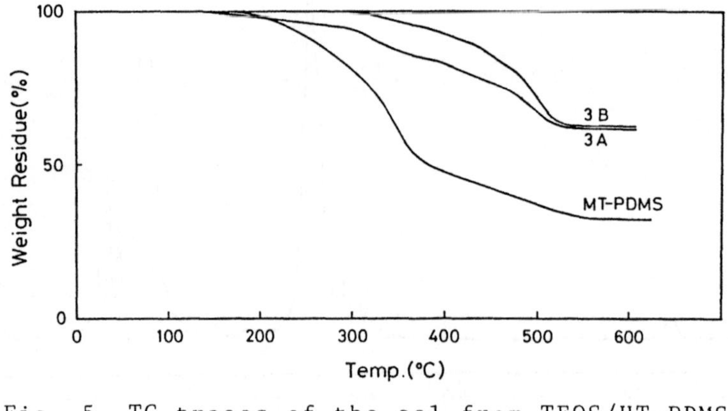

Fig. 5. TG traces of the gel from TEOS/HT-PDMS (See no. 3 in Table 2).

Table 7. Solvent extraction of gel[a] from TEOS/HT-PDMS at 50°C

Sol	Dielectric constant	Sample weight (g)	Wt of extract (g)	Decrease of wt (%)	Residual PDMS (%)
CHX	2.024	0.504	0.048	9.52	86.7
tol	2.381	0.407	0.062	15.2	78.7
CH_2Cl_2	8.93	0.294	0.045	15.3	78.6
THF	7.58	0.363	0.061	16.8	76.5
DOX	2.209	0.349	0.036	10.3	85.6
2-PL	19.92	0.318	0.037	11.6	83.8

a) Sample no. 5 in Table 2.
Sol = solvent, CHX = cyclohexane, tol = toluene, THF = tetrahydrofuran, DOX = dioxane, 2-PL = 2-propanol.

Table 8. The Soxhlet extraction of gels using THF

Sample no.	TEOS/Polymer (molar ratio)	Decrease of wt (%)	Residual polymer (%)
1	1 / 0	0.68	-
2	1 / 0.50 HT-PDMS	16.7	77.5
3	1 / 0.55 PTMG	92.1	0
4	1 / 0.50 ET-PTMO	3.86	94.9
5	1 / 0.52 PPG	93.3	0
6	1 / 0.65 ET-PPO	13.3	84.3
7	1 / 0.57 PEG	95.0	0
8	1 / 0.50 ET-PEO	22.0	73.8

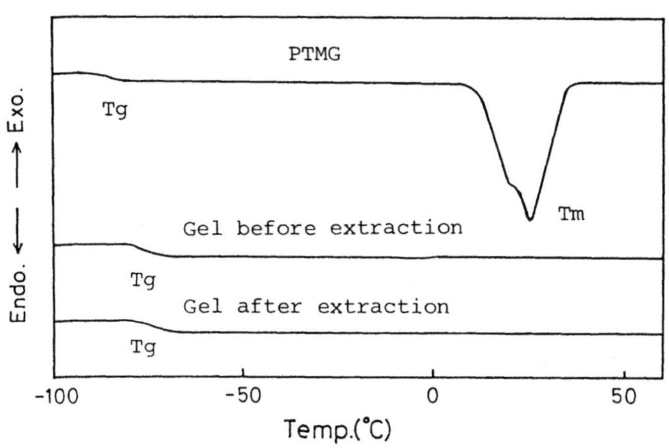

Fig. 6. DSC traces of the gel from TEOS/ET-PTMO (MW 1000).

reaction temperature, i.e., room temperature or $50^{o}C$, did affect the gelation time; several days or several weeks could be shortened to a few tens minutes or a few hours by increasing the temperature to $50^{o}C$. However, the products were almost identical when the feed composition was the same. One drawback of the reaction at $50^{o}C$ was a possibility of bubling, perhaps due to evaporation of solvent.

3.3. Effect of Solvent Extraction of the Gels

In Table 7 are shown the results of solvent extractions of the gels prepared from TEOS and HT-PDMS. About 80 % of PDMS remained in the gel, i.e., a large fraction of PDMS was incorporated into the network formed by TEOS at $50^{o}C$. The extract was found to be PDMS by infrared and proton NMR spectroscopies. The percentages of the insoluble part varied a little from solvent to solvent, but the variation was within 75 and 87 %. This finding is consistent with our assumption that only unreacted polymers were free and solvent-extracted.

Gels from TEOS/PTMG, PPG, PEG or triethoxysilyl terminated polymers of them (ET-PTMO or ET-PPO) were prepared similarly by the sol-gel process. Table 8 shows the results of the Soxhlet extraction on these gels together with that for TEOS/HT-PDMS gel. As already described, only a minor part of HT-PDMS was extracted. However, almost all PTMG, PPG and PEG were removed, that is, they failed to be incorporated into network. There are two possible reasons for the failure of PTMG, PPG or PEG to be incorporated. These polyethers are not reactive enough to silanol groups under the present conditions, or the Si-O-C bonds which are present in the reaction products are not stable enough, leading to bond scission. At present there is not enough evidence to determine which is the case.

The results on TEOS/ET-PTMO, TEOS/ET-PPO and TEOS/ET-PEO systems indicate a drastic decrease of soluble fractions compared with those of TEOS/PTMG, TEOS/PPG and TEOS/PEG systems. Triethoxysilyl groups at the polymer chain ends participated in the network formation during the sol-gel process, and in case of ET-PTMO the efficiencies seemed to be higher than HT-PDMS, i.e., smaller weight losses were observed. It seems that the order of reactivity in the sol-gel process involving TEOS as one component is as follows

$$-Si(OEt)_3 > -Si-OH >> -CH_2-OH$$

The reactivity of the last one, $-CH_2-OH$, may be underestimated because of the lack of stability of the resultant bonds under the sol-gel reaction conditions. Anyway it is very encouraging that organic polymers are found to be useful reacting components in the sol-gel process, if they carry suitable functional groups to react with metal alkoxides or some of the other metallic compounds (metal acetates are now widely tried).

3.4. Thermal Properties of the Gel

Figure 5 shows the TG traces of the gel from TEOS/HT-PDMS before (A) and after (B) solvent extraction together with that of PDMS having trimethylsiloxy terminals (MT-PDMS). The gel from TEOS alone (silica gel) did not show any weight decrease at the temperature range shown in Figure 5, while the gel with PDMS did show the weight loss even after the solvent extraction. The weight loss in the curve A started at about 200°C which coincides well with that for MT-PDMS. The weight decrease of the curve B was shifted to a higher temperature than A. After the extraction the gel is not expected to contain free PDMS, and the rest of the PDMS was resistant to solvent extraction. In other words, the PDMS appears to be bound to the silica network. Therefore, PDMS chains that are incorporated in the silica network seem to be much more heat-resistant than free PDMS.

Figure 6 shows DSC traces for the products containing poly(oxytetramethylene) (MW = 1000) chains. PTMG shows a glass transition (Tg) at -90°C and the melting of crystalline regions (Tm) at 20°C. The gel from TEOS/ET-PTMO did not show melting, and Tg was shifted to -75°C. Poly(oxytetramethylene) chains were bound to the silica network. Thus the shift of Tg towards higher temperature was observed. The immobilization by binding to the network also inhibited the crystallization of poly(oxytetra- methylene) chains in spite of micro-phase separation between inorganic and organic components. Because a major fraction of polymer was incorporated into the network, the solvent extraction had no effect as seen in Figure 6. These results suggest the chemical bonding between the silica network and organic poly(oxytetramethylene) or poly(oxypropylene) chains, i.e., the formation of a new type of inorganic /organic hybrid gel by the sol-gel process, although the clear observation of Tg indicates micro-phase separation between inorganic glass and oraganic polymer portion.

The thermal stability of organic polyether chains was also improved as indicated by TG measurements. The results on TEOS/ET-PPG and TEOS-PTMO gels were essentially the same as those obtained for the gel prepared from TEOS/HT-PDMS (see Figure 5).

3.5. Dynamic Mechanical Properties

The inorganic/organic hybrid gels so far described were obtained as flexible films, except when the TEOS contents in the feed were the highest among the present examples. Additionally, the films were optically clear and transparent, as shown in Figure 7. In cases of ET-PTMOs whose MWs were 2000, the gels were often opaque due to the crystallization, which is seen in Figure 6 for ET-PTMO with MW of 1000.

Dynamic mechanical measurements were carried out in order to elucidate mechanical properties and higher order structures of the elastic gels. Figure 8 shows temperature dispersions of dynamic modulus (E') and dynamic loss (tan δ) of TEOS/ET-PTMO (MW = 2000) measured at 10 Hz. Glass transition (Tg) of PTMO chains is clearly observed at about -60°C. Following this transition, E' displayed a very unusual behavior, i.e., it suddenly increased before the large decrease due to glass transition did not complete. This very unusual increase of E' with temperature is ascribed to the crystallization of PTMO chains. The PTMO chains were amorphous in the gel, and it was freezed by cooling to -100 °C. Upon heating they became to be able to orient to crystalize after the glass transition. In fact, this increase of E' was not observed as shown in Figure 9, if the gel had been annealed before the measurement. The annealing was

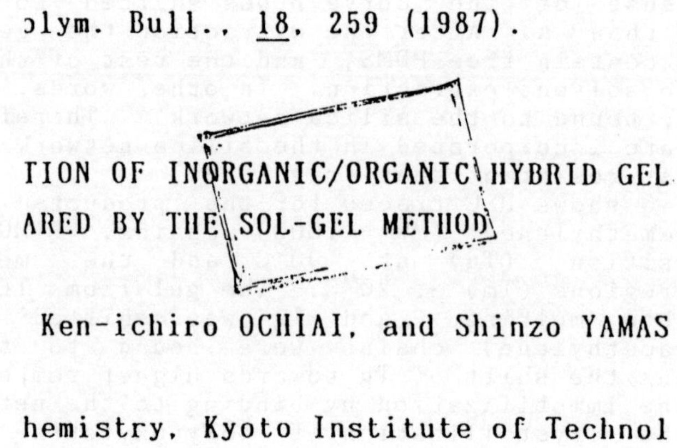

Fig. 7. Transparency of the gel from TEOS/ET-PTMO (MW 2000).

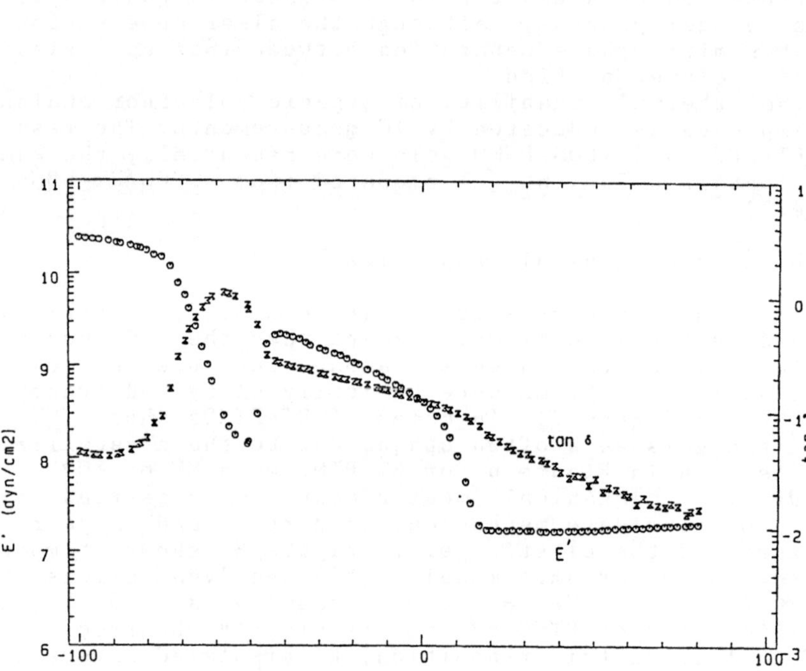

Fig. 8. Temperature dispersion of E' and tan of the gel from TEOS/ET-PTMO (MW 2000) as prepared.

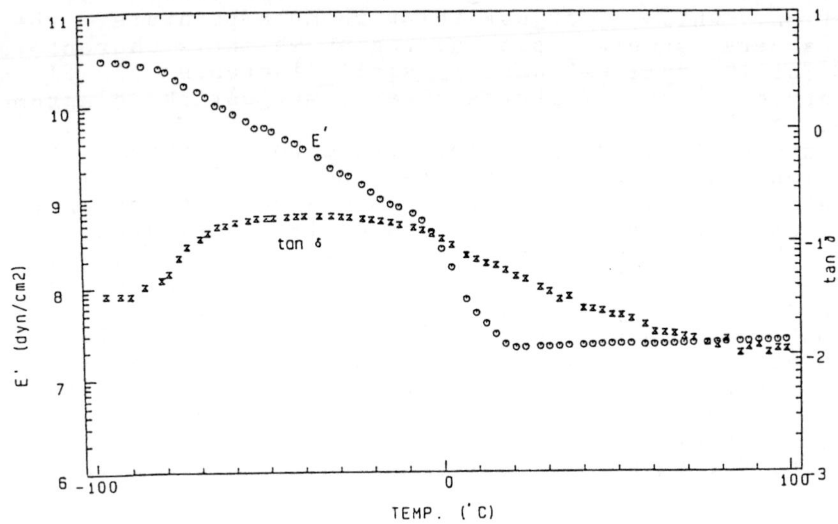

Fig. 9. Temperature dispersion of E' and tan of the gel after annealing.

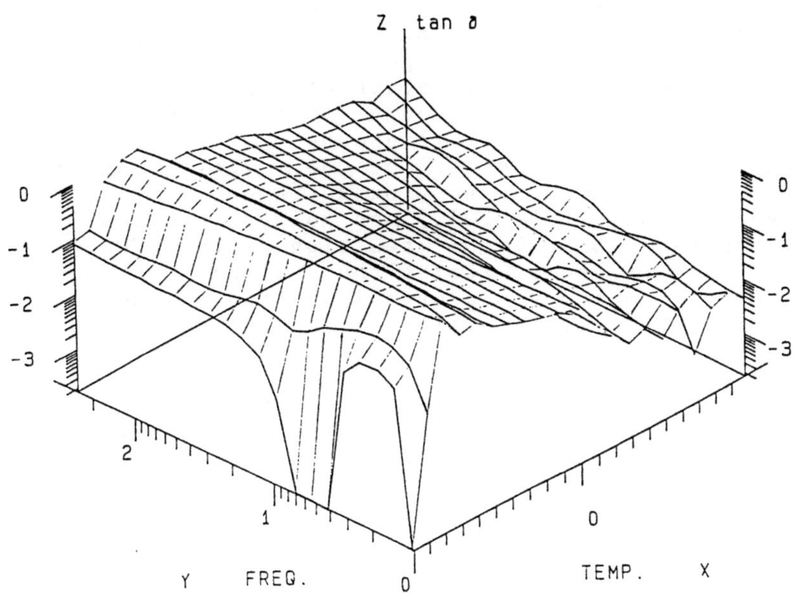

Fig. 10. Dynamic loss of the gel from TEOS/ET-PTMO (MW 2000).

carried out by slowly increasing the temperature up to -30 C or so, and again down to -100°C. Just below room temperature (about 10°C) another decrease of E' was observed, which is probably due to the melting of PTMO crystals. After this transition, which was just below room temperature, the gel shows a very stable plateau region which is characterized by a E' of 10^{7} $dyne/cm^2$ and a small increase of E' with temperature. These features clearly suggest the elastomeric natures of the gel.

Dynamic loss shows a very broad peak between the two transitions as seen in Figures 8 and 9. From these temperature dispersions, it is estimated that the inorganic and organic phases are rather randomly distributed in the gels. The loss was further investigated under various conditions, and the results are displayed in Figure 10. Frequency range was not wide enough, but it is noted that this material shows relatively large and stable loss. Consequently, the gels are supposed to be a candidate of damping materials, i.e., the materials that can attenuate or isolate mechanical vibrations, which are indispensable to protect mechanical and electronics machines of a higher precision. For this purpose further analysis and developments are of course necessary.

4. CONCLUSION

Novel inorganic/organic hybrid gels were prepared using sol-gel process which enabled a low temperature synthesis of inorganic glass. As organic components, polyethers were investigated, and it was found that commercially available hydroxyl terminated polyethers were not incorporated into the silica glass from tetraethoxysilane (TEOS) except hydroxyl terminated poly(dimethylsiloxane). The polyethers were extracted by solvent from the gels. In order to react with TEOS, the chemical modification of organic polyethers was necessary: The hydroxyl groups on the chain ends were converted into triethoxysilyl groups via allyl groups. The modification was found to be very quantitative. Triethoxysilyl terminated poly(oxytetramethylene) (ET-PTMO), poly(oxypropylene) (ET-PPO), and poly(oxyethylene) (ET-PEO) were subjected to sol-gel reaction with TEOS, and the products were found to resist the solvent extraction. In other words, the inorganic/organic hybrid gel seemed to be synthesized. The gels were flexible and transparent. From DSC and temperature dispersions of dynamic modulus (E') and tan, micro-phase separation between inorganic glass and organic polymer was indicated. When the organic component formed a matrix, the materials were elastomeric, and can serve as good damping materials for their future application.

References

1 K.S. Mazdiysni, R.T. Dolloff, J.S. Smith II, J. Am. Ceram. Soc., 52 (1969) 523.
2 H. Dislich, Angew. Chem., 83 (1971) 428.
3 S. Sakka and K. Kamiya, J. Non-Cryst. Solids, 42 (1980) 403.
4 J.D. Mackenzie, J. Non-Cryst. Solids, 48 (1982) 1.
5 S. Sakka, Am. Ceram. Soc. Bull., 64 (1985) 1463.

6 E.M. Rabinovich, J. Mater. Sci., 20 (1985) 4259.
7 L.L. Hench and D.R. Ulrich eds., Science of Ceramic Chemical Processing, John Wiley & Sons, New York, 1985.
8 S. Sakka, Science of Sol-Gel Process, Agune-Shofusha, Tokyo, 1988 (in Japanese).
9 D.R. Ulrich, Chemtech, 18 (1988) 242.
10 T, Yokoo, K. Kamiya, S. Sakka, Nippon Kinzoku Gakkaishi, 27 (1988) 775 (in Japanese).
11 H.H. Huang, B. Orler and G.L. Wilkes, Polym. Bull. 14 (1985).
12 D. Ravaine, A. Seminal, Y. Charbouillot, M. Vincens, J. Non-Cryst. Solids, 82 (1986) 210.
13 H.H. Huang, B. Orler, G.L. Wilkes, Macromolecules, 20 (1987) 1322.
14 H.H. Huang, G.L. Wilkes, Polym. Bull., 18 (1987) 455.
15 Y. Charbouillot, D. Ravaine, M. Armand, C. Poinsignon, J. Non-Cryst. Solids, 103 (1988) 325.
16 R.H. Glaser, G.L. Wilkes, Polym. Bull., 19 (1988) 51.
17 R.H. Glaser, G.L. Wilkes, Polym. Bull., 22 (1989) 527.
18 H.H. Huang, G.L. Wilkes, J.G. Carlson, Polymer, 30 (1989) 2001.
19 M. Fujita, K. Honda, Polym. Commun., 30 (1989) 200.
20 F. P. Baldwin, A. Malatesta (Esso Research and Engineering Co.), U. S. Patent 3366612 (1968); Chem. Abstr., 68, 60386u (1968).
21 H. V. Voight, Kautsch. Gummi Kunst., 29, 17 (1976).
22 S. Yamashita, A. Yamada, M. Ohata, S. Kohjiya, Makromol. Chem., 186 (1985) 1373 and 2269.
23 J.E. Mark, in ref. 7), p.434.
24 J.E. Mark, C.C. Sun, Polym. Bull., 18 (1987) 259.
25 K. Kajiwara, S. Kohjiya, M. Shibayama, H. Urakawa, "Characterization of Gel Structure by Means of SAXS and SANS" in this book.
26 R. Ohta, A. Watanabe, J. Fukunaga, N. Yoshida, Nippon Seramikkusu Kyokai Gakujutsu Ronbunshi, 97 (1989) 213 (in Japanese, formerly Yogyo Kyokaishi).

BIOLOGICALS

EXCITATION PROCESS AND SWELLING OF NERVE FIBERS

Ichiji Tasaki

Laboratory of Cell Biology
National Institute of Mental Health
Bethesda, Maryland 20892

INTRODUCTION

It is now firmly established that nerve fibers swell transiently when they are excited[1-4]. A variety of nerve cells were also shown to undergo rapid swelling when excited[5,6]. Viewed from a colloid chemical standpoint, this transient swelling is a very interesting phenomenon. In my opinion, however, the real significance of this phenomenon lies in the fact that it throws new light on the mechanism of nerve excitation. As is generally known, many present-day neurobiologists employ equivalent electric circuits of the nerve membrane in explaining their experimental data. The physicochemical basis of these equivalent circuits is obscure. Our analyses of non-electrical manifestations of excitation process furnish a basis for an alternative, physico-chemically oriented interpretation of the excitation process.

In this article, we discuss the process of nerve excitation and conduction from the standpoint of polymer and colloid chemistry. This article consists of three parts, I, II and III. In Part I, we describe several basic properties of the impulse carried by nerve fibers. In part II, we describe the methods and the results demonstrating rapid swelling of nerve fibers evoked by electric stimulation. In Part III, we discuss the relationship between the phenomenon of nerve swelling and the mechanism of nerve excitation. This article was written with the understanding that many readers of this book may have limited knowledge of physiology of nerve fibers. Part I is written for the purpose of providing

Polymer Gels, Edited by D. DeRossi *et al.*
Plenum Press, New York, 1991

background information required for discussing the implications of our experimental findings. In Parts II and III, the importance of non-electrical manifestations of the excitation process is emphasized.

Part I. DISTINCTIVE FEATURES OF THE NERVE IMPULSE

(1) *Propagation and collision of nerve impulses*

When an electric shock is delivered to the nerve of a sciatic nerve-muscle preparation of the frog, a twitch is produced in the muscle. It was Helmholtz[7] (1850) who found that the message carried to the muscle by the nerve (i.e. nerve impulse) travels at a velocity far smaller than that of light or sound waves -- namely, at about 25 m/s. When, in the 1930's, single nerve fiber preparations and devices for recording electric signs of the nerve impulse became available, measurements of the velocity of nerve conduction became easy and direct. A simple experiment shown in Fig. 1 shows some of the basic properties of the nerve impulse.

Here, a giant nerve fiber of the squid is immersed in a shallow pool of seawater. To evoke an impulse in the fiber, a pair of platinum electrodes is placed near each end of the fiber (A and B in the figure). A glass pipette electrode (about 2 μm in diameter at the tip) is pushed into the interior of the fiber through the surface membrane. The glass

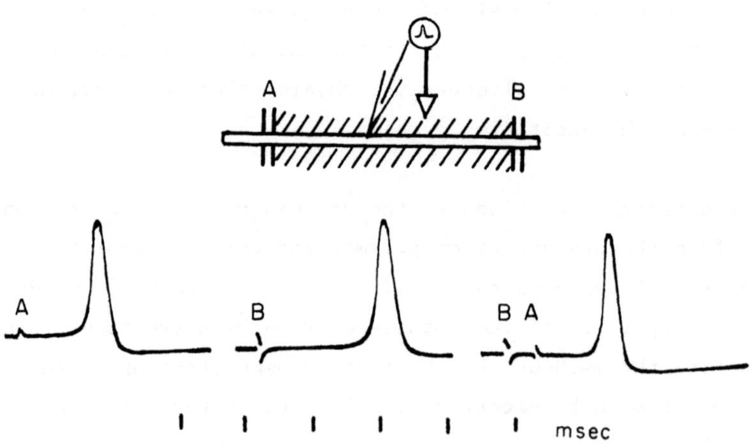

Fig. 1. Action potential of a giant nerve fiber of the squid evoked by a shock delivered by use of electrodes A (left), by electrodes B (center) and by both A and B (left). (Adapted from Tasaki, 1956)

pipette electrode is connected to a device for recording variations of the electric potential associated with conduction of a nerve impulse. When a current pulse is delivered to the fiber with one pair of platinum electrodes, a rapid variation in the potential difference between the fiber interior and the external medium is observed (provided that the pulse is strong enough). The observed potential variation, i.e.the action potential, is evoked by the stimulating current pulse in an *all-or-none* manner.

The interval from the time of delivery of a stimulating pulse to the onset of the action potential varies directly with the distance between the stimulating and recording electrodes. The conduction velocity of the squid giant fiber determined in this manner is about 15 m/s at room temperature. In the nerve fiber with a uniform diameter, the velocity of an impulse propagating in one direction along the fiber is the same as that traveling in the opposite direction. Neither the size nor the shape of the action potential varies along the fiber. Hence, the potential variation, V, at position x along the fiber may be written as

$$V = f(t - x/v) \tag{1}$$

where f(t) denotes the time-course of the action potential observed at x=0 and v the velocity. The negative sign on the right-hand side of equation (1) indicates that the impulse is traveling in the positive direction of x.

Now, we consider what happens when stimulating current pulses are delivered nearly simultaneously at both ends of the fiber. Under these circumstances, an impulse is evoked near each of the two ends and these two impulses propagate toward the middle. Obviously, these impulses are bound to collide at the halfway point of the fiber. The record on the right-hand side of Fig. 1 (marked "B, A") indicates that only one action potential is recorded under these conditions. As long as the second stimulating pulse is delivered at the opposite end of the fiber before the arrival of the nerve impulse, neither the size nor the shape of the recorded action potential is affected by the delivery of the second stimulating pulse. The two impulses are annihilated after head-on collision in the middle.

The phenomenon of annihilation of impulses by collision clearly indicates that the process of nerve conduction is fundamentally different from the propagation of sound or light waves. Whereas sound and light waves behave in accordance with the principle of superposition, this principle can not be applied to the propagation of nerve impulses. The relevance of equation (1) does not indicate the applicability to the nerve of the familiar second-order partial differential equation describing the propagation of a variety of waves. This behavior of the nerve impulse is directly related to the situation that the physical state of the fiber during the rising phase of the action potential is quite different from that during the falling phase.

2) *Ion-antagonism and cation-exchanger property*

The term of "ion-antagonism" was introduced into physiology by Jacques Loeb[8,9]. He noted that excised muscles and nerves lose their ability to carry impulses when they are immersed in an aqueous solution containing a pure salt of monovalent cation, such as NaCl or KCl. Addition of bivalent cation salt, such as $CaCl_2$, to the solution "antagonizes" the poisonous effect of Na or K ions. Ordinarily, solutions containing only divalent cation salts do not sustain the excitability of these tissues. It is important, for the maintenance of excitability, to use proper mixtures, or balanced solutions, of these salts. Loeb had a fairly clear idea about the physicochemical basis of this ion-antagonism.

In 1900, he wrote: "*The salts, or electrolytes in general, do not exist in living tissues as such exclusively, but are partly in combination with proteins or fatty acids. The salts or electrolytes do not enter into this combination as a whole, but through ions. The great importance of these ion-proteid combination (or soaps) lies in the fact that, by the substitution of one ion for another, physical properties of the proteid change.*" In 1906, he said further: "*The normal irritability of animal tissues depends upon the presence in these tissues of Na, K, Ca and Mg ions in right proportion; --- any sudden change in the relative proportion of these ions --- alters the properties of these tissues and gives rise to an activity or inhibition of activity.*"

It is evident from these paragraphs that Loeb had a fairly good understanding of the cation-exchanger property of "irritable" tissues. Extending Loeb's notion, Höber[10] expressed the view that immersion in K-rich media produce "Auflockerung" (loosening) and in Ca-rich media bring about "Verdichtung" (tightening) of the plasma membrane.

More recently, Williams[11,12] examined interactions of alkali and alkali earth metals ions with biopolymers. He states that Na and K ions bind weakly with oxygen anions, RO^-, of biopolymers and that this binding is electrostatic in origin. Ca ions bind also to oxygen anions, but this binding is moderately strong. "*It is becoming clear that metal ion-protein interactions fulfill a great diversity of functions in biology"* (Williams[11], 1966). *"As calcium has a high affinity for many proteins, it can be used as an initiator of structural change, provided that its concentration can be dynamically controlled"* (Williams[12], p.337). *"There is the strong indication that activity owes itself to a cooperative interaction between metal ion and protein"* (p. 333).

It is noteworthy that, based on the consideration of changes in protein hydration, Williams predicted the occurrence of contraction of the nerve fiber in the excited state.

(3) *Local current*

Shortly before World War II, when single nerve fiber preparations became available, a dramatic revival of the old concept of "re-stimulation by local current" took place. Diagram A in Fig. 2 is adapted from Hermann's handbook[13] published in 1879; it illustrates his vision of the lines of current flow in the vicinity of the excited portion (E in the figure) of a nerve fiber. The e.m.f. measured with a pair of reversible electrodes, one inside and the other outside the excited portion of a nerve fiber, is different from that measured in the portion of the fiber at rest. The direction and the intensity of the current generated by this difference

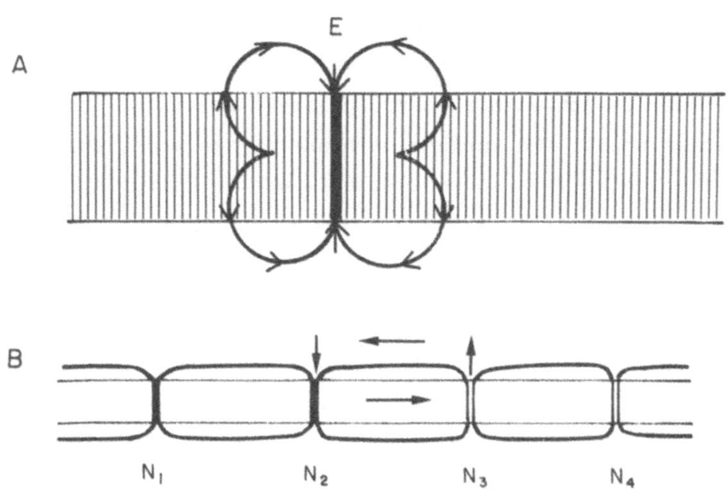

Fig.2. A: Local current in the non-myelinated nerve fiber. E represents the excited area. (Adapted from Fig. 26 in Handbuch der Physiologie, 1879). B: Local current in myelinated nerve fiber. Nodes N_1 and N_2 are in the excited state. N_3 and N_4 are in the resting state.

in e.m.f. are such that the resting portion is thrown also into the excited state. The idea that nerve conduction is mediated by re-stimulation by local current is firmly established in the non-myelinated nerve fiber (see Hodgkin[14] ; Katz and Schmitt[15]).

Large nerve fibers of vertebrate animals (e.g., frogs, rabbits, etc.) are all myelinated. The structure of these fibers is complicated by the existence, outside the protoplasm, of multiple layers of myelin interrupted at more-or-less regular intervals, namely, at nodes of Ranvier. Shortly before World War II, I demonstrated that the myelin sheath has high electric resistivity and, consequently, that electric currents pass predominantly through nodes of Ranvier[16-19]. Diagram B in Fig.2 shows the pattern of the electric current associated with propagation of an impulse. [Note that, in this diagram, the diameter of the fiber (about 12 μm) is drawn disproportionately large as compared with the internodal distance (about 2 mm).]

When a single myelinated fiber is immersed in a large pool of salt solution, the potential variation generated in the external salt solution by the fiber is very small. In the interior of the fiber, however, a large potential variation (about 110 mV in amplitude and 1 - 2 ms in duration) appears when the fiber carries an impulse. When both nodes N_1 and N_2 are in the excited state, there is no appreciable current between these nodes. If node N_3 is at rest, there is a current between N_2 and N_3, its strength being determined by Ohm's law applied to the cylinder of protoplasm between these nodes (roughly 2 nA at the peak). The return path of this current is through the external salt solution. Due to a finite electric resistance of the surface membrane at the node, there is a spread of current to node N_4 (not shown in the figure). There is also a brief current through the myelin sheath between N_2 and N_3 (not shown).

It is to be noted that, in large myelinated nerve fibers, the local current through N_3 is roughly five times as strong as the lowest level required for re-stimulation. Because the nerve fiber carries an impulse with such a large safety factor, it is difficult to block nerve conduction by freezing a short portion of the protoplasm between two neighboring nodes (personal observation). It is possible to to re-stimulate a second nerve fiber by efficiently leading the current developed one fiber (Tasaki[19]).

(4) *Termination of prolonged action potential*

Under a variety of experimental conditions which involve alterations of the monovalent-bivalent cation-concentration ratio outside and/or inside nerve fibers, it is possible to record prolonged action potentials with roughly rectangular time-courses. Fig. 3, left, shows one of the early examples we have encountered. By adding a small amount of the salt of Ni or Co ion to the external medium, we found that the duration of the action potential developed by the node could be markedly prolonged. The end of this prolonged action potential is characterized by a fairly rapid fall of the potential. Since a single node of Ranvier is very narrow (only about 1 μm in width), we can quickly raise the Ca ion concentration outside the node. We found that a prolonged action potential is prematurely terminated when the external Ca ion concentration is suddenly raised. We note, under these conditions, that the nodal membrane tends to assume either one of the two distinct potential levels, the resting level or the excited. A rise in the external Ca ion concentration favors the appearance of the resting potential level.

Spyropoulos[20] applied a brief pulse of heat (either radiant or Joule heat) to the node developing a prolonged action potential. He found that the action potential can be prematurely terminated by suddenly raising the temperature (Fig. 3, right). Based on this and other related experimental

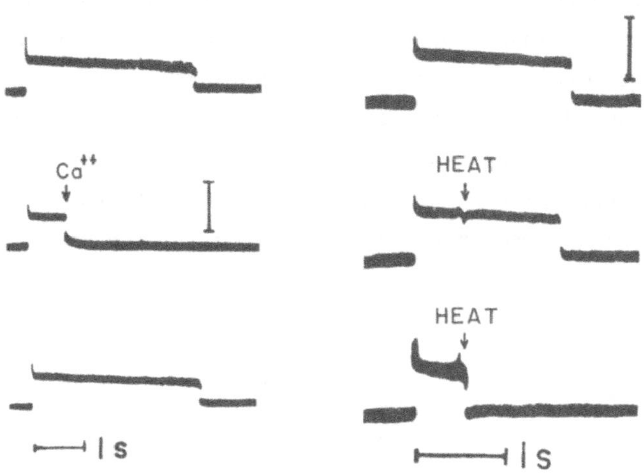

Fig.3. left: Premature termination of the prolonged action potential of a node of Ranvier brought about by a sudden transient rise in the external Ca ion concentration. The current between the excited node and the neighboring resting node was recorded. Right: Premature termination of the action potential of a node by a sudden rise in the temperature. (Adapted from Spyropoulos[20], 1961.)

findings, we arrived at the conclusion that the process of termination of an action potential involves replacement of the monovalent cations in the nerve membrane with Ca ions and that this process is endothermic. A reverse cation-exchange process in the nerve membrane must then be exothermic (see Fig. 6 below).

PART II. RAPID SWELLING IN NERVE FIBERS EVOKED BY STIMULATION

In an early (unsuccessful) attempt to demonstrate rapid mechanical changes in the nerve fiber following stimulation, we employed a simple interferometer (of the Michelson type). With a detector of this type, we were unable to record displacements smaller than the wavelength of the light employed (about 500 nm). Apparently, the detector was not suitable for our purpose. More recently, Hill et al.[21] (1977) employed a laser-beam interferometer and reported that they observed "contraction" of a crayfish nerve fiber.

We noted that the use of a bundle of nerve fibers (instead of a single fiber) greatly facilitates detection of rapid mechanical changes. Fig. 4, left, illustrates an example of our early experiments. Here, a small pebble (about 1 mm in diameter) was placed on a crab nerve. A beam of light from an incandescent lamp, led by means of an optical fiber, was partially interrupted by the pebble. The uninterrupted portion of the light was led to a photodiode. A change in the light intensity, detected by use of a photodiode, was taken as an indication of movement of the nerve surface. This experiment has shown unquestionably that the nerve fibers undergo swelling (not contraction) when an impulse is evoked by an electric shock applied to the nerve.

Later on, we noticed that a Fotonic sensor (Analog Devices, Inc.) is a very useful optical device for measuring movements of the nerve surface. This device consists of two bundles of optical fibers which are mixed randomly and bound together at one end (see Fig. 4, top right). When a small piece of mirror is placed close to the end of the sensor, the intensity of the light transmitted from one (unmixed) end to the other end varies directly with the distance between the mirror and the tip of the sensor. The results obtained by using a Fotonic sensor have also shown that nerve fibers swell when they develop an action potential. The outward displacement of the surface of the crab nerve associated with nerve conduction was found to be 30 - 60 nm.

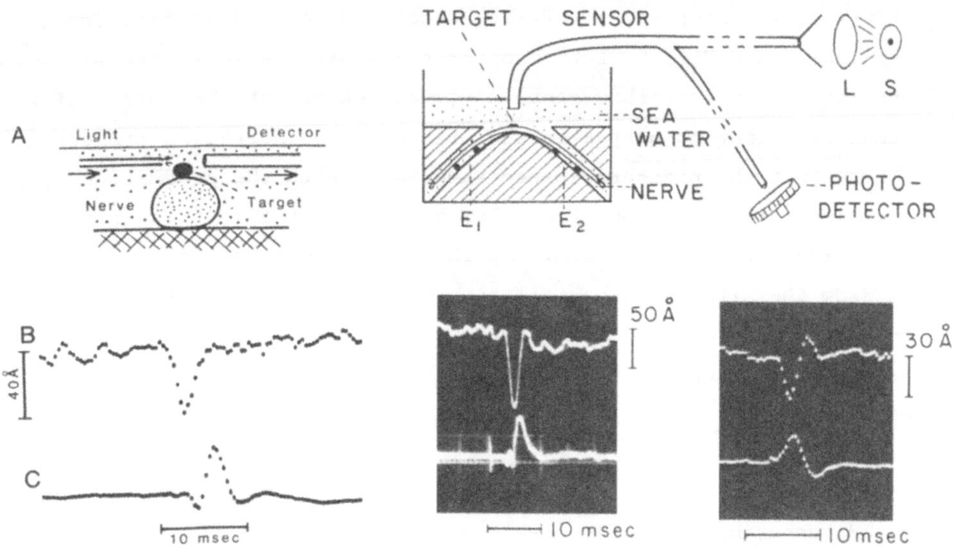

Fig. 4. Left. A: The method of detecting a small movement using a light-obstructing target on a crab nerve. B: Transient decrease in the light intensity evoked by stimulation. C: Externally recorded action potential. A brief electric shock was delivered at the beginning of the time marker. (Reproduced from Science, 210:338, 1980.)

Right. Top: Sensor (Fotonic) represents two bundles of optical fibers mixed at one end. E_1 and E_2 are stimulating and recording electrodes, respectively. Bottom right and left: oscillograph records indicating a transient decrease in the distance between the light-reflecting target on the crab nerve and the tip of the sensor (upper trace) and externally recorded action potentials (lower traces). (Reproduced from J. Physiol., Paris 77:1156, 1981)

The displacement of the surface of the squid giant fiber was also measured by use of a Fotonic sensor. The outward displacement of the surface at the peak of the action potential was found to be roughly 10 A. The initial phase of swelling in the giant fiber was followed by a phase of pronounced shrinkage of the fiber.

The temporal relationship between the action potential and the mechanical changes in the squid giant fiber was examined by using a piezoelectric transducer (piezoceramic bender, Gulton Industries, Inc.). This transducer (see Fig. 5, left) converts a small change in the pressure developed by the nerve fiber into a voltage change at the output of an operational amplifier connected to the transducer. The record presented in Fig. 5 shows that the pressure change associated with the production of an action potential is also diphasic. The peak of swelling (i.e. pressure rise) coincides fairly accurately with the peak of the action potential recorded from the site of mechanical recording. This record indicates also that the large shrinkage occurs during the phase of large hyperpolarization (i.e. intracellular negativity) which is unique to the giant nerve fiber of the squid.

In a recent experiment using olfactory nerve of the garfish, we have re-examined the temporal relationship between the action potential and the mechanical change[22]. The garfish olfactory nerve consists of a large number of small non-myelinated nerve fibers with a highly uniform diameter (about 0.25 μm). We found that, in the vicinity of the site of stimulation, the nerve fibers start to swell simultaneously with the beginning of the action potential. Again, the peak of swelling was found to coincide with the peak of the action potential recorded from the site of mechanical recording. In this nerve, no distinct shrinkage (contraction) was observed following the phase of swelling.

The occurrence of mechanical changes in association with excitation is quite common in the nervous system[5, 6]. We have seen large swelling also in sensory cells in the squid retina evoked by photic stimulation[23].

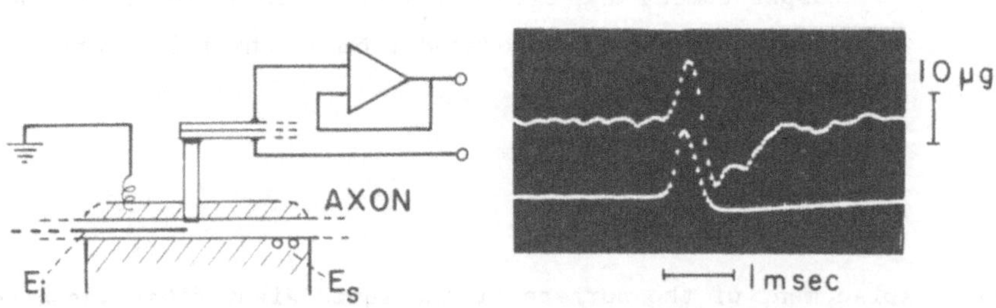

Fig. 5. Left: diagram of the setup used for detecting a mechanical response of a giant nerve fiber of the squid. A Gulton bender is placed on the surface of the fiber. Right: Mechanical response (upper trace) and action potential recorded internally from the site of mechanical recording (lower trace). (Reproduced Biochem. Biophys. Res. Commun. 95:1328, 1980.)

PART III. MACROMOLECULAR APPROACH TO NERVE EXCITATION

In a colloid chemical point of view, a nerve fiber is a macromolecular elastic gel. The problem confronting us now is to clarify the relationship between the rapid physicochemical changes taking place in this gel and the process of nerve excitation. We note that the structure of a nerve fiber is very complex. Electron microscopic studies reveal that the lipid bilayer covering the axoplasm is penetrated by intrinsic proteins and that these proteins are connected to various fibrous elements in the cortical zone of the axoplasm (Metuzals and Tasaki[24], Matsumoto in this book). Within this complex structure, the process of nerve excitation proceeds with a great rapidity. In the giant fiber of the squid, the physicochemical changes occurring in the superficial layer of the fiber do not affect the state of the protoplasm located deep inside the fiber (away from the surface). In garfish olfactory nerve fibers (roughly 0.25μm in diameter), however, the entire population of macromolecules inside the fiber are expected to be exposed to the same, time-dependent changes in the ionic environment. The methods we employ for studying physicochemical changes in nerve fibers do not provide any information concerning the behavior of individual components of the microscopic structures of the nerve fibers.

In anticipation of the following discussion, we have already described several basic properties of the nerve impulse. Several additional well-known facts are to be taken into consideration in the following discussion. At the outset, we stress the following points.

(1) In artificial polymer gels containing numerous anionic groups, replacement of bivalent counter-ions for monovalent cations brings about a pronounced shrinkage (see Katchalsky and Zwick[25]; see Tanaka[26] and his chapter in this book). The mechanical changes observed in the nerve fiber are considered not to be fundamentally different from those seen in these animate colloidal materials.

(2) In nerve fibers immersed in Ca-rich media, the electric resistance of the surface membrane is high. An increase in the monovalent cation concentration in the medium lowers the membrane resistance.

(3) The production of an action potential is associated with a rapid, drastic fall in the membrane resistance. This fall in resistance is considered to be a direct consequence of a rise in the water content,

or swelling, of the superficial layer of the nerve fiber. The potential difference across this layer is altered by an enhancement of the ion mobilities associated with the rise in the water ccontent.

(4) There is a sudden rise in the rate of heat production when a nerve is stimulated. In the garfish olfactory nerve[22], the peak of heat production coincides with the peak of action potential (see Fig. 6). Our recent determination of the quantity of the heat generated by this nerve suggests that roughly 10^{-9} equivalent of Ca ion per gram of wet nerve weight is involved in the Ca-Na exchange[27].

In Part I, we have emphasized that the generation of an action potential by a stimulating current pulse is all-or-none. When only a portion of the surface membrane of a node of Ranvier is excited by a stimulating current pulse of the threshold strength, the remaining portion of the membrane of the same node is thrown into the excited state after a short delay. The occurrence of such a cooperative process may be visualized as arising from the spread of swelling of the membrane elements (domains or macromolecules) by virtue of mechanical and electric coupling between

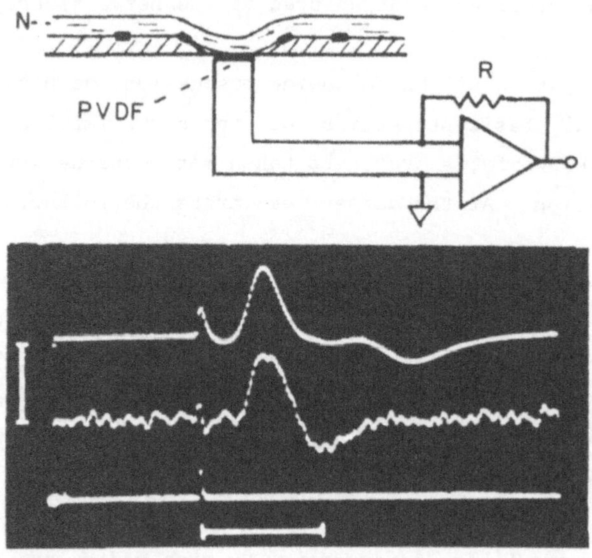

Fig. 6. Comparison of the time-course of heat production (lower trace) and the action potential of the garfish olfactory nerve (upper trace). The rate of temperature rise was taken as the measure of heat production. The action potential was recorded from the site of thermal recording. Time marker, 20 ms. (Reproduced from Biophys. J. 55:1038, 1989.)

different elements. These elements can assume only one of the two discrete, stable states, either the swollen state or the shrunk state.

A long time ago, we proposed a hypothesis that the existence of two discrete, stable states of the nerve membrane derives from its cation-exchanger property (Tasaki[28]). The S-shaped curve in Fig. 7 is a theoretical cation-exchange isotherm for the nerve membrane immersed in a medium containing both monovalent and bivalent cations. In calculation, it was assumed that the occupancy of two neighboring anionic sites by cations of different valency is energetically unfavorable. In the state represented by the portion A--C of the S-shaped curve, the membrane is Ca-rich; D--F represents a monovalent cation-rich state. These two states are stable. The portion C--D represents an unstable, unrealizable state. When the fraction of the monovalent cation is raised to the critical level (C in the figure), either by changing the medium or by the delivery of an outwardly-directed membrane current, the nerve membrane undergoes a *cooperative transition* from its Ca-rich (resting) state to the monovalent-cation rich (excited) state. A reverse transition is associated with an increase in the fraction of the bivalent cation in the nerve membrane. The termination of an action potential is considered to be associated with such a reverse transition.

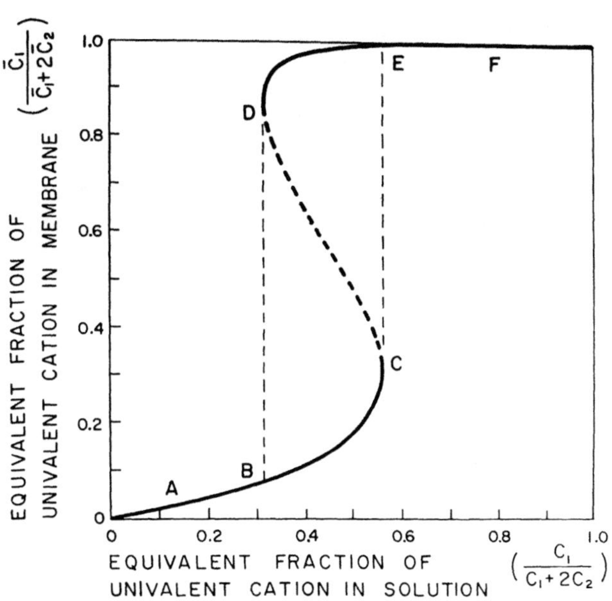

Fig. 7. Theoretical cation-exchange isotherm of the nerve membrane immersed in a salt solution containing C_1 mole of univalent cation and C_2 mole of bivalent cation. (Reproduced from J. Gen. Physiol. 46:770, 1963.)

Quite recently, we have expanded our arguments along this line (Tasaki[19,27]). In addition, we have discussed the process leading to termination of the excited state toward the end of an action potential. [This topic is beyond the space of this article.]

Finally. we make a brief, critical comparison between the present macromolecular approach and the well-known equivalent-circuit approach[29] to the process of nerve excitation.

(1) An electric circuit consisting of two batteries with fixed e.m.f.s, and variable internal resistances, connected in parallel, is equivalent to a single battery with a time-dependent e.m.f. and a variable internal resistance (Fig. 14.5 in Tasaki[19]). This time-dependent e.m.f. is uniquely determined by the choice of the two variable resistances of the equivalent circuit. Hence, the success in evaluating the membrane e.m.f. by use of the voltage-clamp data does not prove that there are separate, independent ion-pathways in the nerve membrane. In other words, the voltage-clamp data are not in conflict with the present macromolecular approach which assumes the e.m.f. to be time-dependent.

(2) Voltage-clamp measurements indicate the existence of a range of membrane potential which gives rise to a negative membrane resistance. Viewed from the macromolecular standpoint, this is a reflection of an increase in the fraction of the membrane area in the excited state with a rise in the membrane potential. In this range, the membrane potential, V, is unstable, in the sense that the rate of its change, dV/dt, increases with a rise in V. The behavior of the nerve membrane at threshold and the rising phase of the action potential is governed by this instability of V.

(3) The dependence of the conduction velocity and the fiber diameter and on the membrane resistance is adequately described by an equation derived by assuming two stable states of the membrane (Matsumoto and Tasaki[30]). This equation adequately describes the rates of impulse conduction in squid giant fibers internally perfused with dilute solutions of various monovalent cation salts.

(4) The process of nerve excitation is a cooperative phenomenon. It is possible that tetrodotoxin, a powerful nerve poison, manifests its effect at extremely low concentrations by interfering with the cooperativity of the nerve membrane. Large hydrophobic cations, such as

tetraethylammonium, are expected to affect the cation-exchange process by binding to the anionic sites in the nerve membrane and suppressing the degree of swelling (or the rise in the water content) of the nerve membrane in its excited state. The effects of these agents on the process of excitation do not prove the existence in the nerve membrane of separate pathways for different monovalent and bivalent cation species.

CONCLUSIONS

(1) Nerve fibers undergo rapid, reversible swelling when they generate an action potential. The swelling starts at about the same time as the action potential. The peak of swelling coincides with the peak of the action potential.

(2) The production of an action potential is accompanied by a rise in heat production. The peak of heat production coincides roughly with the peak of the action potential.

(3) Both swelling and heat production is considered as a consequence of replacement of monovalent cations, Na and K, for Ca ions bound to the anionic sites in the nerve membrane.

(5) The production an action potential is considered to involve a cooperative transition of the membrane macromolecules from its compact state to the swollen state.

REFERENCES

1. K. Iwasa and I. Tasaki. Biochem. Biophys. Res. Commun. 95:1328 (1980).
2. I. Tasaki, K. Iwasa and R. C. Gibbons. Japan. J. Physiol. 30:897 (1980).
3. I. Tasaki and K. Iwasa. Japan. J. Physiol. 32:69 (1982).
4. I. Tasaki and K. Iwasa. Japan. J. Physiol. 32:505 (1982).
5. I. Tasaki, T. Nakaye and P. M. Byrne. Brain Res. 331:363 1985.
6. I. Tasaki and P. M. Byrne. Brain Res. 475:173 (1988).
7. H. Helmholtz. Arch. Anat. Physiol. 1850:276 (1850).
8. J. Loeb. Amer. J. Physiol. 3:327 (1900).

9. J. Loeb. "The Dynamics of Living Matter". Columbia University Press. New York. (1906).

10. R. Höber. "Physikalische Chemie der Zelle und der Gewebe". Wilhelm Engelmann, Leipzig. (1926).

11. R. J. P. Williams. The selectivity of metal-protein interaction, Vol.14 in "Protides of the Biological Fluids," H. Peeters, ed., Elsevier Pub. Co., Amsterdam (1966).

12. R. J. P. Williams. Quart. Rev. 24:331 (1970).

13. L. Hermann. "Handbuch der Physiologie". 1st. Theil. F. C. W. Vogel, Leipzig (1879).

14. A. L. Hodgkin. J. Physiol. (London) 94:560 (1939).

15. B. Katz and O. H. Schmitt. J. Physiol. (London) 97:471 (1940).

16. I. Tasaki. Amer. J. Physiol. 127:211 (1939).

17. I. Tasaki. "Nervous Transmission". Thomas, Springfielkd, Illinois (1953).

18. I. Tasaki. "Handbook of Physiology". Vol.1, Chap.III. J. Field, H. W. Magoun and V. E. Hall. Amer. Physiol. Soc. Bethesda (1958).

19. I. Tasaki. "Physiology and Electrochemistry of Nerve Fibers". Academic Press. New York. (1982).

20. C. S. Spyropoulos. Amer. J. Physiol. 200:203 (1961).

21. B. C. Hill, E. D. Schubert, M. A. Nokes, and R. P. Michelson. Science 196:428 (1977).

22. I. Tasaki, K. Kusano, and P. M. Byrne. Biophys. J. 55:1033 (1989).

23. I. Tasaki, and T. Nakaye. Science 223:411 (1984).

24. J. Metuzals, and I. Tasaki. J. Cell. Biol. 78:597 (1978).

25. A. Katchalsky, and M. Zwick. J. Polymer Sci. 17:221 (1055).

26. T. Tanaka. Sci. Amer. 244:110 (1981).

27. I. Tasaki. Physiol. Chem. Phys. Med. NMR 20:251 (1988).

28. I. Tasaki. J. Gen. Physiol. 46:755 (1963).

29. A. L. Hodggkin, and A. F. Huxley. J. Physiol. (Lond.) 117:500 (1952).

30. G. Matsumoto, and I. Tasaki. Biophys. J. 20:1 (1977).

CYTOSKELETONS ARE FUNCTIONALLY DIFFERENTIATED IN THE SQUID GIANT AXON

Gen Matsumoto, Michinori Ichikawa and Takao Arai*

Electrotechnical Laboratory, Tsukuba, and *Institute of Basic Medical Sciences, University of Tsukuba, Tsukuba, Ibaraki 305 JAPAN

The neuron is a highly differentiated cell for receiving, processing, conducting and transmitting signals. Morphologically distinguishable portions, such as the cell body, dendrites and the axon, are functionally differentiated for this task. The squid giant axon has long been used for physiological and biochemical studies of the various functional sites in the neuron because of its exceptional large size. Most studies have concentrated on the axonal transport and impulse conduction. The central part of the giant axon, the central axoplasm, can easily be extruded with a tiny roller without loss of its electrical excitability and large amounts of transported vesicles are observed in the central axoplasm. These facts suggest that (1) the central axoplasm is functionally differentiated from the peripheral axoplasm, the axoplasm remaining in the axon after the extrusion, and (2) the both parts have distinct rheological properties (for review see Matsumoto *et al.*, 1989).

Classical biologists imagined that (1) the cytoplasm was composed of organella such as mitochondria, endoplasmic reticulum and Goli Apparatus, and the cytosol, a high concentrated solution of water-soluble proteins and (2) spaces among organella were occupied with homogeneous cytosol. But recently, not a few cell biologists become to think that the so-called cytosol is not a homogeneous protein solution but the soluble proteins interact with a semisolid scaffold, cytoskeleton, and organization of the cytoskeletons and their interactions with the soluble proteins affect the rheological properties of the cytosol and the cytoplasm (Porter, 1976; Condeelies, 1983). Three cytoplasmic fibers, actin filaments (microfilaments), microtubules and intermediate filaments, are well known as major elements of the cytoskeletons. These fibers are not a rigid but dynamic structures that dynamic equilibrium is present between the fibers and their constituent proteins. Various investigators reported that depolymerzed and polymerized states of actin filaments contributed the sol- and gel-states of cytoplasm, respectively (Condeelies, 1983).

In order to investigate significance of the cytoskeletons for the regional differentiation and the rheological distinction of the squid giant axon, we compared axonal transport activity, structural organization of the cytoskeletons and localization of the cytoskeletal proteins using video-enhanced differential interference contrast (AVEC-DIC) microscopy, electron microscopy and monoclonal antibodies. The present article will describe these results and discuss significance of the cytoskeletons for the function and rheological properties of the cytoplasm.

Polymer Gels, Edited by D. DeRossi *et al.*
Plenum Press, New York, 1991

MATERIALS AND METHODS

Materials

Nervous tissues (fin fibers, giant axons, stellate ganglia and optic lobes) were collected from squid (*Doryteuthis bleekeri*) maintained in a laboratory tank under healthy conditions (Matsumoto & Shimada, 1980). The tissues were homogenized in buffer A [10 mM 2-(N-morpholino)ethanesulfonic acid (MES) buffer, pH 6.5; 0.5 M NaCl; 1 mM EGTA; 1 mM dithiothreitol; 0.1 mM phenylmethylsulfonyl fluoride; 0.01 mg/ml pepstatin; and 0.01 mg/ml leupeptin] and the supernatants of the centrifugation at 100,000 g for 30 min at 4°C were used as the high-salt extracts of the tissues. In the case of the giant axon, the axoplasm was extruded using a tiny roller. The extruded axoplasm was homogenized in buffer B (0.1 M MES, pH 6.5; 2 mM EGTA and 0.5 mM magnesium acetate). The homogenate was centrifuged at 100,000 g for 30 min at 4°C. The supernatant and the precipitate were used as the labile and the stable fractions of the central axoplasm, respectively. Microtubules were prepared from optic lobes and the labile fraction of the central axoplasm according to the method of Arai (1983).

Video Microscopy and Electron Microscopy

Transported vesicles and microtubules in the intact axoplasm of the giant axon were observed using video-enhanced differential interference contrast (AVEC-DIC) microscopy (Allen *et al.*, 1981). The isolated giant axon was immersed in natural sea water which was bubbled sufficiently with air until use. The aeration is very important for obtaining constant results. Various regions of the axoplasm were examined as a function of distance from the axolemma by changing the focal plane (Ichikawa & Matsumoto, 1986).

The peripheral axoplasm of the giant axon (the remaining axoplasm in the sheath after extrusion with the roller) was chemically prefixed with chemical fixative (cacodylate buffer, pH 7.4; 2.5% glutaraldehyde, 2% formaldehyde, 0.5% tannic acid and 1 mM EGTA) after an intraaxonal perfusion with standard internal solution [20 mM N-2-hydroxyethylpiperazine-N'-2-ethanesulfonic acid (HEPES) buffer, pH 7.25; 380 mM KF and 4% glycerol] or a solution containing 20 mM colchicine. Then, the sample was cut into small pieces and postfixed with 1% OsO_4, embedded in Epon 812, thin-sectioned with a diamond knife and examined as described previously (Tsukita *et al.*, 1986).

The rapid-freeze, deep-etch replica method are performed according to Tsukita *et al.*, (1986). The opened axon on a specimen holder was rapidly frozen by touching its inner face against a pure copper block cooled to 4°K by liquid helium using an Eiko freezing apparatus, RF-10 (Eiko Engineering, Mito, Japan). The frozen samples on the holder were transferred to a cryokit (FTC/LTS-2) fitted to a Sorval MT-2 ultramicrotome and cut with glass knives at -120°C to remove the outermost metal-contact surface of the sample. After cutting, the sample with a mirror-smooth face was transferred to liquid nitrogen in a small cup and put into an Eiko freeze-etch device, FD-3 (Eiko Engineering) together with the specimen holder using a carrier with a cooled cover. In the device the exposed surface was deeply etched in vacuo (1×10^{-7} mmHg) at -95°C for 8-10 min followed by rotary-shadowing with platinum (30°) and carbon (90°). The sample was then removed from the freeze-etch device and immersed in bleach (almost 100% sodium hypochloride). The replicas floating off the sample were washed three times with distilled water and picked up on formvar-filmed grids. The replicas were examined in a JEOL 1200EX electron microscope equipped with a tilting stage at an accelerating voltage of 100 kV.

Production of Monoclonal Antibodies and Immunochemical Procedures

Monoclonal antibodies against MAPs, tubulins and neurofilament proteins were produced by the fusion of mouse myeloma (Ag.8-x63-6.5.3. or

PAI) cells with mouse spleen cells immunized with optic lobe microtubules or the stable fraction of the central axoplasm by the *in vivo* or *in vitro* immunization procedure according to Galfre *et al.* (1977) with some modifications (Arai & Matsumoto, 1988a, b). Monoclonal antibodies against chick (DM1A) and rat (Tub 2.1) β-tubulins were obtained from Amersham (code 357), and BioMakor (code 6068), respectively.

Quantitative measurement of reactivities with monoclonal antibodies was performed by two enzyme-linked immunosorbent assay (ELISA) procedures, the microplate and dot-immunobinding procedures, as described previously (Arai & Matsumoto, 1988b). Specific reactivity with monoclonal antibodies was examined using the Western blotting method. Proteins separated on 3-12% gradient gels were electrophoretically transferred to a double layer of nitrocellulose membrane filters according to Towbin *et al.* (1979). The nearest filter to the gel was used for immunoblotting and the other for protein staining by Aurodye (Janssen).

RESULTS AND DISCUSSION

Regional Differentiation of Axoplasm in the Squid Giant Axon

We investigated axonal transport activity in the various regions of the intact giant axon at the different distances from the axolemma using AVEC-DIC microscopy. We found that the axoplasm can be divided into three regions, the subaxolemmal, cortical and central axoplasm. In the subaxolemmal axoplasm, within 2 μm from the axolemma, only a small number of transported vesicles were observed, although numerous microtubules were present in this region. In contrast, large numbers of vesicles were transported in the cortical axoplasm, 2-20 μm from the axolemma. In the inner region more than 20 μm from the axolemma (central axoplasm), we observed numerous small vesicles transported at a much faster rate (2-3 μm/sec). The slower and faster rates correspond to the medium and fast axonal transport rates, respectively (Table 1).

Table 1. Sensitivity of microtubules and axonal transport in the various regions of the squid giant axon.

Axoplasm (Distance away from axolemma)	Axonal transport: size of vesicles	Axonal transport: transport velocity	Axonal transport: sensitivity to colchicine	Microtubules: sensitivity to colchicine
subaxolemmal (0 - 2 μm)	few transported vesicles			resistant
cortical (2 - 20 μm)	~0.5 μm	0.1 - 0.3 μm/sec	resistant	resistant[a)]
central (20 μm-)	~0.1 μm	2 - 3 μm/sec	sensitive	sensitive

5 mM colchicine was added to the solution in which the axon was immersed. a) The cortical microtubules were sensitive to an intraaxonal perfusion of 20 mM colchicine but were resistant to the treatment.

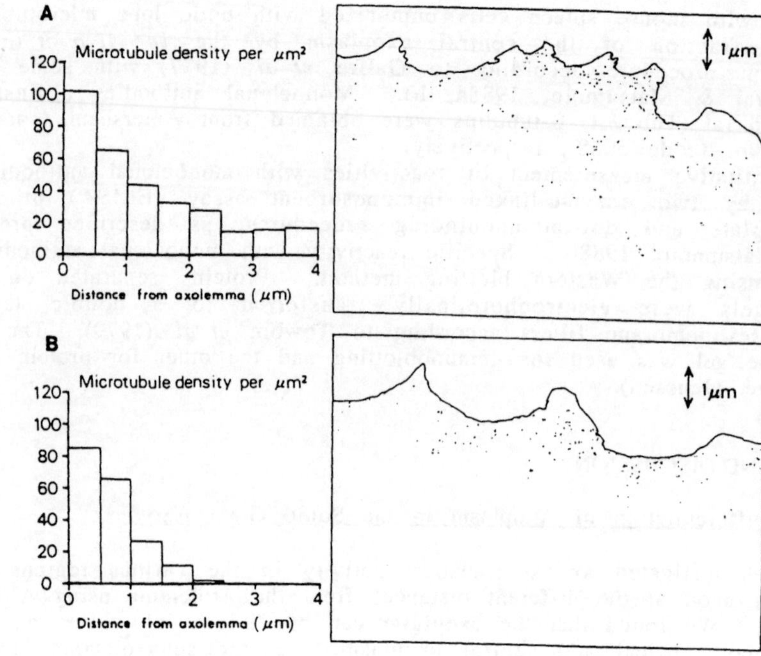

Fig. 1. Distribution of microtubules in the peripheral region of the squid giant axon. The axon was intraaxonally perfused with standard internal solution without (A) or with (B) 20 mM colchicine as described in "MATERIALS AND METHODS".

The medium transport and the cortical microtubules were insensitive to 5 mM colchicine whereas the fast one and the central microtubules were sensitive (Table 1). These results indicate that: (1) the subaxolemmal microtubules are functionally differentiated from axonal transport microtubules localized in the cortical and central axoplasm; and (2) the cortical and central microtubules are involved in the medium and fast axonal transports, respectively.

The difference between the subaxolemmal and cortical microtubules were demonstrated by electron microscopy (Fig. 1). In the transversely cut plane of the axon, the density of the microtubules at the region near the axolemma (subaxolemmal axoplasm) was very high and these microtubules were resistant to colchicine. This density rapidly decreased towards the interior of the axon. In the region more than 2 µm away from the axolemma (cortical axoplasm), most of the microtubules were sensitive to colchicine. This finding suggests that these two types of microtubules, subaxolemmal and cortical, are functionally and structurally differentiated.

Structural Organization of Axonal Cytoskeletons

Subaxolemmal and cortical axoplasm. The subaxolemmal cytoskeleton is characterized by parallel bundles of longitudinally oriented microtubules, which appear to be interconnected by slender strands (Fig. 2). When the cortical axoplasm of the rapidly frozen giant axon is fractured tangentially at a distance ~2 mm away from the axolemma, the replicas obtained are

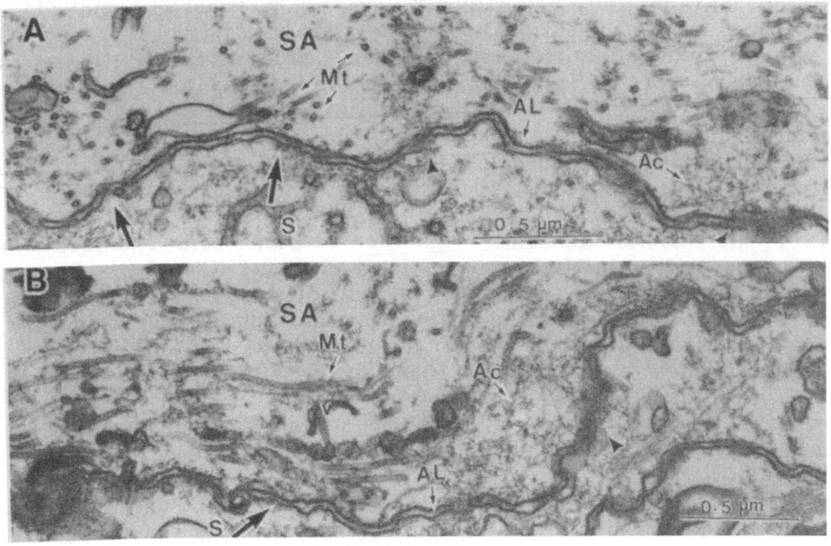

Fig. 2. Transverse (A) and longitudinal (B) sections of the subaxolemmal cytoskeleton of a squid giant axon. The axolemma (AL) of a giant axon is divided into two domains: microtubule-associated (large arrows) and microfilament-associated (arrowheads). Microtubules (Mt) are oriented longitudinally, and some are associated laterally with axolemma. One microtubule runs along the cytoplasmic surface of the axolemma and attaches laterally to the axolemma through slender strands in a side-to-membrane fashion. Microfilaments ~4-5 nm in thickness form spot-like clusters beneath the axolemma. S, schwann cell. SA, subaxolemmal axoplasm.

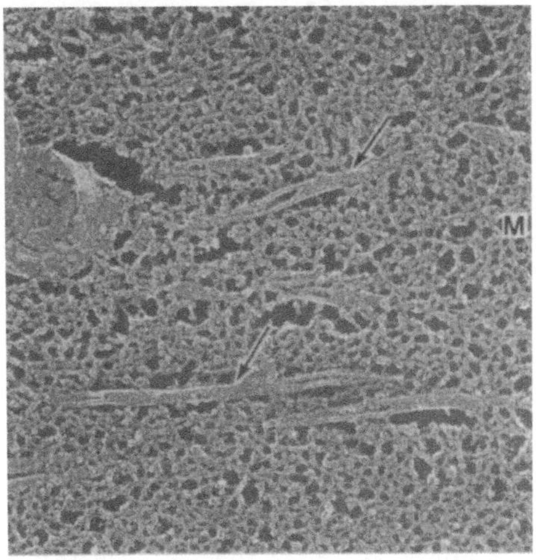

Fig. 3. Freeze-etch replicas of the subaxolemmal axoplasm fractured tangentially at a distance ~2 μm from the axolemma. Before the rapid freezing, the specimen was chemically fixed by perfusing the fixative intracellularly. The fixative was washed out with distilled water containing 10% methanol, then the specimen was rapidly frozen. The replica is characterized by the longitudinally oriented microtubules (arrows) embedded in a well developed, three-dimensional fine meshwork (M).

characterized only by the longitudinally oriented microtubules embedded in a well developed, three-dimensional fine meshwork of the axolinin (Fig. 3). Since axolinin, one of microtubule-associated proteins (MAPs), (see Fig. 7), induces bundling of microtubules *in vitro* in the absense of other MAPs (Murofushi *et al.*, 1983), the axolinin meshwork may arrange the subaxolemmal microtubules into groups.

The axolemma is divided into a microtubule-associated domain and an actin filament-associated domain. The axolemma is covered over 10% ± 5% of its area with the actin filament-associated domains (Tsukita *et al.*, 1986). The bundles of microtubules asociated with the axolemma are excluded from the axolemma near the actin filament-associated domains (Fig. 2). In the actin filament-associated domain, the actin filament cluster is a complicated three-dimensional network in which actin filaments are intertangled (Fig. 4). Some actin filaments exhibit a ~5-nm period on their surface, which may correspond to the native helical pitch of the filaments. Within the network of actin filaments, thin cross-linking strands and globular structures are uniformly distributed. Occasionally, some microtubules are observed to run into the actin filament clusters and to be intermingled with actin filaments.

Central axoplasm. The cytoskeleton of the central axoplasm is characterized by longitudinally oriented microtubules and neurofilament lattice, which are interconnected by slender strands, as shown in Figure 5. Morris and Lasek (1982) found that neurofilaments are a prominent feature of axoplasm both before and after extraction of soluble proteins. Before extraction, neurofilaments have irregular profiles and interconnect to form a three-dimensional network extending through the axoplasm (Fig. 5). After extraction, the cross-linking network is preserved and the neurofilaments in general have smooth profiles. The size of the spaces between adjacent neurofilaments does not appear to change greatly, suggesting that the density of the neurofilament network is unaltered during the loss of the soluble components of axoplasm. Mitochondria and smooth endoplasmic reticulum are

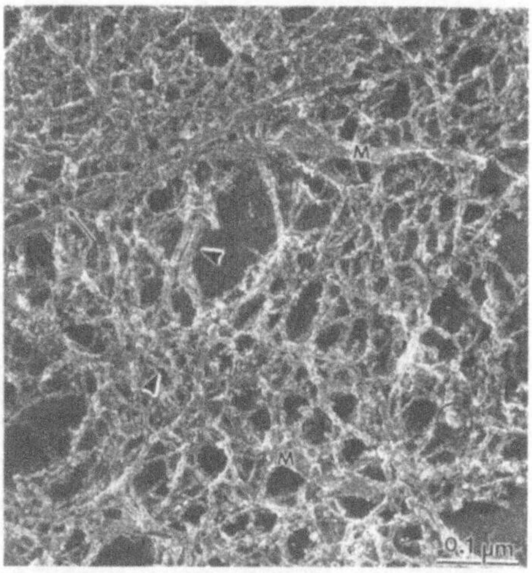

Fig. 4. Freeze-etch replicas of microfilament clusters beneath the axolemma. Some microfilaments show ~5-nm period on their surface (arrowheads), which may correspond to the helical pitch of actin filament. Microfilaments are intertangled, forming a complicated three-dimensional network, held together by thin cross-linking strands. Globular structures ~30-35 nm in diameter showing a horseshoe-like appearance are circled. Some microtubules (M) are observed to run into the microfilament clusters and to be associated with microfilaments through short, slender strands (arrow).

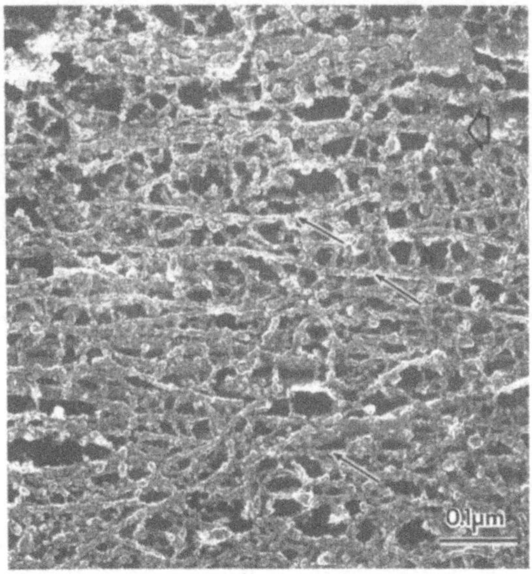

Fig. 5. Freeze-etch replica of the extruded axoplasm of a squid giant axon. The axoplasm is chemically fixed and washed with distilled water before freezing. Note that the longitudinally oriented neurofilaments (arrows) are connected with each other by thin strands, resulting in a ladder-like appearance.

retained in the insoluble fraction as well as a small number of (stable) microtubules. The microtubules either can act as nucleating sites for equilibrium microtubules. Thus it can be concluded that stable polymers constituted mainly of neurofilaments and/or stable microtubules maintain the structural pattern within the cytoskeleton. Actin filaments are difficult to preserve and identify in ultrastructural studies (Lehrer, 1981; Small, 1981; Maupin & Pollard, 1983; McDonald, 1984).

Morphological differentiation of axonal cytoskeletons. From the previous results, the architecture of the cytoskeleton in the squid axoplasm can be shematically drawn in Fig. 6. In the subaxolemmal axoplasm, the plasmalemma (PL) can be divided into two domains, microtubule-associated (RA) and microfilament-associated (RB). The subaxolemmal cytoskeleton is composed mainly of spot-like clusters of microfilaments (AF) and longitudinally oriented microtubules (MT), which are embeded in the fine meshwork consisting mainly of axolinin (AL). The horseshoe-shaped structures (small arrows) on the microfilaments may be the morphological counterparts of the 255 kD protein (see Fig. 7) (Kobayashi *et al.*, 1986). Note three different types of cross linkers between MT and AF (A), MT and PL (b), AF and PL (c).

In the central axoplasm, the cytoskeleton is constituted of longitudinally oriented microtubules (Mt) and a neurofilament (NF) lattice, both of which are interconnected by slender strands. Note the four types of cross-linker strands: 1) between microtubules, 2) between neurofilaments, 3) between microtubule and neurofilament, and 4) between microtubule and transported vesicle (V). In the figure, microtubule and neurofilament territories are indicated by large white and black arrows, respectively.

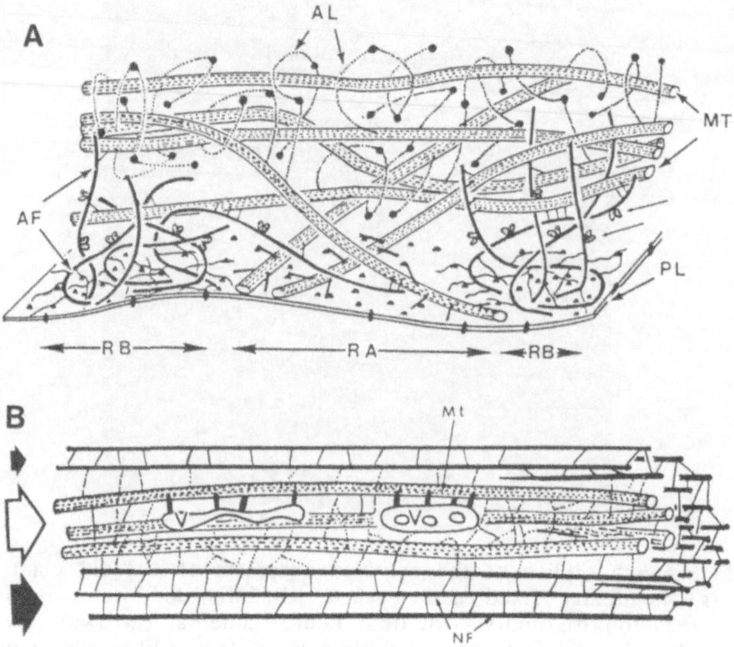

Fig. 6. Schematic drawing of the architecture of the subaxolemmal (A) and central (B) cytoskeletons of a squid giant axon.

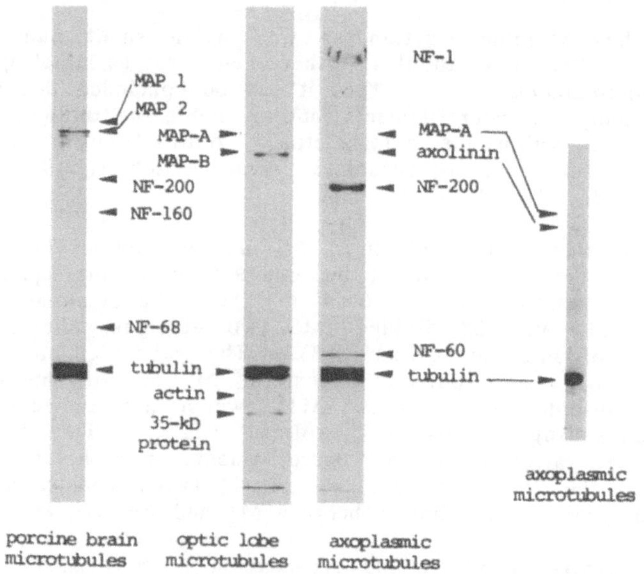

Fig. 7. Analysis of protein species contained in microtubule preparations from squid nervous tissues. Microtubule preparations shown in a and b were prepared as described in "MATERIALS AND METHODS" and were asembled in the presence of 0.45 M sodium glutamate (Sakai & Matsumoto, 1978), respectively. Protein species were analysed by SDS-polyacrylamide gel electrophoresis using 3-12% gradient gel.

Localization of Cytoskeletal Proteins in the Differentiated Regions

Cytoskeletal proteins in the squid nervous tissues. Protein species contained in microtubules prepared from optic lobes and the central axoplasm of the giant axon were examined by SDS-polyacrylamide gel electrophoretic analysis according to Laemmli (1970) (Fig. 7). The optic lobe microtubules were composed of MAP A (300 kD), MAP B (280 kD) (Arai & Matsumoto, 1988a) and α- and β-tubulins whereas the axoplasmic microtubules contained axolinin (280 kD) (Kobayashi *et al.*, 1986; Tsukita *et al.*, 1986) instead of MAP B. In addition to these microtubule proteins, neurofilament proteins [NF60 (60 kD), NF200 (200 kD) and NF1 (>400 kD)] (Morris & Lasek, 1982) and actin were present in these preparations.

Production of monoclonal antibodies. Monoclonal antibodies against these cytoskeletal proteins were produced from spleen cells immunized with the optic lobe microtubules or stable fraction of the central axoplasm by the *in vivo* or *in vitro* immunization procedure as described in "MATERIALS AND METHODS". The properties of monoclonal antibodies used in this study are summarized in Table 2.

Localization of microtubule proteins. The squid giant neuron was divided into 4 parts: (1) cell body (stellate ganglion), (2) peripheral axoplasm, (3) labile fraction of the central axoplasm, and (4) the stable fraction of the

Table 2. Properties of monoclonal antibodies against cytoskeletal proteins of the squid neuron used in this study.

Monoclonal antibodies	Squid protein recognized	Antigen immunized	Parent myeloma	References
NC4B4	axolinin	axolinin[a)]	PAI	Arai & Matsumoto (1988b)
F6F10	MAP A			
O1A2	MAP A & B	optic lobe microtubules[a)]	X63.6,5,3	Arai & Matsumoto (1988a)
E6B6	β_1-tubulin			
DM1A	β_1 & β_2-tubulins			
Tub 2.1	β_1* & β_2*-tubulins		commercially available	
TA143		porcine brain microtubules[a)]	X63.6.5.3	
E5115	NF200 &	stable fraction of central		Arai & Matsumoto (to be submitted
E1024	NF1	axoplasm[a)]	PAI	to Cell Motil. Cytoskeleton)
IV2B4		stable fraction of central		
IV2D9	NF200	axoplasm[b)]		

Antigens a) and b) were immunized by *in vivo* and *in vitro* procedures, respectively. β_1* and β_2*-tubulins are modified forms of β_1 and β_2-tubulins, respectively.

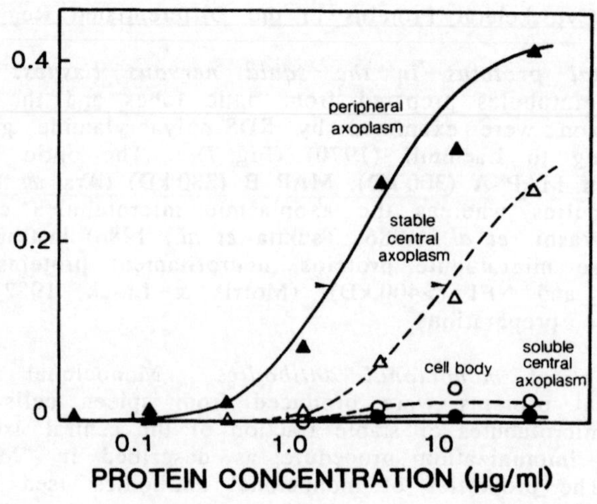

Fig. 8. Subcellular localization of axolinin in the squid giant neuron. Distributions of axolinin in the cell body (O), peripheral axoplasm (▲), and labile (●) and stable (Δ) fractions of the central axoplasm were measured by the microplate procedure using anti-axolinin monoclonal antibody, NC4B4. The relative contents of the protein were determined using the method described by Arai & Matsumoto (1989) where the optical densities adopted are indicated by the arrow heads.

central axoplasm. MAPs contained in these parts were examined by immunochemical procedures using monoclonal antibodies, F6F10 (anti-MAP A), O1A2 (anti-MAP A and MAP B) and NC4B4 (anti-axolinin). Fig. 8 shows that the relative contents of axolinin in these parts were <1/100, 1, <1/100, and 1/6, respectively. In contrast, MAP B was localized in the cell body. MAP A was mainly localized in the cell body and labile fraction of the central axoplasm.

There were significant differences between the chemical properties of MAP B and axolinin. Axolinin was insoluble in low NaCl solution and soluble in 0.5 M NaCl (Kobayashi *et al.*, 1986). To solubilize the protein in an SDS solution, >4 M urea was required. Axolinin bound numerous sugar chains recognized by Concanavalin A. In contrast, MAP B was soluble in a low NaCl solution and in the SDS solution without urea. The different subcellular localization of axolinin and MAP B and the difference between the chemical properties of the two proteins suggest that cell body microtubules consist of tubulin isotypes different from the axoplasmic microtubules.

To study the presence of tubulin isotypes in the squid neuron, reactivities of anti-tubulin monoclonal antibodies with optic lobe (mainly composed of cell body of neurons) and fin fibers (bundles of axons of small neuron) were compared by the microplate procedure. It was found that the reactivities were not always in parallel in both the optic lobe and fin filber preparations. This suggests that these monoclonal antibodies recognize different tubulin isotypes and the isotypes are distributed differently in the cell body and axon. The results of the β-tubulin isotypes can be reasonably explained by hypothesizing that (1) the squid neuron posseses two kinds of β-tubulin isotypes, β_1 and β_2; (2) DM1A recognizes both isotypes while E6B6 reacts only with β_1; (3) Tub 2.1 recognizes modification forms of the isotypes; and (4) the optic lobe contains both isotypes while fin fiber contains only β_1-tubulin.

Table 3. Subcellular distribution of β-tubulin isotypes in the squid giant neuron.

Subcellular regions	Reactivity with E6B6	DM1A	Tub 2.1	β-tubulin isotypes localized
cell body	0.42	0.90	0.09	β_1 and β_2
peripheral axoplasm	0.72	0.99	0.17	β_1
stable central axoplasm	0.35	0.55	0.12	β_1
labile central axoplasm	0.80	1.00	0.50	β_1 *

Reactivity was measured by the microplate procedure and was indicated by OD450. Properties of these antibodies are summarized in Table 2.

The above hypothesis was verified by examining the reactivities of these monoclonal antibodies with tubulin fragments using the Cleveland digestion methods (Cleveland *et al.*, 1977) (Arai & Matsumoto, 1988a; Arai *et al.*, 1988). As expected from the hypothesis, (1) the optic lobe and fin fiber contained two and one β-tubulin fragments recognized by DM1A, respectively; (2) E6B6 recognized only one fragment of both nervous tissues; and (3) Tub 2.1 reacted with the same optic lobe tubulin fragment as DM1A and did not react with the fin fiber tubulin fragment.

To investigate the subcellular localization of the β-tubulin isotypes in the squid giant neuron, reactivities of these three monoclonal antibodies with various parts of the giant neuron were examined by the microplate procedure (Table 3). Cell body microtubules consist of β_1- and β_2-tubulin isotypes whereas microtubules of the peripheral and central axoplasm consist of β_1 isotype and its modified form, respectively. This and other results shown in tables 1 and 3 indicate that microtubules localized in the different regions of the giant neuron are functionally and structurally differentiated. Their localization and possible function are shown in Fig. 9.

Neurofilaments and actin filaments. Protein species of the central axoplasm of the giant axon were examined by SDS-polyacrylamide gel electrophoresis (Fig. 10). The figure suggests the presence of structurally differentiated actin filaments in the central axoplasm. Two actin binding proteins, 255-kD protein (Kobayashi *et al.*, 1986) and spectrin (Morris & Lasek, 1982; Gilbert & Sloboda, 1986), are reported. We found that 255-kD protein and spectrin were associated with labile (lane 2) and stable (lane 3) actin filaments, respectively.

To investigate the structural differentiation of the neurofilaments, we studied their stability and protein species in the cell body, and the peripheral and central axoplasms using anti-NF200 monoclonal antibodies which recognize phosphorylated forms (E5115, IV2B4, and IV2D9), a nonphosphorylated form (E1024), and a nonphosphorylated site (TA143) of NF200 and NF1 (see Table 2). We found that (1) cell body neurofilaments are sensitive to a high NaCl solution whereas those in the peripheral and central axoplasm are relatively and very resistant, respectively; (2) the main component of the cell body neurofilaments is NF60 and the high molecular weight components, NF200 and NF1, are not phosphorylated; (3) the axoplasmic neurofilaments contain

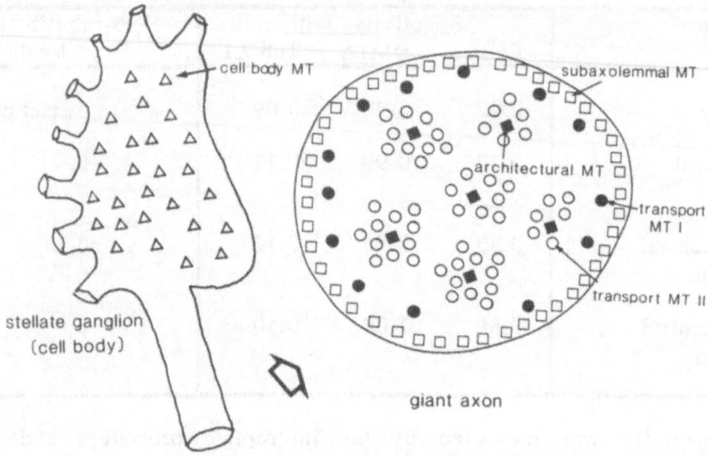

Fig. 9. Localization of differentiated microtubules in the squid giant neuron. Transported microtubules I and II are involved in the medium and fast axonal transports, respectively. Architectural microtubules support the structure of the axon. Functions of microtubules localized in the cell body and the subaxolemmal region are not known yet.

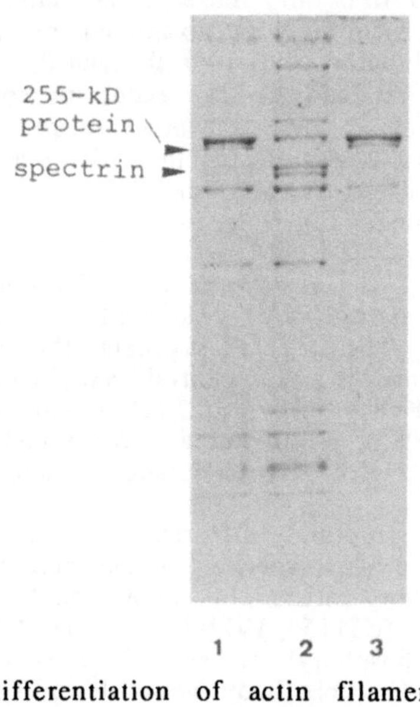

Fig. 10. Structural differentiation of actin filaments present in the central axoplasm of the squid giant axon. 1, stable central axoplasm; 2, resistant fraction of 1 to 0.5 M NaCl; and 3, labile fraction of 1 to 0.5 M NaCl.

Table 4. Structural differentiation of neurofilaments in the squid giant axon.

Subcellular regions	Reactivity with E5115	TA143	E1024	IV2B4	IV2D9	NF200 localized
Cell body	0.03	0.31	0.59	0.01	0.04	nonphosphorylated from
peripheral axoplasm	2.3	2.5	1.4	0.41	0.22	phosphorylated form I
central axoplasm	3.1	2.8	1.5	1.2	1.8	phosphorylated form II

Reactivity was indicated by OD450. Properties of these antibodies are summarized in Table 2.

large amounts of phosphorylated high molecular weight proteins; and (4) NF200 of the central neurofilaments might be further modified as compared to that of the peripheral ones. The last conclusion was based on the fact that monoclonal antibodies produced by *in vitro* immunization procedure recognize specifically the central proteins. The modified ones are thought to be very labile during *in vivo* immunization. These results are summarized in Table 4.

Functional and structural differentiation of cytoskeletons, and its significance for regional differnetiation. Microtubules are involved in a wide variety of biological phenomena such as the determination of cell shape during morphological differentiation, maintenance of the differentiated cell shape, chromosome movement during mitosis, axonal transport, regulation of membrane functions, and movement of cilia and flagella (Schliwa, 1986). One cell contains several kinds of MAPs and tubulin isotypes and the different subcellular distribution of these proteins is demonstrated in some cells (McKiethan & Resenbaum, 1984). Miller *et al.*, (1987) showed that the lobster axon contained two distinct populations of microtubules, transport and architectural microtubules. These facts led us to consider the functional and structural differentiation of individual microtubule (Fulton & Simposon, 1976; Cleveland & Sullivan, 1985). This study clearly shows the presence of at least four functionally differentiated microtubules in the squid giant neuron and the close relationship between the function and structure of these microtubules.

We further demonstrated the presence of structurally differentiated nerofilaments and actin filaments in the axon. Taken together with above findings, these facts strongly suggest that these cytoskeletons play important role in the regional differentiation. Electron microscopic observation showed that the subaxolemmal cytoskeleton formed gel (see Fig. 3). This indicates that the cytoskeleton contributes the gel-like properties of the subaxolemmal axoplasm.

SUMMARY

Video-enhanced microscopic observation of transported vesicles reveald that the squid giant axon was composed of three functionally differentiated regions, the subaxolemmal axoplasm for membrane functioning, the cortical axoplasm for intermediate axonal transport of larger vesicles and the central axoplasm for fast axonal transport of smaller vesicles. By electron microscopic observation and immunochemical study using monoclonal antibodies, we found that organization of cytoskeletons was specialized in these regions and cytoskeletons localized in the different regions were composed of distinct protein components. These findings strongly suggest that cytoskeletons in these regions are functionally and structurally differentiated and that the distnict cytoskeletons play important role in the regional differentiation.

ACKNOWLEDGMENTS

The authors thank Drs. Shoichiro Tsukita (Natl. Inst. of Physiol.), Takaaki Kobayashi (Jikei University, School of Medicine) and Yoshio Hanyu (Keio University, Dept. of Physics) for stimulating discussions and for critical reading of the manuscropt.

REFERENCES

Allen, R.D., Allen, N.S., & Travis, J.L. (1981) Cell Motil. 1, 291-302.

Arai, T. (1983) FEBS Lett. 155, 273-276.

Arai, T., Ichikawa, M., & Matsumoto, G. (1988) in Dynamics of Microtubules, (Hotani, H., ed.), pp.260-282, The Taniguchi Fundation, Kyoto.

Arai, T. & Matsumoto, G. (1988a) J. Neurochem. 51, 1825-1838.

Arai, T. & Matsumoto, G. (1988b) Hybridoma 7, 583-593.

Arai, T. & Matsumoto, G. (1989) J. Neurochem. 52, 93-100.

Cleveland, D.W., Fisher, S.G., Kirschner, M.W., & Laemmli, U.K. (1977) J. Biol. Chem. 252, 1102-1106.

Cleveland, D.W., & Sullivan, K.F. (1985) Ann. Rev. Biochem. 54, 331-365.

Condeelies (1983) in Modern Cell Biology, Vol.2, (Satir, P., ed.) pp.225-240, Alan R. Liss, New York.

Fulton, C., & Simpson, P.A. (1986) in Cell Motility, (Goldman, R., Polland, T., & Rosenbaum, J., eds), pp.987-1005, Cold Spring Harber Laboratory, New York.

Galfre, G., Howe, S.C., Milstein, C., Butcher, G.W., & Howard, J.C. (1977) Nature, 266, 550-552.

Gilbert, S.P., & Sloboda, R.D. (1986) J. Cell Biol. 103, 947-965.

Ichikawa, M., & Matsumoto, G. (1986) in Proceedings of the First International Symposium for Science on Form (Ishizaka, S. ed.), pp.85-90. KTK Scientific Publishers, Tokyo.

Kobayashi, T., Tsukita, S., Tsukita, S., Yamamoto, Y., & Matsumoto, G. (1986) J. Cell Biol. 102, 1699-1709.

Laemmli, U.K. (1970) Nature 227, 680-685.

Lehrer, S.S. (1980) J. Cell Biol. 90, 459-466.

Matsumoto, G., & Shimada, J. (1980) Biol.Bull. 159, 319-324.

Matsumoto, G., Tsukita, S., & Arai, T. (1989) in Cell Movement, Vol.2: Kinesin and Microtubule-Associated Proteins (Warner, F.D., McIntosh, J.R., eds.). pp.335-356, Alan R. Liss, New York.

Maupin, P. & Pollard, T.D. (1983) J. Cell Biol. 96, 51-62.

McDonald, K. (1984) J. Ultrastruct. Res. 86, 107-118.

McKiethan, T.W., & Rosenbaum, J.L. (1984) in Cell and Muscle Motility (Shay, J.W., ed.). pp.255-288, Plenum Press, New York.

Miller, R.H., Lasek, R.J., & Katz, M.J. (1987) Science 235, 220-222.

Morris, J.R., & Lasek, R.J. (1982) J. Cell Biol. 92, 192-198.

Murofushi, H., Minami, Y., Matsumoto, G., & Sakai, H. (1983) J. Biochem. 93, 639-650.

Porter, K.R. (1976) in Cell Motility, (Goldman, R., Pllard, T., & Rosenbaum, J., eds.) p.1, Cold Sprin Harber Laboratory, New York.

Sakai, H., & Matsumoto, G. (1978) J. Biochem. 83, 1413-1422.

Schliwa, M. (1986) in Cell Biology Monographs, Vol.13: The Cytoskeleton, An Introductory Survey. Springer-Verlag, Heidelberg.

Small, T.V. (1981) J. Cell Biol., 91, 695-705.

Towbin, H., Staehelin, T., & Gordon, J. (1979) Proc. Natl. Acad. Sci. USA 76, 4350-5354.

Tsukita, S., Tsukita, S., Kobayashi, T., & Matsumoto, G. (1986) J. Cell Biol. 102, 1710-1725.

SYNTHETIC VITREOUS BODY OF PVA HYDROGEL

Aizo Yamauchi

Research Institute for Polymers and Textiles

1-1-4 Higashi, Tsukuba, Ibaraki 305 Japan

INTRODUCTION

Among optical diseases of the eye, those diseases concerned with the cornea and the crystalline lens can be treated by transplanting of human cornea and by extraction of cloudy lens and implantation of plastics intraocular lens (IOL). So far, however, there has been no method developed for curing diseases related to the anomaly of the vitreous body, such as the clouding of vitreous and detachment of retina. In recent years, replacement of diseased vitreous has been of clinical interest and the use of synthetic materials as silicone oil has been investigated for this purpose. But these materials have not yet sufficient properties for clinical use.

At the Research Institute for Polymers and Textiles we have long studied the crosslinking of poly(vinyl alcohol) (PVA) in aqueous solution, forming a hydrogel or a network polymer highly swollen by water. Then, jointly with doctors of Nara Medical University, we have studied its application as a synthetic vitreous body and performed experiments with albino rabbits, in vivo and in vitro conditions (see references).

PREPARATION OF SYNTHETIC VITREOUS BODY

As shown in Fig. 1, the hydrogel was prepared by irradiating gamma-ray to 7 % aqueous solution of PVA, sealed in ampoules, and then extracting the unreacted portion of the polymer. The polymer used had an average degree of polymerization of 2,000 and a degree of saponification of almost 100. The gel obtained in this way was a transparent, fluid material which was easily handled by means of a syringe through needle. After sterilizing by autoclave heating, the gel was replaced in the eyes of rabbits, from which a portion of vitreous had been extracted.

The compatibility with vitreous was examined from ophthalmological and histological findings as well as the measurement of infrared spectrum, refractive index and ultraviolet absorption.

An image for the nature of the gel is given in the following photograph (see Fig. 2). We should mention that this method became possible only since we developed fluid hydrogel. We can now perform operations with almost no failures.

As shown in Fig. 3, the plot of refractive indices against swelling rations is an almost linear relation and the line just crosses the shadowed area covering the data for natural vitreous, suggesting that PVA gels could

Polymer Gels, Edited by D. DeRossi *et al.*
Plenum Press, New York, 1991

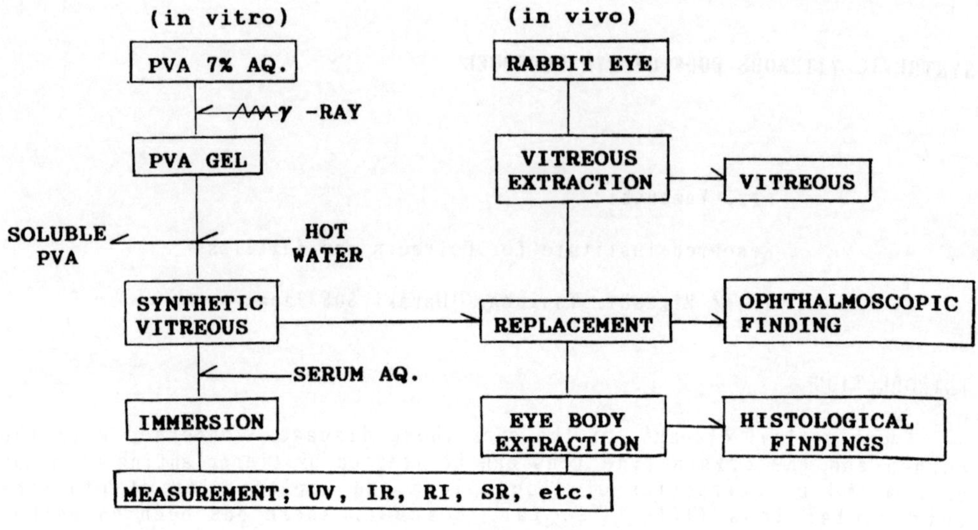

Fig.1 FLOW SHEET OF EXPERIMENT

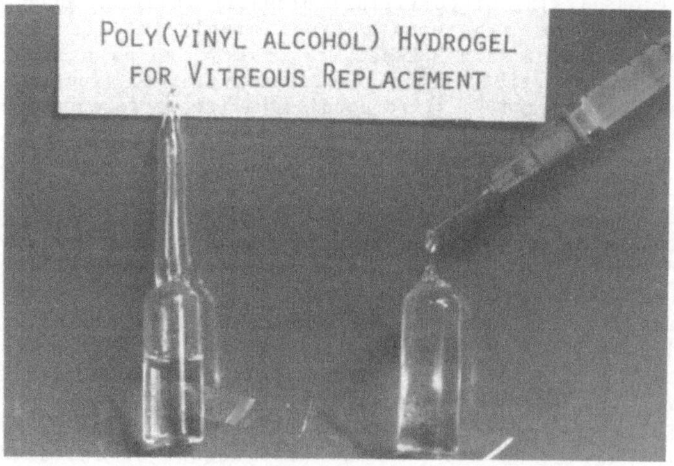

Fig. 2 PVA Hydrogel for Vitreous Replacement

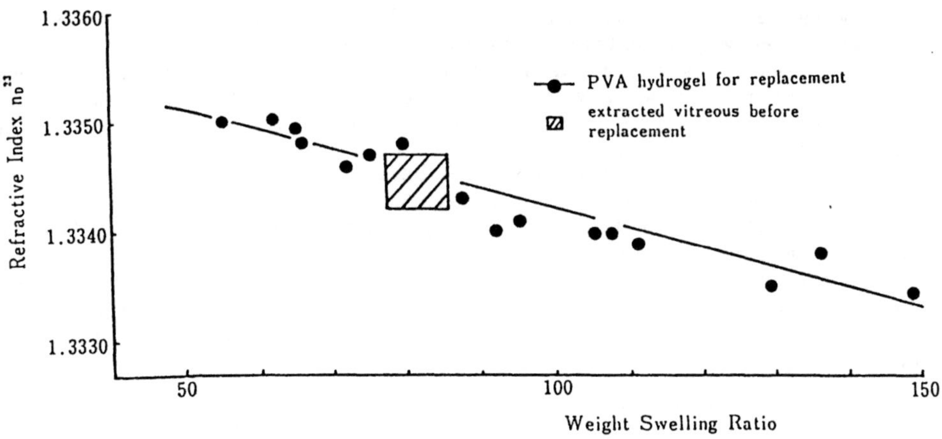

Fig. 3 Relation between Refractive Index and Weight Swelling Ratio of PNA Hydrogel

Table 1 Some Properties of Vitreous Body and PNA Hydrogel

	Vitreous		PVA Hydrogel
	Albino Rabbit	Human*1	
Weight of Vitreous Body (g)	1.1	4.0	
Water Content (%)	98.8	99	98.3 - 99.3
Swelling Ratio*2	81.3		60 - 150
Refractive Index*3 (n_D^{23})	1.3345	1.334	1.3334 - 1.3350

*1 : O'. W. Richards and E.B.Hague ; Amer. J. Ophth., 55,151 (1963)
*2 : Swelling ratio = natural or swollen weight / dried weight
*3 : Water : n_D^{23} = 1.3327

be a useful material as a substitute for the vitreous body. The degree of swelling, about eighty has been achieved by the dose, 0.6 Mega Rad.

Table 1 summarizes some properties of vitreous in comparison with corresponding properties of PVA gel, used for replacement. As a whole, the synthetic gel has similar values to natural vitreous and in particular the refractive index is very close with each other.

SUBSTITUTION OPERATIONS

The amount of rabbit vitreous is so small that operation for it has to be made with special concentration by skilled hands.

Fig.4 illustrates the cross-section of eye. The vitreous body, numbered by 6, comprises a large part of eye filling up the whole space between the lens, 4 and the retina, 7. The vitreous body should always be transparent but a serious disease results, when opacity develops. Also, if the volume of retina shrinks for some reason, the contact between the retina and the choroid is detached, causing another diseases.

The substitution operation is made by means of a syringe, as shown in the Fig.4, through sclera, choroid and retina. First, a portion of vitreous, ranging from 0.1 to 1 ml, is extracted through a needle and replaced by the same or a smaller amount of PVA gel. Before and after these operations the injection points were treated by diathermy to cure the damage.

EVALUATION OF SYNTHETIC VITREOUS BODY

The clinical evaluation was made by ophthalmoscopic observation, ocular tension measurement, electroretinography, etc., and a group of results is shown in Table 2-1 and 2-2. The table 2-2 includes the histological finding in the last column too. In these cases, ophthalmoscopic and histological findings were almost normal.

The compatibility between PVA hydrogel, replaced gel and vitreous body was studied by infrared spectra. As you see in spectrum (see Fig.5), a big difference lies between the specter of vitreous and the original gel. However, after the gel was embedded 6 days in rabbit eye, the spectrum of extracted PVA gel turned to be very similar to the spectrum of vitreous itself, implying the components of vitreous transferred and merged very well into the synthetic polymer gel.

Fig.6 shows the refractive indices itself of the gel and surrounding vitreous after substitution operation. As you see the values are very close with each other at every measuring time showed a tendency to increase initially and comes down to the original level with in a half day time. After a day or two, they tended to go up significantly but recovery to the normal level was observed within a fortnight or month, though it not shown in this figure. Such a long range change is a medical problem and doctors generally say it is the process of physiological phenomenon which frequently occurs after various kind of operations.

In order to evaluate the effect of PVA hydrogel on the retina of Albino rabbits. Gel was substituted to left eyes and saline solution to right eyes each 10 eyes a group. Clinical and histological findings of groups were evaluated 1, 2, 3 and 4 months of after the operation. In Table 3, the results were summarized. From the table, during 1 and 2 months there are no differences between right and left eyes. However after 3 months the tissues surrounding PVA gel is slightly different from saline.

In the Fig.7, photograph 1 is retina of ocular fundus before operation and 2 is that one month after PVA gel substitution operation. In No.2 the photograph shows the capillaries on the surface of retina, observed through vitreous body containing PVA gel. We can see the bottom of the eyes clearly and transparently without deflection, and maybe rabbit can see us clearly without any deformation, also.

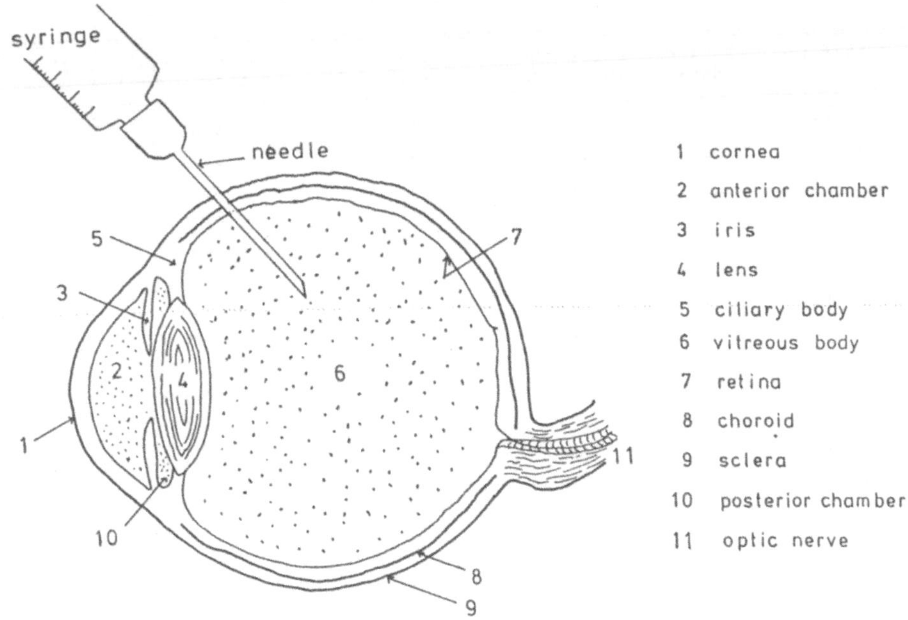

Fig. 4 Cross Section of Eye

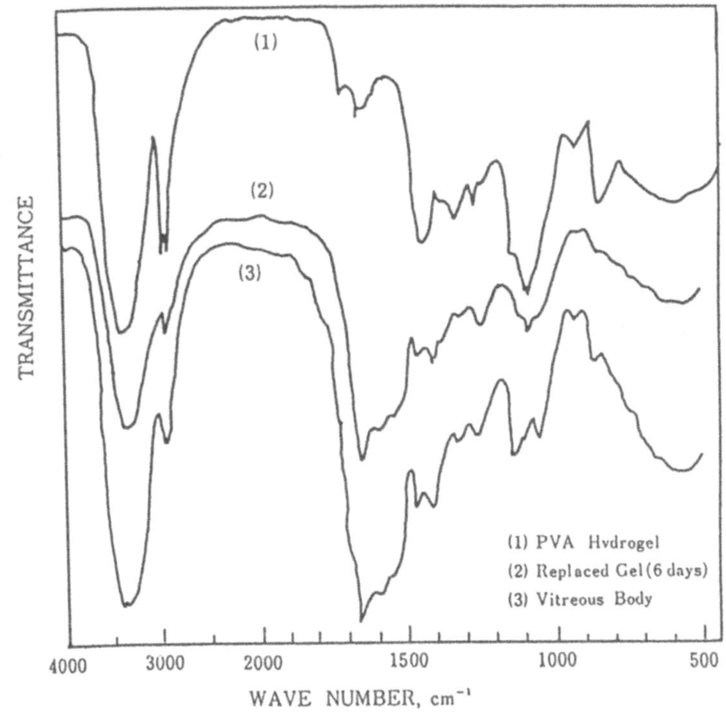

Fig. 5 Infrared Spectra of Vitreous Body and Replaced PVA Hydrogel

Table 2-1 Ophthalmological Findings (1)

Rabbit No.	Period (day)	Extracted volume (ml)	Replaced volume (ml)	Change of interocular tension (mm Hg)
19 (R)	62	0.6	0.5	- 5.88
19 (L)	21	1.0	0.5	- 5.87
20 (R)	62	0.8	0.5	- 3.68
20 (L)	18	0.2	0.2	- 2.95
21 (L)	18	0.8	0.5	- 7.09
22 (R)	49	0.7	0.4	+ 1.10
22 (L)	18	0.5	0.3	+ 2.21
23 (R)	49	0.5	0.5	+ 2.20
24 (R)	49	0.8	0.6	- 3.30
24 (L)	21	1.0	0.6	- 4.41
25 (R)	49	0.3	0.3	- 3.67
25 (L)	18	0.7	0.3	- 2.21
26 (R)	21	0.5	0.4	- 1.11
26 (L)	49	0.3	0.3	- 0.44
28 (R)	22	0.6	0.4	- 2.64
28 (L)	15	0.2	0.2	------
Mean (16)	34	0.6	0.4	- 2.52

Table 2-2 Ophthalmological Findings (2)

Rabbit No.	Electro-retinagram	Ophthalmoscopic finding	Histological finding
19 (R)	normal	(-)	(-)
19 (L)	weak	(-)	(-)
20 (R)	weak	(-)	(-)
20 (L)	normal	(-)	(-)
21 (L)	normal	(-)	(-)
22 (R)	normal	(-)	(-)
22 (L)	normal	(-)	(-)
23 (R)	weak	(-)	(-)
24 (R)	weak	(-)	(-)
24 (L)	normal	(-)	-----
25 (R)	negative	(-)	(-)
25 (L)	normal	(-)	(-)
26 (R)	normal	(-)	(-)
26 (L)	normal	(-)	(-)
28 (R)	normal	(+)	-----
28 (L)	weak	(+)	-----
Mean (16)	normal 10 weak 5 negative 1	(-) 14 (+) 2	(-) 13

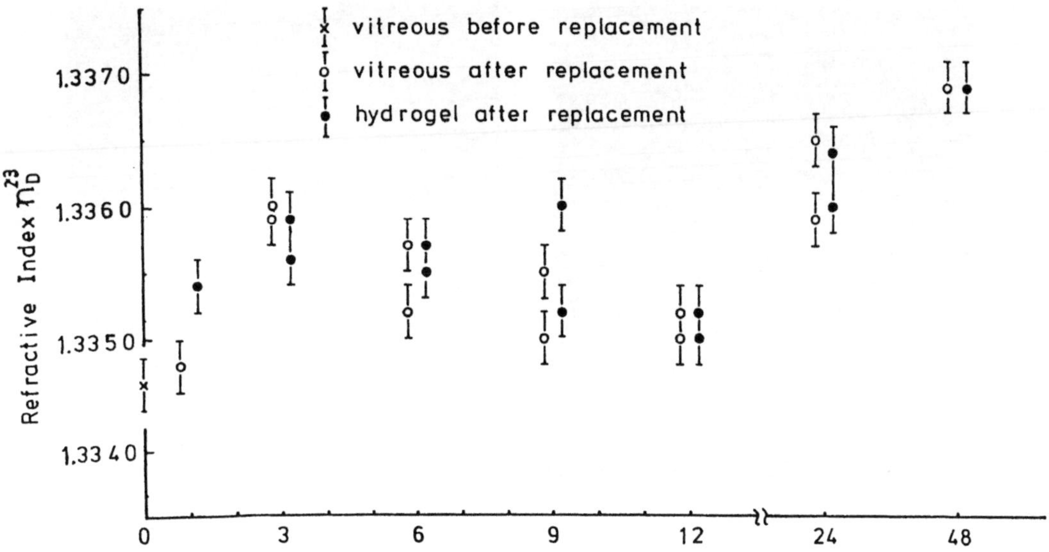

Fig. 6 Refractive Index Value for Hydrogel and Vitreous Body after Substitution Operation

Table 3 Clinical and Histological Findings of Rabbit Eyes Treated with PNA Hydrogel (Month after Operation)

	1		2		3		4	
treatment	PVA	Saline	PVA	Saline	PVA	Saline	PVA	Saline
conjunctival injection	0	0	0	0	0	0	0	0
iris congestion	0	0	1	0	0	0	0	0
cataracta	2	1	2	3	2	1	1	1
vitreous opacity	4	2	1	1	3	1	2	1
retina	0	0	0	0	0	0	0	0
Iris								
edema	0	0	1	0	0	0	0	0
congestion	1	0	0	1	0	0	0	0
cell infiltration	0	0	1	0	0	0	0	0
Ciliary body								
edema	1	0	3	0	4	0	3	2
congestion	1	0	1	0	2	0	1	0
cell infiltration	3(1f)	0	1	0	2	0	2	0
Choroid								
congestion	1	0	0	0	2	1	2(1f)	1
cell infiltration	5(1f)	4(2f)	0	0	4(3f)	1(1f)	1(1f)	0
granulation	1(1f)	2(2f)	0	0	3(3f)	1(1f)	1(1f)	0
Retina								
defect	1(1f)	3(3f)	0	0	3(2f)	1(1f)	2(2f)	0
detachment	5(1f)	2(1f)	0	0	1(1f)	0	2	0
granulation	1(1f)	3(3f)	0	0	2(2f)	1(1f)	2(2f)	0
Sclera								
cell infiltration	2	2(1f)	0	0	4(4f)	0	0	0

PVA: polyvinyl alcohol

f: focal change

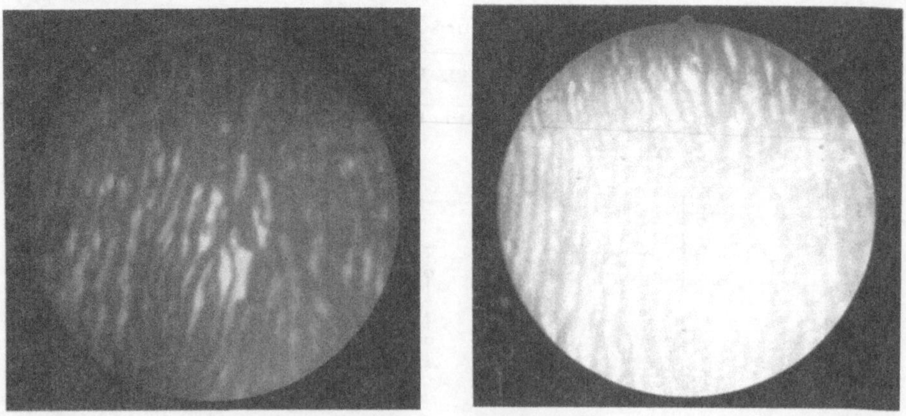

Photo. 1 Before Operation　　　Photo. 2 One Month After Operation

Fig. 7 Photograph of Ocular Fundus

CONCLUSION

To conclude briefly, we have studied the possibility of using PVA gel as a substitute for diseased vitreous body and the result looked to be promising at least when experiments were made for rabbit.

There were two particular points in this experiments. Firstly, the application of gel to eye is not cause any problem of so-call blood compatibility. Secondly, we were able to observe the results even in vivo.

Between PVA hydrogel and saline solution there are not heavy differences. It concluded from the experiments of rabbits that PVA hydrogel will be suitable material for vitreous substitution.

References

Yamauchi, A., Matsuzawa, Y., Nishioka, K., Hara, Y. and Kamiya, S., 1977, PVA hydrogel for Vitreous Replacement, Kobunnshi Ronnbunnshu, 34:261

Hara, Y., Nishioka, K., Kamiya, S., Yamauchi, A. and Matsuzawa, Y., 1977, Replacement of Vitreous by PVA Hydrogel for Albino Rabbit, J. Ophthalmol. Jpn., 28:576

Hara, Y., Nishioka, K., Kamiya, S., Matsuzawa, Y. and Yamauchi, A., 1977, Behaviour of Poly(vinyl alcohol) Hydrogel in Vitreous Body of Rabbit, ibid., 28:1522

Yamauchi, A., Matsuzawa, Y., Hara, Y., Matsushima, S., Nishioka, K., Nakao, S. and Kamiya, S., 1978, Behaviour of Poly(vinyl alcohol) Hydrogel in Vitreous Body of Rabbit (II), ibid., 29:1922

Yamauchi, A., Matsuzawa, Y., Hara, Y., Kamiya, S., Nishioka, K., Saishin, M. and Nakao, S., 1979, Behaviour of Poly(vinyl alcohol) Hydrogel in Vitreous Body of Rabbit (III), ibid., 30:385

Hara, Y., Kamiya, S., Nishioka, K., Saishin, M., Nakao, S. and Yamauchi, A., 1979, Behaviour of Poly(vinyl alcohol) Hydrogel in the Vitreous Body of Albino Rabbits (IV), Acta Soc. Opthalm. Jap., 83:1478

Hara, Y., Hara, T., Hatanaka, O., Hirai, H., Ichiba, S., Kamiya, S., Nakao, S., Saishin, M., Hiasa, Y. and Yamauchi, A., 1984, The Effects of PVA Hydrogel on the Retinas of Albino Rabbit, J. Ophthalmol. Jpn., 35:1340

IMMOBILIZATION OF BIOACTIVE SUBSTANCES WITH PVA SUPPORTS

Hisao Ichijo, Kunihiro Ichimura, Hatsuho Uedaira, Tetsuro Suehiro, Jun'ichi Nagasawa, Aizo Yamauchi, Noboru Aisaka

Research Institute for Polymers and Textiles
1-1-4 Higashi, Tsukuba, Ibaraki 305 Japan

INTRODUCTION

A variety of methods and materials for immobilizing biocatalysts have been proposed and tested. All of the different methods or materials has some advantages and disadvantages. No one technique or support seems to have a clear edge over the others. In any case, the technique or support, which can make the most of the advantage and minimize the disadvantage, should be selected.

The structural network of cross-linked poly(vinyl alcohol) (PVA) gels consists mainly of $-CH_2-CHX-$ segments with alternate backbone carbon atoms bearing alcohol groups (X=-OH). The great abundance of these groups is responsible for the hydrophilic character of the polymers. Goldstein and Manecke stated[1] that supports rich in hydrophilic groups bound on the average considerably larger amounts of protein; in such preparations the bound enzyme retained a higher proportion of its activity and usually exhibited higher activity. It has also been demonstrated in a number of cases that the presence of hydrophilic groups on a support can mitigate the deleterious effects of a hydrophobic polymeric environment. Since many kinds of functional groups are easily incorporated into PVA by acetalization and/or acylation, the polymer can be modified in various ways. So far, immobilized biocatalysts have been used in food and pharmaceutical industries and "not toxic" is very important for industrial application. PVA is really safe, and its physical and chemical properties have well been characterized.

Since fibrous texture generally has a large surface area exposed to solution, they would immobilize a great amount of bioactive substances on their surfaces. There are various kinds of fiber available such as porous fibers, ultra fine fibers, hollow fibers, fibers with a cross section in unusual form, *etc.* Moreover, fibers can be formed into a variety of shapes such as string, cloth, knitted fabric[2], filter paper[3], braid, suitable for each intended application. Therefore, various types of bioreactor can be designed for each intended application by the use of fibrous supports.

Photochemical technique to entrap bioactive substances has the advantages due to milder network formation of polymeric matrix. The technique do not require any chemical modification and any change in pH, temperature *etc.*

In this chapter, we review the characteristics and applications of

Polymer Gels, Edited by D. DeRossi *et al.*
Plenum Press, New York, 1991

both PVA super fine fiber and photocrosslinkable PVA bearing styrylpyridinum group.

PVA SUPER FINE FIBERS (SFF)

As extremely fine fibers have very large surface area per unit mass, a large amount of biocatalysts would be immobilized on their surfaces. If the functional groups incorporated into a fibrous support have an ionic interaction with biocatalysts, the procedure for immobilization and regeneration does not require any chemical modification of biocatalysts.

Preparation of SFF Incorporating Functional Groups

The solution of completely saponified PVA (C-PVA) is mixed with the solution of partially saponified PVA (P-PVA) or poly(ethylene oxide) (PEO) in a 5:5 to 3:7 weight ratio. The mean degree of polymerization (DP) of the P-PVA ranges from 800 to 1700. When DP of the P-PVA used is too low or too high, it is very difficult to spin a fiber. The saponification value of the P-PVA ranges from 85% to 90%. P-PVA with the lower value lessens resistance to heat treatment and C-PVA is not well separated from P-PVA with the higher value. The mixture is extruded through the spinneret, which has plurality of orifices into a vertical cell where jets of spinning solution leaving the spinneret come into contact with a stream of hot air in which water vaporizes. The continuous drawing and heat treatment directly follows the spinning. Those treatments give a fiber crystallinity and orientation. They are accomplished by the combined effects of heat treatment and mechanical stretching of the extruded fibers. When the fiber is washed with water, P-PVA or PEO dissolves in water and only C-PVA exists in a shape of fiber because P-PVA or PEO is not well oriented. After washing, the filament is split into several tens to some hundreds of ultra fine filaments. Although the swelling ratio of the fine fiber is controlled by the degree of drawing and the temperature of heat treatment, the ratio from 2 to 3.5 is desirable. The fiber with the lower ratio does not immobilize enzymes well, while the physical properties of the fiber with the higher value is deteriorated. The process to make SFF [4,5] is illustrated in Fig.1.

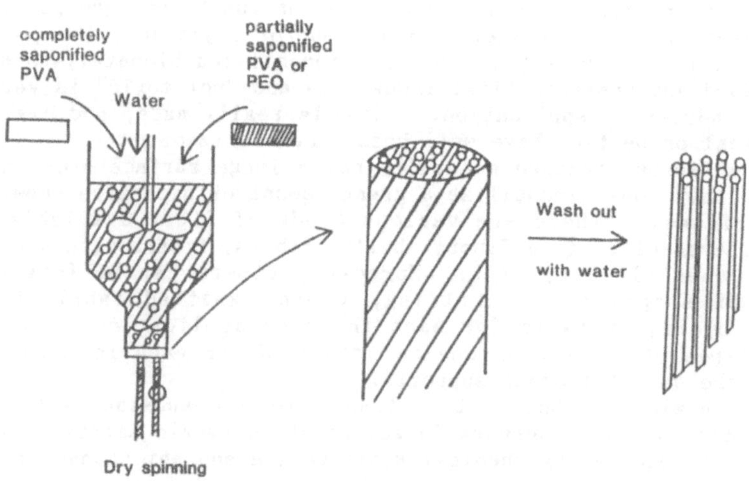

Fig.1 Preparation of PVA super fine fibers

SFF is first acetalized with bromoacetaldehyde to incorporate functional groups. It is then treated with formaldehyde solution not to dissolve under the following process. It is finally treated with amine, sulfite, *etc.* Aminated SFF could be made by another method. After PVA powder is directly acetalized, the PVA solution mixed with an additive is then extruded and drawn[6].

Pore Size Distribution

The pore size distribution of three materials determined by methanol adsorption method is shown in Fig.2[7]. The surface area of SFF is larger than 200 m^2/g and is as large as 10-20% of that of activated charcoal. Besides, relatively large sized pores form a large proportion of pore size distribution of SFF. An ion exchange resin commercially available also has larger surface area than SFF, but the most of their surfaces consists of small sized pores. Since enzyme molecules are generally large, the surface of an immobilization support should consist of large sized pores.

Immobilization of Enzymes

Effect of surface area on immobilization. When both an ordinary PVA fiber and SFF are observed through a scanning electron microscope, SFF is found to be much finer than the ordinary fiber[8]. Since SFF has a larger surface area than an ordinary PVA fiber, it is expected that SFF has much greater ability for immobilizing biocatalysts than an ordinary PVA fiber. There is a great difference in the immobilization rate and immobilization yield of invertase between SFF and the ordinary fiber (see Fig.3)[5].

The amounts of lipase immobilized on three different cation exchangers

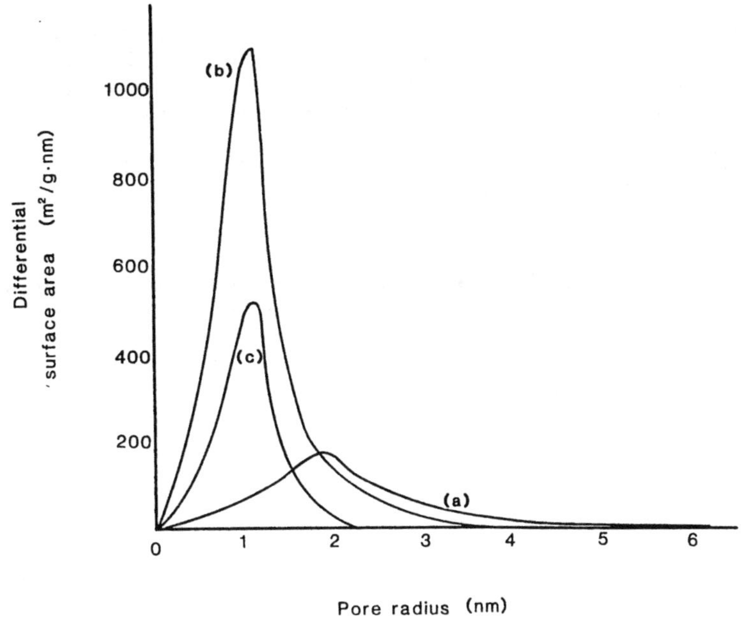

Fig.2 Pore size distribution of three materials[7]

(a) PVA super fine fiber (217 m^2/g)
(b) activated charcoal (1400 m^2/g)
(c) ion exchange resin (410 m^2/g)

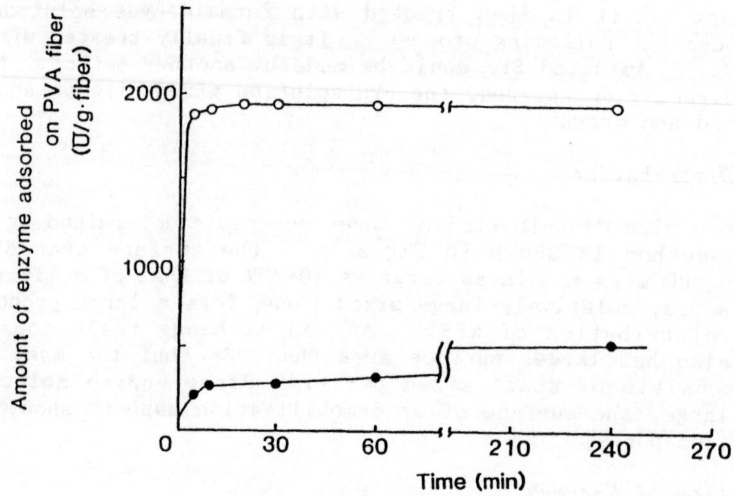

Fig.3 Immobilization rate of enzyme by PVA fibers[8]

○ SFF, ● ordinary PVA fiber

are compared in Table 1[9]. The lipase immobilized on SFF is much greater than those on an ion exchange fiber (IEF) and an ion exchange resin (IER), although the values of weight and an ion exchange capacity of SFF are smaller than those of IEF and IER. It is due to the difference in a surface area effective for enzyme immobilization. The lipase consists of low molecular weight lipase and high molecular weight lipase[10]. The diameter of low molecular weight lipase is estimated at 4 nm by assuming that a lipase molecule is spherical in shape. The surface consisting of pores larger than 4 nm in diameter would be effective for lipase immobilization. The surface areas composed of pores larger than 4 nm in diameter are then estimated and listed in the same table.

The value for SFF is almost twice as large as that for IEF and is about ten times that for IER.

Table 1 Immobilization of lipase on various supports

Support	SFF	IEF	IER
Weight [mg]	111	252	221
Ion exchange capacity [meq/g]	0.94	1.3	4.24
Total surface area [m²/g]	217	515	410
Surface area composed of pores larger than 4 nm in diameter [m²/g]	182	94	18
Amount of immobilized enzyme [mg/g]	66	5	5

Table 2 Enzymes immobilized on SFF carrying a different functional group

Enzyme	Isoelectric point	Yield(%) D-SFF[1]	T-SFF[2]	S-SFF[3]
Invertase	3.8	90	73	0
Glucose oxidase	4.3	98	98	27
β-galactosidase[4]	4.6	92	92	15
β-glucosidase	7	16	7	98
Lipase	7.3	65	2	100
β-galactosidase[5]	8.4	30	14	100

1) Dimethylaminated SFF
2) Trimethylaminated SFF
3) Sulfonated SFF
4) from *E. coli*
5) from jack bean

SFF also immobilizes β-galactosidase from jack bean 3 to 8 times as much as that bound on IEF and IER, and the enzyme-SFF conjugate shows even higher observed activity than the enzyme bound on the other two supports[11].

A great amount of yeast invertase is immobilized on SFF, though the enzyme molecule is very large and its diameter is estimated at 8.55 nm[8]. Those results may be due to the large surface effective for immobilization of large molecules.

Ionic interaction between supports and biocatalysts. As shown in Table 2, the acidic enzymes (invertase, glucose oxidase and β-galactosidase from *E. coli*) combine with the anion exchanger such as D-SFF and T-SFF, whereas the basic enzymes (β-glucosidase, lipase and β-galactosidase from jack bean) combine with the acidic form of SFF[9,11,12]. Mitz et al. stated[13] that the anion exchanger like diethylaminoethyl-cellulose adsorbed the acidic enzymes such as pepsin and acylase while the cation exchangers like carboxymethylcellulose bound the basic proteins such as chymotrypsin and trypsin. Our results are in agreement with the observations by them. The results suggest that enzymes are immobilized on SFF mainly by electrostatic force of attraction.

Fig.4 illustrates the relationship between the total nitrogen content of SFF and the amount of invertase immobilized on the SFF[8]. The immobilized invertase increases with the nitrogen content and reaches almost the maximum at 1.0 wt%. The result also demonstrates that an ionic interaction is a main force for the enzyme immobilization by SFF.

The enzyme immobilized on SFF is released from the support, when it is immersed in concentrated salt solution. Almost no enzyme detaches from the support in 0.01 M sodium chloride, while around 90% of immobilized enzyme detaches in 0.5 M sodium chloride. These results support that SFF binds enzymes by ionic interaction. That is another advantage of the support, because SFF is easily regenerated after immobilized biocatalysts lose their activity.

In general, acidic enzyme begins to lose its negative charge and to detach from an ion exchanger, when pH approaches the isoelectric point of enzyme[14]. However, it is confirmed that the invertase immobilized on SFF does not detach from the support through the sucrose hydrolysis performed at its optimal pH[8]. The β-galactosidase bound on SFF is not released from the support during the enzymic hydrolysis of a synthetic substrate, either[11].

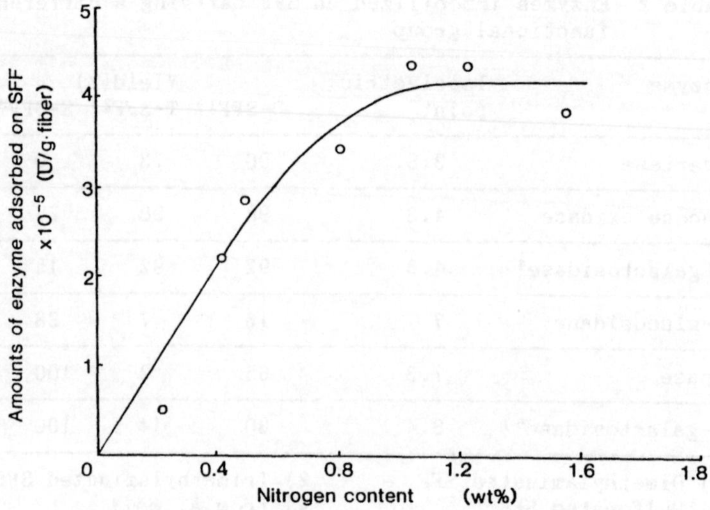

Fig.4 Relationship between immobilized enzyme and nitrogen content of SFF[8]

Immobilization rate of enzyme. The isotherm for the immobilization of invertase on SFF has the shape characteristic of the Langmuir adsorption isotherm[15].

Assuming that agitation is strong enough to mix enzyme solution and a SFF sample completely, an immobilization rate equation is given by

$$-\frac{V\,dC}{W\,dt} = k_a C(1-\theta) - k_d \theta \qquad (1)$$

$$\theta = V(C_o - C)/NW \qquad (2)$$

C : enzyme concentration (mg-Enz./dℓ)
C_o : initial enzyme concentration (mg-Enz./dℓ)
t : elapsed time (min)
θ : fraction of surface covered by adsorbed enzymes (-)
k_a : forward adsorption rate constant (dℓ/min · g-fiber)
k_d : reverse adsorption rate constant (mg-Enz./min · g-fiber)
V : volume of enzyme solution (dℓ)
N : maximum value of adsorbed enzyme per unit mass of SFF (mg-Enz./g-fiber)
W : weight of SFF (g-fiber)

At equilibrium condition, eq.(1) is equal to zero and eq.(2) becomes

$$\theta_\infty = V(C_o - C_\infty)/NW = K_a C_\infty/(1+K_a C_\infty) \qquad (3)$$

K_a : $=(k_a/k_d)$ adsorption equilibrium constant (dℓ/mg-Enz.)

where the subscript ∞ indicates equilibrium condition. The parameters, N, K_a and k_a, evaluated by least squares method are substituted into eq.(1) and the time course of enzyme adsorption is simulated. The comparison of the calculated enzyme concentration with the observed values gives excellent agreement in the initial enzyme concentration range lower

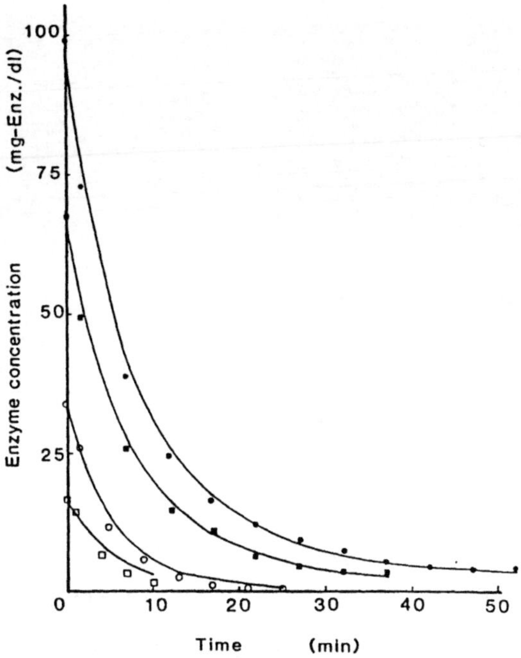

Fig.5 Comparison of calculated enzyme concentration with observed values[15]

than 100 mg-enz./dℓ (see Fig.5)[15]. The immobilization of invertase on SFF hence follows the Langmuir adsorption mechanism. However, in the higher concentration, there is not good agreement between them. After immobilization proceeds and most of the support surface exposed to the enzyme solution is covered, free enzyme in bulk solution have to diffuse through small spaces among twisted filaments in order to reach vacant sites. The apparent immobilization rate hence seems to decrease with the progress of immobilization.

Immobilization of Microorganism

After SFF is put in yeast solution and stirred vigorously, a lot of yeast cells are observed on aminated SFF surfaces[16]. However, no yeast is found on sulfonated SFF surface. Most of microorganisms dispersed in water generally have a net negative charge on their surface. The charge is thought to be produced by the dissociation of carboxylic and phosphonic acids in cell membranes[17]. Morisaki et al.[17] stated that strongly basic anionic resins had positive charges on their surfaces and were able to adsorb cells more than 10^{10} by electrostatic force of attraction.

A gram of SFF immobilizes more than 200 mg of dried baker's yeast and is expected to immobilize more yeast. Compared with other results in the literature [18,19], SFF has a higher capability to immobilize microorganisms on its surface.

After a small amount of yeast was immobilized on aminated SFF, the complex was incubated in glucose solution. The scanning electron micrographs of the immobilized yeast before and after incubation are shown in Fig.6[7,16]. As the considerable growth of the immobilized

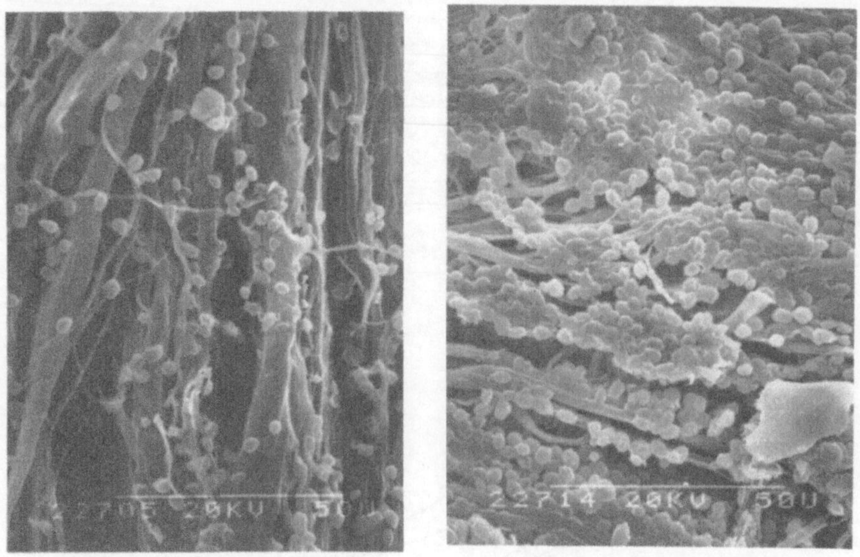

(a) before incubation (b) after incubation

Fig.6 Scanning electron micrographs of bound baker's yeast[7,16]

yeast is recognized, SFF may have a potential to be applied to other cell culture. When free and adsorbed yeast were left in glucose solution, there was no big difference observed in ethanol production. Immobilized yeast have the activity as high as free one.

When the SFF-yeast complex is charged into a glass column and glucose solution is fed, the solution containing ethanol continuously flows out of the column. The immobilized yeast cell itself grows during continuous alcohol fermentation and some of the proliferated cells detaches from SFF.

An attempt was also made to apply SFF to waste water treatment. The microorganisms acclimated to biodegrade glucose, acrylonitrile and cyanic compounds are well bound on SFF. Compared with free activated sludge, the activated sludge bound on SFF uses the energy liberated in living organisms to further decompose BOD components[20]. It hence suggests that they decomposes pollutants effectively.

The durability of a variety of supports is tested in anaerobic fermentation broth[21]. Although natural polymers are completely biodegraded within 20 d, synthetic polymers including aminated SFF are not biodegraded for the period longer than 100 d. Besides, the weight of aminated SFF increases after the long durability test, which suggests that the microorganisms in anaerobic fermentation broth attaches to SFF.

Thermal Stability of Immobilized Enzyme

The effects of immobilization on thermal stability have widely been studied, but it has so far been evaluated by measuring enzyme activities. An overall thermal behavior of various enzyme preparations can be obtained easily by differential scanning calorimetry (DSC). Therefore, the thermal stability of free and immobilized enzyme is investigated by both activity assay and calorimetry.

Influence of heat treatment on enzyme denaturation. The DSC analysis for free invertase and the invertase immobilized on SFF shows that the immobilization shifts the denaturation temperature, T_d, to higher temperature by about 5°C[22]. The alteration of higher-order structure starts at lower temperature for the free enzyme than for the immobilized one, and the transition is completed at lower temperature for the free enzyme than for the immobilized one.

Fig.7 shows the thermal stability of free and immobilized invertase in a plot of relative activity and relative denaturation enthalpy against the temperature of heat treatment[7,22]. When the free enzyme is heated over 60°C, enzyme activity decreases rapidly with the increase in temperature. Sixty percent of the original activity remains in the sample heated to 70°C and 6% in the sample heated to 72.6°C. The immobilized enzymes heated to 75°C and 77°C maintain 50% and 25% of its original activity, respectively. At 70°C, 3.5% of the immobilized enzyme changes its conformation, but this change seems to be reversible, because the immobilized enzyme keeps its original activity even after heating.

The invertase immobilized on SFF is more thermostable than that in solution. An increase in thermal stability is also observed when invertase is entrapped in PVA film[23]. It is well known that polyhydric alcohols and carbohydrates have a stabilizing effect on enzymes[24,25]. PVA is a kind of polyhydric alcohol and invertase from *C. utilis* contains about 50% carbohydrate[26,27]. Consequently, the increase in thermal stability may be due to the interaction among water, the PVA support and the carbohydrate contained in the enzyme.

Glucose oxidase containing about 16% carbohydrate[28] is thermally stabilized by the immobilization on SFF[29], while β-galactosidase having no carbohydrate[30] is not stabilized. Those results also support the effect of the carbohydrate in enzymes.

The T_d value of the immobilized invertase increases with enzyme loading, which suggests the effect of carbohydrate, but is slightly influenced by pH change. On the other hand, the thermal stability of the immobilized glucose oxidase is more dependent on pH than enzyme loading[29].

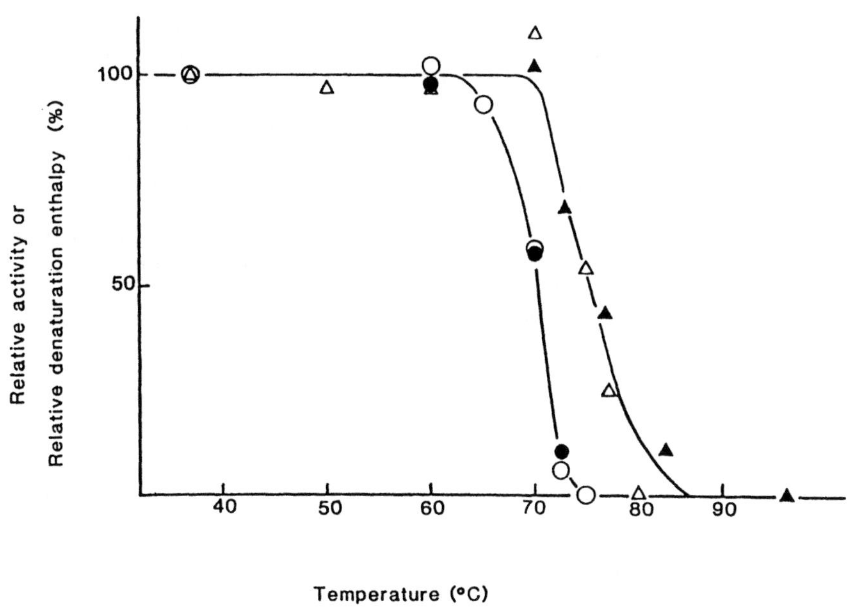

Fig.7 Thermal stability of free and immobilized invertase[7,22]

relative activity : ○ free, △ immobilized
relative denaturation enthalpy : ● free, ▲ immobilized

PHOTOCROSSLINKABLE PVA

PVA substituted by styrylpyridinum group (SbQ) shows very high photosensitivity even when the content of SbQ is extraordinarily low. Furthermore, as the polymer demonstrates excellent mechanical properties and is also formed into various shapes such as membrane, particle, gel-fiber, *etc.*, it is easily applied to photoresist and a support for immobilizing biocatalysts.

Preparation of PVA-SbQ

SbQ group is easily and efficiently incorporated into PVA in an acidic aqueous solution by acetalization. The lower reaction temperature gives rise to the higher conversion; only less than about 15% of formylstyrylpyridinum salt is recovered unchanged when the reaction is performed at room temperature, whereas about 30% of the salt is recovered at 60°C. PVA carrying SbQ (PVA-SbQ) is isolated by precipitating in acetone and washing thoroughly with methanol. The purified polymer is dissolved in distilled water to afford a viscous, pale yellow solution that is stored for a long period in the dark. The efficiency of photoinsolubilization is very dependent on both the content of SbQ and the irradiation period. The PVA having more than about 1.0 mol% of SbQ is suitable for the immobilization of enzyme since more than 90% of the polymer is insolubilized[31]. Besides, PVA-SbQ can be modified to be more hydrophobic and more ionic by incorporating various functional groups at hydroxyl groups of PVA.

Immobilization of Enzymes

Since a film of PVA-SbQ is highly photosensitive to light of wavelength shorter than ca. 460nm, conventional light sources, such as a mercury arc, a xenon arc and a fluorescent lamp, are all suitable for the immobilization.

Enzymes are easily entrapped in photocrosslinked gelatine-like material when the mixture of enzyme and water soluble PVA-SbQ is insolubilized by the cyclodimerization of SbQ caused by the irradiation of visible light being not harmful to bioactive substances (see Fig.8)[31]. Besides, this reaction occurs specifically without free radicals which may attack biocatalysts. It is very advantageous to irradiate the PVA in the air dried film state, because the irradiation time is considerably reduced.

The network formation takes place exclusively between the photosensitive groups, which means that enzymes are solely entrapped in the physical manner without the formation of any enzyme-enzyme covalent bonding. Coulombic interaction may play a negligible role for the immobi-

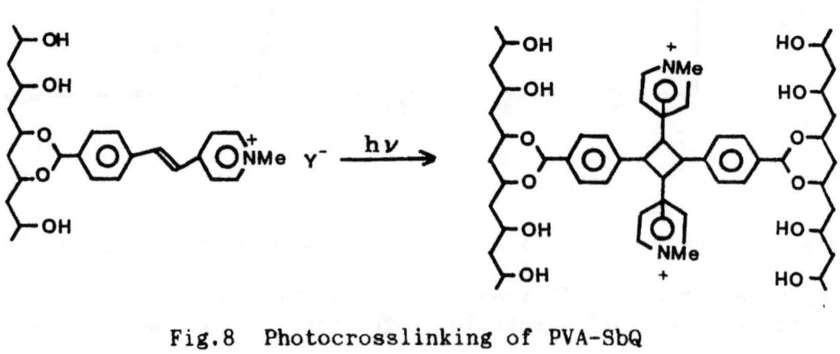

Fig.8 Photocrosslinking of PVA-SbQ

lization because there are few cationic sites on the PVA. Various enzymes are completely and efficiently entrapped in the network of the polymer and no escape of enzyme is observed, when the content of the photosensitive group is more than about 1.0 mol%. The size of the polymer network is controlled by SbQ content and molecular weight of PVA.

The addition of poly(ethylene glycol) to PVA-SbQ causes the phase separation of the two water-soluble polymers, resulting in the formation of a porous structure of PVA-SbQ[32]. Therefore, the entrapment of invertase in the porous PVA-SbQ facilitates the diffusion of substrate molecule into the polymer network.

Effect of Gel Hydrophilicity on Product Separation

The cultured cells of *L. vera* entrapped in PVA-SbQ grow well inside gel matrices and synthesize blue pigments in the presence of L-cysteine as an inducer[33]. The entrapped cells are much better than free or calcium alginate entrapped cells judging from cell growth and total pigment productivity. However, the pigments are poorly released from PVA-SbQ partially saponified, because the product is scarcely soluble in medium under the conditions. While, the pigments are well excreted into the medium from the free and calcium alginate entrapped cells. Those results suggest that release of the pigments may be remarkably affected by hydrophilicity-hydrophobicity balance of the cell entrapping gel materials. Hydrophilicity of PVA-SbQ gels is regulated by controlling the degree of saponification of PVA. As shown in Fig.9, pigment release is greatly enhanced by increasing the degree of saponification, that is, gel hydrophilicity. More than 50% of the pigments produced by the entrapped cells is released from the gel into the medium when PVA is completely saponified. The partition coefficient for the pigments in the gel against that in the medium decreases with increasing degree of saponification. Recovery of the pigments produced by PVA-SbQ entrapped cells is thus improved an increase in the gel hydrophilicity.

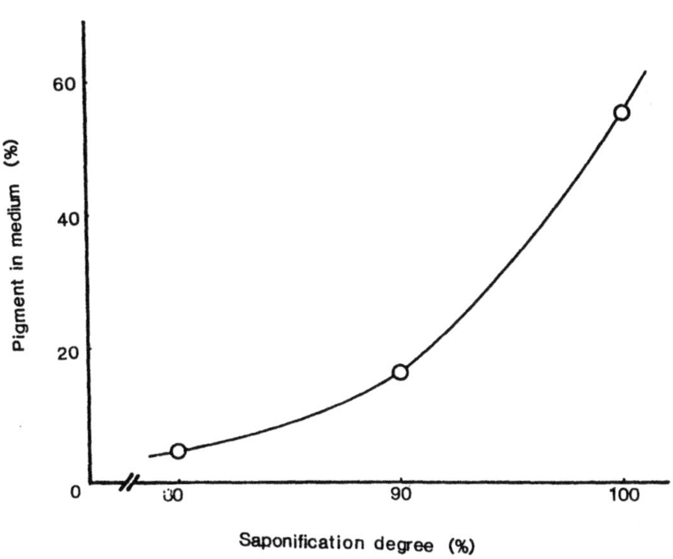

Fig.9 Effect of gel hydrophilicity on pigment production by PVA-SbQ entrapped *L. vera* cells

REFERENCES

1) L. Goldstein and G. Manecke, The Chemistry of Enzyme Immobilization, in: "Immobilized Enzyme Principle" ,L. Goldstein and E. Katchalski Eds., Academic Press, N.Y., (1976).
2) H. Ichijo, T. Suehiro, J. Nagasawa, A. Yamauchi and M. Sagesaka, *Biotechnol. Bioeng.*, 27:1077 (1985).
3) H. Ichijo, T. Suehiro, J. Nagasawa, A. Yamauchi and M. Sagesaka, *Sen-i Gakkaishi,* 42:T636 (1986).
4) M. Takashio, A. Shirasaka and I. Nagashima, Japanese Patent, No.5477720 (1979).
5) A. Yamauchi, T. Suehiro, M. Suzuki, M. Uzumaki, M. Takashio and N. Fujii, Japanese Patent, No.5547130 (1980).
6) H. Toyama, M. Uzumaki and H. Haioka, Japanese Patent, No.3310663 (1958).
7) H. Ichijo, J. Nagasawa and A. Yamauchi, *J. Biotechnol.*, in press.
8) H. Ichijo, T. Suehiro, A. Yamauchi, S. Ogawa, M. Sakurai and N. Fujii, *J. Appl. Polym. Sci.*, 27: 1665 (1982).
9) H. Ichijo, T. Suehiro, J. Nagasawa, A. Yamauchi and M. Sagesaka, *Se-i Gakkaishi*, 39:T532 (1983).
10) Y. Horiuchi and S. Imamura, *J. Biochem.*, 81:1639 (1977).
11) H. Ichijo, T. Suehiro, J. Nagasawa, A. Yamauchi and M. Sagesaka, *Sen-i Gakkaishi*, 41:T303 (1985).
12) H. Ichijo, T. Suehiro, J. Nagasawa, A. Yamauchi, and M. Sagesaka, *Sen-i Gakkaishi*, 42:T115 (1986).
13) M. A. Mitz and R. J. Schlueter, *J. Am. Chem. Soc.*, 81:4024 (1959).
14) H. E. Klei, D. W. Sundstrom and R. Gargano, *Biotechnol. Bioeng.*, 20: 611 (1978).
15) H. Ichijo, *J. Appl. Polym. Sci.*, 28:1447 (1983).
16) H. Ichijo, T. Suehiro, J. Nagasawa, A. Yamauchi and M. Sagesaka, *Sen-i Gakkaishi*, 43:271 (1987).
17) H. Morisaki and R. Hattori, "Kaimen to Biseibutsu", Gakkai Syuppan Center, Tokyo, (1986).
18) A. J. Daugulis, N. M. Brown, W. R. Clue and D. B. Dunlop, *Biotechnol. Lett.,* 3:651 (1981).
19) T. A. Krug and A. J. Daugulis, *Biotechnol. Lett.*, 5:159 (1983).
20) K. Kubota, K. Urano and H. Ichijo, Suishitsuodakugakkai koen Yokosyu, 243, (1986).
21) H. Sano, Y. Souma and I. Toyama, Osakakogyogijutsushikensyo Kiho, 37: 99 (1986).
22) H. Ichijo, H. Uedaira, T. Suehiro, J. Nagasawa and A. Yamauchi, *Agric. Biol. Chem.*, 49:3591 (1985).
23) H. Uedaira, A. Yamauchi, J. Nagasawa, H. Ichijo, T. Suehiro and K. Ichimura, *Sen-i Gakkaishi*, 40:T317 (1984).
24) H. Uedaira, *Sen-i Gakkaishi*, 37:P436 (1981).
25) K.Gekko, *J. Biochem.*, 91:1197 (1982).
26) M. Iizuka, H. Chiura and T. Yamamoto, *Agric. Biol. Chem.*, 42:1207, (1978).
27) N. P. Newmann and J. O. Lampen, *Biochem.*, 16:2105 (1977).
28) B. E. P. Swoboda and V. Massey, *J. Biol. Chem.*, 240:2209 (1989).
29) H. Ichijo, H. Uedaira, T. Suehiro, J. Nagasawa, A. Yamauchi and N. Aisaka, *Agric. Biol. Chem.*, 53:833 (1989).
30) A. V. Fowler and I. Zabin, *Proc. Natl. Acad. Sci U.S.A.*, 74:1507 (1977).
31) K. Ichimura, *J. Polym. Sci.*, 22:2817 (1984).
32) K. Ichimura, *Makromol. Chem.*, 188:763 (1987).
33) H. Nakajima, K. Sonomoto, H. Morikawa, F. Sato, K. Ichimura, Y. Yamada and A. Tanaka, *Appl. Microbiol. Biotechnol.*, 24:266 (1986).

ELECTRIC MODULATION OF POLYMER GEL CONTRACTION AND ANIMAL CELL PROLIFERATION

Hiroaki Shinohara, Masashi Yaoita
Yoshihito Ikariyama, and Masuo Aizawa

Department of Bioengineering, Faculty of Engineering
Tokyo Institute of Technology
Ookayama, Meguro-ku, Tokyo 152, Japan

ABSTRACT

Modulation of gel morphology of synthetic polymer and biopolymer upon low potential application was investigated. Morphology of polymethacrylic acid (PMAA)-Ca^{2+} chelate microgels has been electrically modulated on a conducting polymer film-coated electrode. The electrode induced a drastic change of ion flux in the vicinity of the conducting polymer film without any water electrolysis, which caused the PMAA-Ca^{2+} microgels to change rapidly their shape. Electric effects on animal cells cultured on an electrode were also investigated in this line of study. Morphology and proliferation of HeLa cell was dependent on electrode potential during culture. The modulation mechanism is interpreted in relation to the structural change in cytoskeletal gel composed of F-actin.

INTRODUCTION

Polymer gels are widely employed in biomedical fields since these polymers are relatively inexpensive and simple to use. Their use is, however, mostly limited to passive purpose, which may include the uses as a matrix for immobilizing enzymes and living cells.

In contrast to such passive uses, an increasing interest has arosen in developing polymer gels as active elements. Stimuli-responsive motions of polymer gels are attracting an intensive attention in a wide area. An expected application concerns with the development of an actuator system modeled on muscular system[1-3]. It may also be exemplified by the modulation of biopolymer gel structures. The morphology of living cells may be regulated by such a modulation, because cells contain gel phase biopolymers.

This paper primarily concerns with the electric modulation of synthetic polymer gel contraction and the designing of an actuator system modeled on the neuromuscular system. Although electrically controlled contraction of polymer gels has been reported by several researchers, these systems need high voltage application in the range from several Volt to several tens of Volt to induce gel contraction, because these electrically induced gel contraction is based on electrophoresis and electro-osmosis[4-9]. High voltage application may result in water electrolysis. In order to avoid water electrolysis, gel contraction should be preformed by low voltage application.

On the other hand, tissue contraction by a small electric stimulation is carried out in the neuromuscular system which may suggest the most sophisticated actuator system. In this system, an impulse, small membrane potential change through a neuron, is transduced to molecular signal, a release of neurotransmitters. The release induces a Ca^{2+} flux from the sarcoplasmic reticulum in the muscular cell. This ion flux promotes the contraction and relaxation of myofibril, which is regarded as a gel actuator element. This mechanism suggests that polymer gel morphology may be modulated by controlling ion flux upon low potential application.

We have taken notice of an electrochemical incorporation and release of ions onto conducting polymers to modulate contraction of ion responsive polymer gel by law potential application. An electrochemical incorporation and release of ions onto a polypyrrole film-coated electrode were reported by our group[10]. Furthermore, the drastic pH change due to OH^- transfer in the vicinity of a polypyrrole film by small shift of potential was also described in a previous paper[11]. Based on these findings, a polypyrrole electrode was coupled with a pH responsive mechanochemical polymer gel. In this study, polymethacrylic acid (PMAA)-Ca^{2+} chelate microgel was employed to achieve rapid modulation of gel morphology by an electrically induced pH change.

In the further study, the electric modulation has successfully been extended to control morphology of living cells. This modulation is expected to be promising to control cellular morphology and cell function such as proliferation.

Electrical techniques have been attracting strong interest in cell manipulation, since it has been shown that cells suspended in a non-electrolytic solution are easily fused upon high d.c. voltage application. However, few studies on cellular behavior under low d.c.potential have been reported to date. In our previous publications we reported the electrochemical effects on erythrocytes as an animal cell model. We employed an optically transparent electrode (OTE) for d.c. potential application to induce the changes in morphology and membrane fluidity of the cells. By taking advantage of the transparency of an indium trioxide (In_2O_3) sputtered OTE, we were able to observe directly the lysis of erythrocytes at a potential of ca. 1.5V vs. Ag/AgCl[12].

In this paper, HeLa cells, a human cancer cell, was cultured on an In_2O_3 OTE to know changes in cellular morphology and proliferation rate under low potential application. These electrochemical effects are discussed in relation to the change of cytoskeleton, a biopolymer gel.

EXPERIMENTAL

Preparation of PMAA-Ca^{2+} Microgels

An aqueous solution containing 25% polymethacrylic acid (PMAA) sodium salt (molecular weight of 15000), calcium chloride and sodium chloride were purchased and used without further purification. Pyrrole was used after vacuum distillation. All reagent solutions were prepared with ion exchanged and distilled water.

PMAA-Ca^{2+} microgel suspension was prepared by mixing of 1% PMAA Na aqueous solution with 0.1M $CaCl_2$ aqueous solution in a volume ratio of 1 : 1. When the two solutions were mixed with a mixer, the solution was turbid due to the formation of microgels. Spherical shape of PMAA-Ca^{2+} chelate microgels in this suspension was observed by a microscope. The average diameter of gels was 3μm.

Preparation of a Polypyrrole Film-coated OTE

A Pt thin layer was prepared on a glass plate (1cm × 3cm) by an ion sputter coater (Hitachi, E-102). A polypyrrole thin film was electrochemically prepared on this Pt sputtered optically transparent electrode (OTE). A Cu wire with Teflon theath was connected to the Pt sputtered glass plate with a silver past. The connecting part was covered with epoxy resin to prevent leakage of electricity. A polypyrrole thin film was electrochemically prepared on this Pt OTE to observe directly the morphological change of microgels which are in contact with the polypyrrole film-coated electrode. An aqueous solution containing 0.1 M pyrrole and 0.1M NaCl was prepared as the electrolyte solution for polymerization. The Pt OTE, a Pt plate counter electrode and a salt bride connected to an Ag/AgCl electrode were positioned in the electrolyte solution. Oxidative polymerization of pyrrole was carried out at a constant potential of 0.8V vs. Ag/AgCl using a potentiostat (Hokuto Denko Ltd., HA-301). Thickness of the synthesized film was controlled at ca. 200nm by terminating electrochemical oxidation at a passed charge of 80mC/cm^2. After thorough rinse with distilled water, the polypyrrole film-coated Pt OTE was used to promote the morphological change of microgels by low potential application.

Electric Stimulation and Observation of Microgel Morphology

Figure 1 shows the experimental setup for the observation of morphological change of PMAA-Ca^{2+} microgels which are in contact with a polypyrrole film-coated Pt OTE under potential control. A polypyrrole film-coated Pt OTE, a Pt counter electrode and a salt bridge connected to an Ag/AgCl electrode were positioned in a flat dish (4cm in diameter) on the stage of an inverted microscope (Olympus, IMT-2) as shown in Fig.1. The microgel suspension was then poured into the dish.

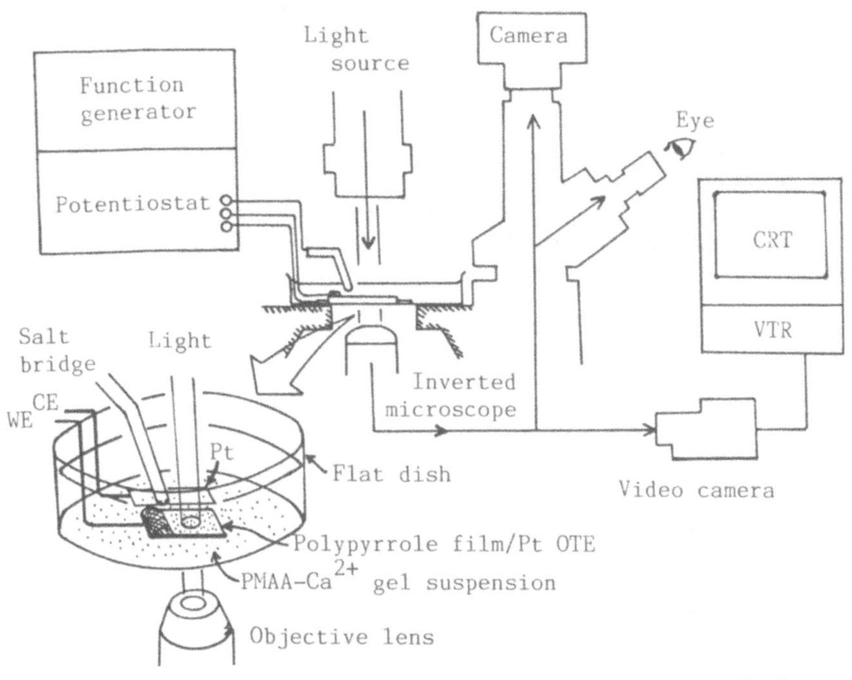

Fig.1 Experimental setup for observation of morphological change by potential control of PMAA-Ca^{2+} microgels which are in contact with a polypyrrole film-coated Pt OTE.

Microgels contacted with the polypyrrole film-coated Pt OTE by sedimentation. Potential of the polypyrrole film-coated Pt OTE was controlled to modulate ion flux with a potentiostat and a function generator (Hokuto Denko Ltd., HB-105). Morphological change of gels which were in contact with the electrode was observed with an inverted microscope, and monitored with a video system. The size and the number of microgels were determined from the minitored display.

Characterization of Electrically Controlled Ion Transfer in the Vicinity of the Electrode

Electrochemical oxidation and reduction of a polypyrrole film electrode was characterized by cyclic voltammetry and differential pulse voltammetry with a polarographic analyzer (Yanaco, P-1100). pH change in the vicinity of a polypyrrole film electrode by potential step was measured *in situ* using a microfablicated pH probe (Microelectrode Co. Ltd., MI-410) which was contacted with the electrode surface and connected to a pH meter (DKK Ltd., HG-3).

Culture of HeLa Cells on an In_2O_3 OTE

HeLa cells were cultured in a Petri dish with built-in electrodes. An In_2O_3 OTE was used as a working electrode (1cm × 2.5cm). In_2O_3 sputtered glass plates were purchased from Matsuzaki Sinku Co.Ltd. An Ag/AgCl reference electrode (1mmφ × 10mm) and a Pt plate counter electrode (1cm × 1cm) were positioned with an In_2O_3 OTE in the Petri dish. McCoy's 5A mixed with 10% fetal calf serum was used as the cultivation medium[13]. The cultivation was performed in a CO_2-controlled incubator at 37°C. The CO_2 concentration was regulated at 5%. Electrode potential was controlled with a potentiostat to apply electric effect on the cells which are cultured on the electrode. The rest potential of the In_2O_3 OTE was about 50 mV vs. Ag/AgCl.

Observation of Cell Proliferation under Potential Controlled Culture

HeLa cells were cultured on the In_2O_3 OTE under constant potential application. Cells cultured on the electrode were counted everyday to investigate the electric effect on proliferation rate. Observation was done using an inverted microscope. The potential application was suspended for 30min per day for cell counting.

Observation of Morphological and Cytoskeletal Changes of HeLa Cells by Potential Application

Morphological change of HeLa cells by potential application was observed using an inverted microscope in a similar manner as the observation of microgel morphology. The changes were recorded by taking pictures.

Cytoskeletal proteins such as F-actin take an indispensable role of supporting membrane structure. Rhodamine phalloidin(Rh-Ph), a fluorescent dye, is here exploited to investigate the changes of cytoskeleton during the application of electrode potential. HeLa cells were cultured on In_2O_3 OTEs for 3 days, and then potential application was initiated. After potential applications of 0V or 1.0V vs. Ag/AgCl for 1hr in saline, the cells were fixed with 3.7% formaldehyde solution and then stained by rhodamine phalloidin. However, this procedure was not done when negative potential was applied, because cell was detached from the electrode after negative potential application. Therefore, after staining with rhodamine phalloidin, HeLa cells on an OTE were subjected to the potential application of -0.6V vs. Ag/AgCl in saline. Staining of HeLa cells with Rh-Ph was performed by the method of Amato et al[14]. HeLa cells cultured on an In_2O_3 OTE was incubated in a Rh-Ph PBS solution containg palmitoyl ribophosphatidyl choline at 37°C for 20 to 30 mins. After rinse, the electrode was settled in Petri dish which was filled with a saline solution, and then potential application was carried out. Cytoskeletal changes by the potential application was observed using an inverted microscope with a fluorescence observation system (Olympus, IMT2-RFL).

RESULTS AND DISCUSSION

ELectrochemical Characteristics of a Polypyrrole Film-coated Electrode

Cyclic voltammetry and differential pulse voltammetry of a polypyrrole film-coated Pt electrode was carried out in 0.1M NaCl aqueous solution in the potential range from -0.8V to 0.4V vs. Ag/AgCl. On the positive potential scanning, anodic current due to the oxidation of polypyrrole arose at -0.5V vs.Ag/AgCI. In the reverse scanning, cathodic current due to the neutralization of oxidized polypyrrole was observed until the potential shifted to -0.5V vs.Ag/AgCl. No distinctive gas evolved from the polypyrrole electrode in this potential region. This result suggests that polypyrrole is oxidized above a potential of -0.5V with anion incorporation and is reversibly neutralized below the potential with the release of the incorporated anions.

Electric Modulation of PMAA-Ca^{2+} Microgels on a Polypyrrole Electrode[15]

In order to induce a morphological change of PMAA-Ca^{2+} microgels by anion transfer, the electrode potential of a polypyrrole film-coated Pt OTE was controlled between -0.6V and 0.4V vs. Ag/AgCl. Figure 2 shows photographs of the morphological change of PMAA-Ca^{2+} microgels on the polypyrrole film upon the potential shift. When the potential was set at -0.6V vs. Ag/AgCl, many spherical microgels (about 3μm in diameter) were observed, and no apparent morphological change was observed (Fig.2a). When the potential was stepped to 0.4V vs. Ag/AgCl, dissolution of the gels begins within 5 sec, and the dissolution was completed within 15 sec(Fig.2c). Reformation of microgels was demonstrated within 10sec, when the potential was stepped to -0.6V vs. Ag/AgCl (Fig.2e). Figure 3a shows the time-course of a gel diameter when the potential shifted. Figure 3b shows the relative number of the gels on the polypyrrole eletrode during the potential shift. Initial gel number was referred as the standard. This dissolution and gelation upon the small potential shift within 1V was reversibly continued over 1 hour without gas evolution due to water electrolysis.

On the other hand, no apparent morphological change was observed for PMAA-Ca^{2+} microgels which were in contact with a bare Pt OTE under the same potential step. This result supports that the anion is transfered in the vicinity of the polypyrrole film by electrochemical oxidation and neutralization. In order words, anion doping and undoping promote the morphological change of PMAA-Ca^{2+} microgels.

Furthermore, potential dependence on the rate of morphological change was investigated. Rapid and reversible gelation and dissolution was demonstrated by stepping the potential between -0.6V and 0.4V vs. Ag/AgCl.

A Postulated Mechanism for Electric Modulaton of Microgel Morphology

Dissolution and gelaration of PMAA-Ca^{2+} microgels may be due to the suppresion and accerelation of chelating of PMAA by Ca^{2+}. And this chelation seems to be controlled by pH change. PMAA-Ca^{2+} microgels dissolve in a solution of low pH..

On the other hand, we have found a drastic pH change in the vicinity of a polypyrrole film upon the potential step. Figure 4 shows the time-course of the pH change in the neighborhood of a polypyrrole film-coated Pt electrode upon a potential step between -0.6V and 0.6V vs.Ag/AgCl. High pH at -0.6V seems to be attributed to mainly OH^- production by electrochemical reduction of dissolve oxygen. The drastic pH decrease by the potential shift to 0.6V is considered to be attributed to OH^- doping with Cl^- into the oxidized polypyrrole film. This OH^- incorporation might induce the increase of H^+ concentration in the vicinity of the film. The drastic pH change was reversibly measured upon the potential step between -0.6V and 0.4V vs. Ag/AgCl. The pH response upon the potential step corresponds to the morphological change of microgels in contact with the polypyrrole film electrode.

From these results, we conclude that morphological change of PMAA-Ca^{2+} microgels is induced by the pH change around the gels as shown in Figure 5.

Electric Modulation of Cell Morphology and Proliferation[16]

In the further study, electric modulation has been extended to control the morphology of biopolymer gels. The effects of low potential application on living animal cells were investigated to demonstrate the possible regulation of cell morphology and ability.

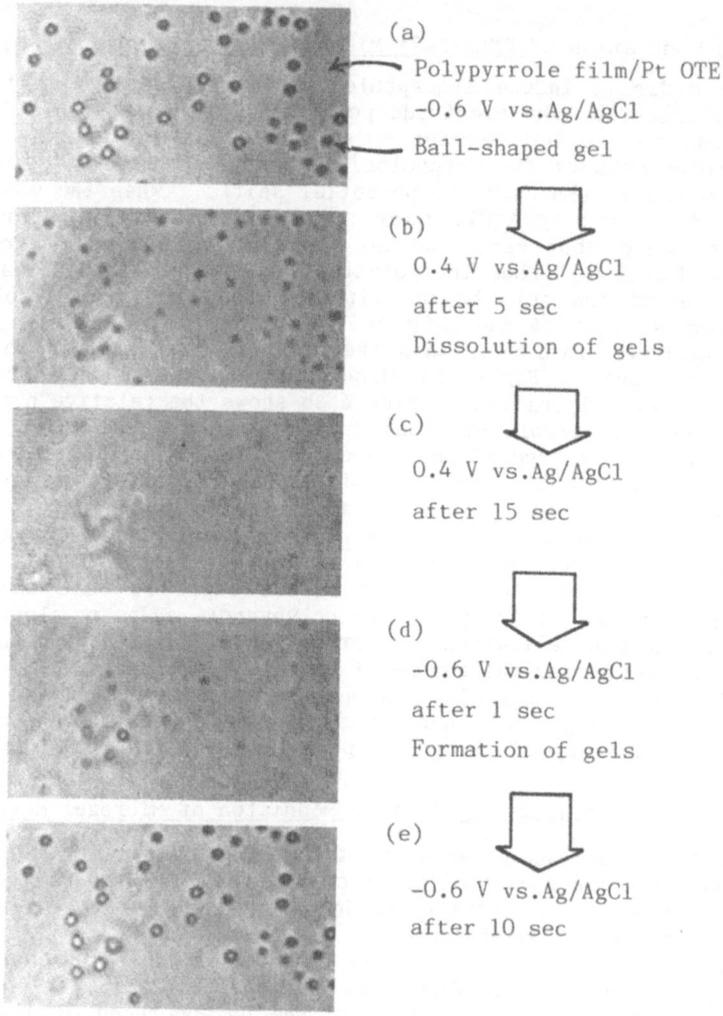

Fig.2 Morphological change by potential step of PMAA-Ca^{2+} microgels which were in contact with the polypyrrole film-coated Pt OTE. Pulse width : 1 min.

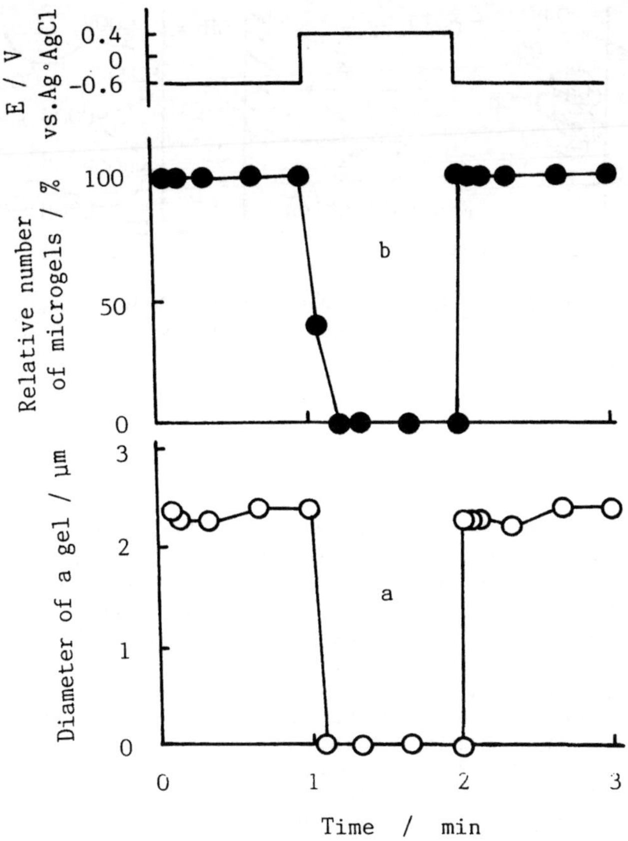

Fig.3 Change of gel diameter (a) and relative gel number (b) for potential shifts between -0.6 V and 0.4 V vs.Ag/AgCl. Pulse width : 1 min.

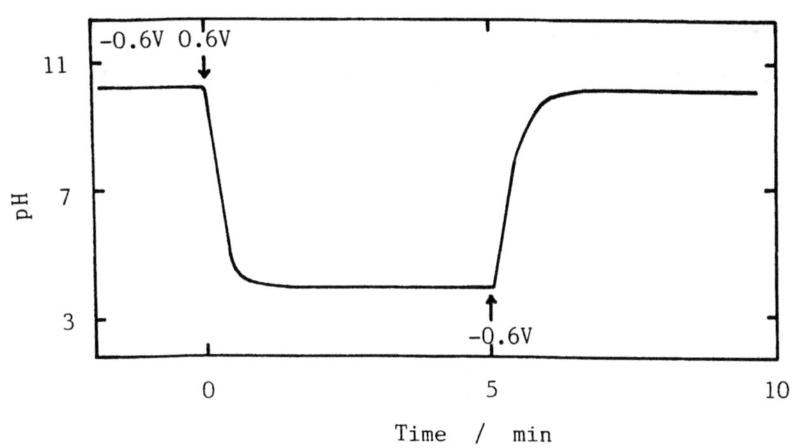

Fig.4

Time-course of pH change near the polypyrrole film-coated Pt electrode upon the potential step between -0.6 V and 0.6 V vs.Ag/AgCl. Electrolyte : 0.1M NaCl aqueous solution, Pulse width : 5 min.

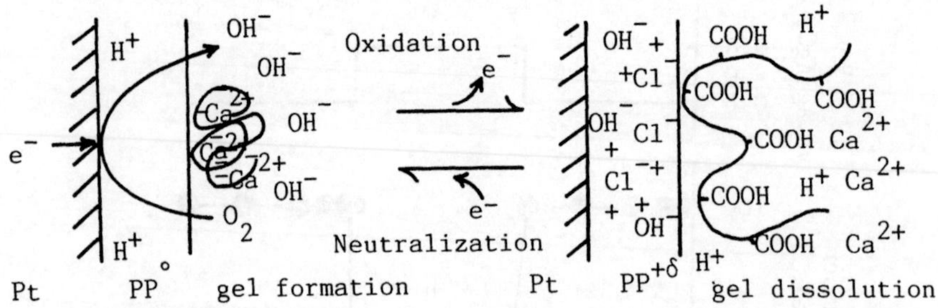

Fig.5 Postulated mechanism of morphological change by potentail control of PMAA-Ca^{2+} microgels which were in contact with a polypyrrole film-coated Pt electrode.

At first, cell morphology change and proliferation of HeLa cells were investigated under a long term potential application. No appreciable change were observed for the cell morphology and proliferation rate, when the potential of the In_2O_3 on which HeLa cells were cultured was controlled in the range between 0 V and 0.5 V vs. Ag/AgCl. On the other hands, the cells drastically changed morphology when they were cultured at a potential region from 0.6V to 0.65V vs. Ag/AgCl. Cells changed their morphology from the usual spindlical shape to circular shape within 3 days, and aggregated into colony form. Furthermore, the rate of proliferation is decreased under the potential application at 0.65V vs. Ag/AgCl, as shown in Figures 6 and 7. Interestingly enough, these morphological change and depression of proliferation were recovered by the termination of the potential application. On the other hand, the application of a potential higher than 0.7V vs. Ag/AgCl results in a lethal effect on the cells.

From these phenomena we conclude that it is possible to control the cell morphology and proliferation rate by low potential application. These results have been motivated us to study the changes in cytoskeleton of HeLa cells, which are mainly constructured by F-actin molecules.

Electric Effects on Cellular Cytoskeletal Changes

At 0V vs. Ag/AgCl, neither morphological nor the fluorescence intensity change (corresponding to F-actin and its distribution) were observed, as shown in Figures 8a and 8b. After an application of 1.0V vs. Ag/AgCl for 1h, viable cells spread over the OTE surface and the corresponding change of intracellular F-actin distribution was observed, as shown in Figures 8c and 8d.

Almost all the HeLa cells became globular and underwent burst after a continuous application of -0.6V V for 30min, which resulted in the fluorescence decrease in the cells, as shown in Figures 8e and 8f. Under a potential application of -0.6V vs. Ag/AgCl, total fluorescence decreased gradually as HeLa cells underwent swelling and burst.

The rhodamine fluorescence was extremely strong in the surrounding of nucleus, of which vicinity the intracellular particles seemed to be coexisted. On the other hand, less intense fluorescence of the rhodamine was seen in the nucleus. This suggests that intracellular particles contains considerable amount of F-actin molecules, and the filament molecules seemed to stay even after the cell burst. However, at the subcellular parts other than the nucleus, the F-actin distribution was considerably decreased at -0.6V vs.Ag/AgCl. At a potential of

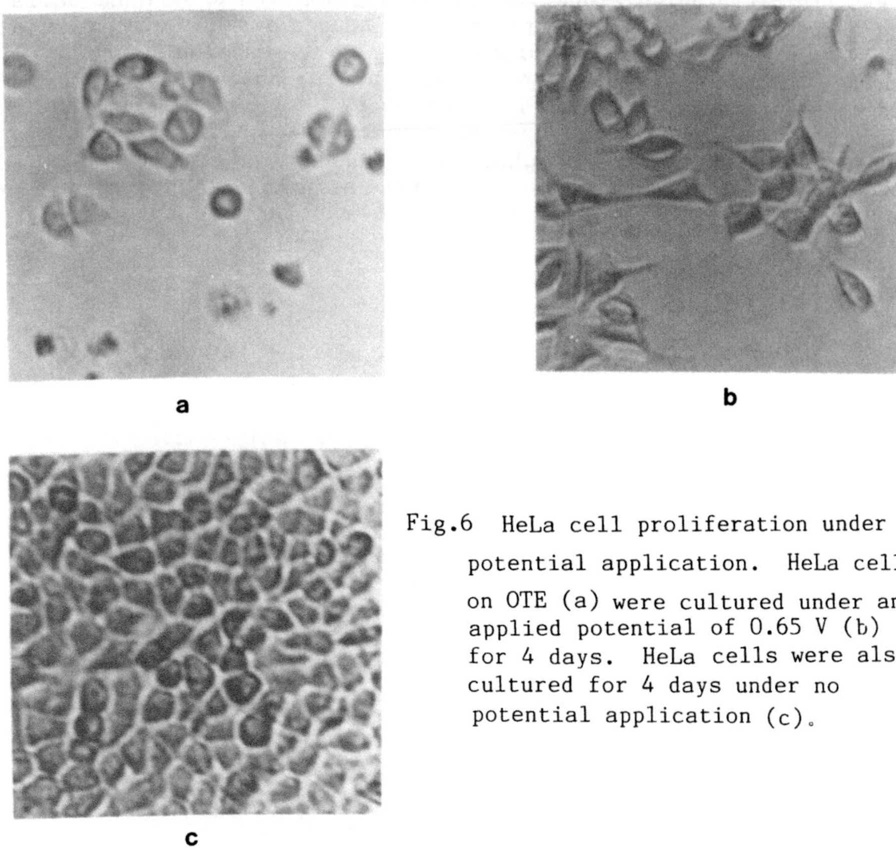

Fig.6 HeLa cell proliferation under potential application. HeLa cells on OTE (a) were cultured under an applied potential of 0.65 V (b) for 4 days. HeLa cells were also cultured for 4 days under no potential application (c).

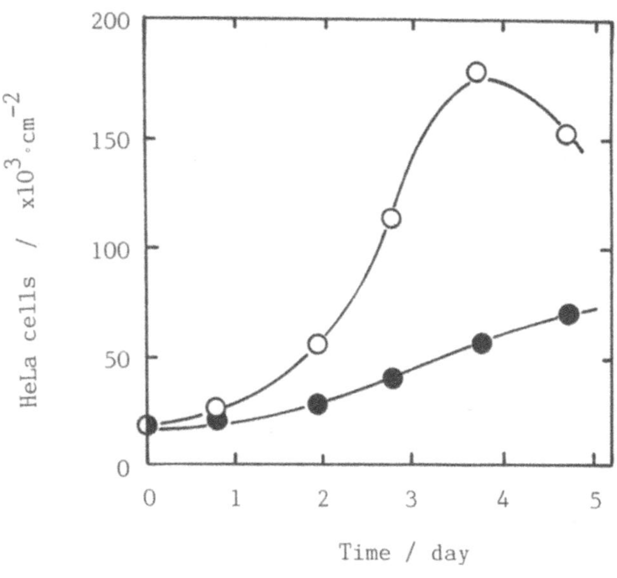

Fig.7 Proliferation curves of HeLa cell cultured on OTEs under potential application. HeLa cells were cultured under no potential application (-○-) or under a potential application of 0.65 V vs.Ag/AgCl (-●-).

1.0V, F-actin distributed along the inside circumference of HeLa cells. This suggests that F-actin localization is induced inside HeLa cells at a positive potential, which resulted in the cell adhesion over an OTE surface. On the other hand, the breakdown of F-actin network may be induced at a more negative potential than -0.6V.

Electrically induced pH decrease and increase over an OTE may be related to these cellular morphological and cytoskeletal changes. Furthermore, different mechanisms of electric effects on cells should be proposed for the positive and the negative potential application, respectively.

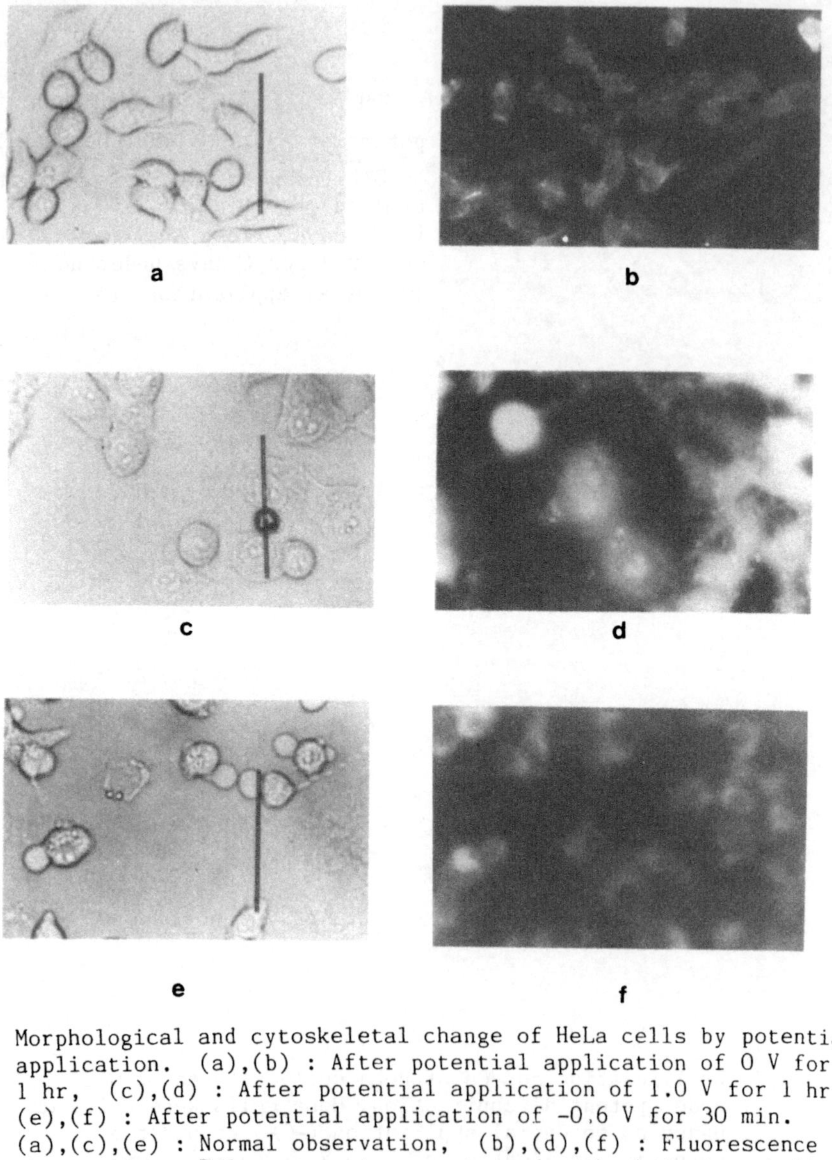

Fig.8 Morphological and cytoskeletal change of HeLa cells by potential application. (a),(b) : After potential application of 0 V for 1 hr, (c),(d) : After potential application of 1.0 V for 1 hr, (e),(f) : After potential application of -0.6 V for 30 min. (a),(c),(e) : Normal observation, (b),(d),(f) : Fluorescence observation, Filament actin was stained with rhodamine phalloidin.

CONCLUSIONS

In this paper, the morphological change and their modulation of synthetic polymer and biopolymer gels were described.

In the first half of the study, rapid and reversible morphological change of polymer microgels was performed on a conducting polymer electrode by small potential shift, which induced ion transfer . The result supports that it is possible to make an actuator system modeled on a neuromuscular system.

In the second half of the study, the effects of low d.c. potential application on living cells were investigated in details. It was found that cellular morphology and proliferation were affected under low potential application. These phenomena were attributed to the structual change of cytoskeleton constructed with F-actin molecules. Though the mechanisms of the present electric effect are not necessarily clear, these phenomena support the possibility of the electric modulation of membrane functions such as cell permeability, mass transport, adhesion, and excitability.

REFERENCES

1. W.Kuhn, B.Hargitay, A.Katchlsky, and H.Eisenberg, Nature, 165, 514 (1950).
2. M.V.Sussman and A.Katchalsky, Science, 167, 45 (1970).
3. Y.Osada and Y.Takeuchi, J.Polym. Sci., Polym.Lett.Ed., 19, 303 (1981).
4. R.Kishi and Y.Osada, J.Chem.Soc., Faraday Trans., 1, 85, 655 (1989).
5. A.Fragala, J.Enos, A.Lalone, and J.Boyack, Electrochim.Acta, 17, 1507 (1972).
6. D.E.DeRossi, P.Chiarell, G.Buzzigoli, and C.Domenici, Trans.Am.Soc. Artif.Intern.Organs, 32, 157 (1986).
7. T.Tanaka, I.Nishino, S-I.Sun, and S.Ueno-Nishino, Science, (Washington,D.C), 218, 467 (1982).
8. Y.Osada and M.Hasebe, Chem.Lett., 1285 (1982).
9. I.V.Yannas and A.J.Grodzinsky, J.Mechanochem.Cell Motility, 2, 113 (1973).
10. H.Shinohara, M.Aizawa, and H.Shirakawa, J.Chem.Soc., Chem.Commun., 87 (1986).
11. H.Shinohara, J.Kojima, and M.Aizawa, J.Electroanal.Chem., 266, 297 (1989).
12. M.Aizawa, T.Toyoshima, M.Yaoita, and Y.Ikariyama, Nihon Kagaku-Kaishi (J.Chem.Soc.Jpn.), 1985, 1302 (1985).
13. T.A.McCoy, M.Mcxwell, and P.F.Kruse, Proc.Soc.Exp.Biol., 100, 115 (1959).
14. P.A.Amats, E.R.Unanue, and D.L.Taylor, J.Cell Biol., 96, 1032 (1977).
15. H.Shinohara and M.Aizawa, Kobunshi Ronbunshuu, 46, 703 (1989).
16. M.Yaoita, Y.Ikariyama, H.Shinohara and M.Aizawa, DENKI KAGAKU, 56, 1086 (1988).

SEPARATION

SEPARATION OF ORGANIC SOLVENT/WATER MIXTURES BY ACRYLAMIDE GEL MEMBRANE SUPPORTED BY SILICA-ALUMINA POROUS MEMBRANE

Shuji Sakohara, Fujikazu Muramoto, Shigeo Sakai,
Masaharu Yoshida and Masashi Asaeda

Department of Chemical Engineering
Hiroshima University
Shitami, Higashi-Hiroshima 724, Japan

INTRODUCTION

Membrane separation of organic solvent/water mixtures, most of which have azeotropic points, has come to attract much attention from view point of energy conservation. Various types of membranes have been used and studied for separation of such mixtures. Most of them are nonporous organic polymer membranes, through which water permeates selectively. Such organic membranes have been improved by incorporating hydrophilic groups into the membrane to obtain high selectivity for water permeation[1]. But, fluxes through such membranes are small, in general. On the other hand, porous inorganic membranes such as silica-alumina porous membrane prepared in our laboratory have quite large flux of water due to the capillary suction pressure[2,3]. But, the selectivity of such porous membranes depends on the pore size and structure or largely on some pin holes left in the membranes[4].

A composite membrane of organic and inorganic porous membranes can be expected to have a high separation factor keeping large flux of water. As an organic material to make such composite membranes, polymeric hydrogels which usually shrink or swell reversibly in aqueous organic solvent solutions corresponding to the solvent concentration[5] are considered to be promising. If the permeation of organic solvent molecules through such composite membranes is blocked by the shrunk network of gel, extremely high separation factor can be expected.

In the present paper, a composite membrane of porous ceramic membrane and acrylamide gel membrane is prepared, and separation properties for various kinds of organic solvent/water mixtures are examined by the pervapouration method. Furthermore, some considerations on separation mechanism are discussed through the shrinking properties of acrylamide gel in such mixtures.

PREPARATION OF MEMBRANE

In preparing a gel membrane, one of the significant problems is that the mechanical strength of gel is too small to make a thin membrane. In this work, acrylamide gel membrane is prepared in the pores of thin ceramic membrane of silica-alumina by copolymerising acrylamide of primary monomer and N,N'-methylenebisacrylamide as crosslinking agent. The prepa-

Polymer Gels, Edited by D. DeRossi *et al.*
Plenum Press, New York, 1991

ration techniques of such porous thin membrane of silica-alumina have been established in our laboratory[2,3]. Most of the pores of this ceramic membrane are considered to have constrictions[3], which is considered to be convenient for preparing the gel in it.

Preparation of silica-alumina thin membrane

The pore size of silica-alumina membrane is very important factor in preparing such composite membrane. If it is too large, it is difficult to make a thin stable film of gel. On the other hand, if it is very small like that prepared in the previous works, far less than 3nm[2,3], it may be difficult to make the gel in it. In this work, although the preparation procedures are basically similar to those in the previous works[2,3], the pore diameter was made larger. The procedures are as follows;

As a supported substrate of silica-alumina thin membrane was used a cylindrical coarse porous ceramic (0.01m in outer diameter, 0.001m in thickness and 0.15m in length) which has high porosity (about 50%) and an average pore diameter of 1μm. Near the surface of such porous substrate was formed a thin membrane using alumina sol which was prepared by hydrolysing aluminium isopropoxide at about 80°C with quite small amount of nitric acid. The procedures are as follows; soaking the sol, drying it and firing it at about 400°C. The pore size of thin alumina membrane can be controlled by the repetition times of above procedures. And then alumina membrane was treated with sodium silicate by similar manner as in the previous papers[2,3]. Finally it was washed in boiling water.

The schematic figure of the membrane module is shown in Fig.1. The thickness of silica-alumina thin membrane is considered to be less than 10μm estimating from the photograph of cross section of membrane module shown in the previous paper[3].

Preparation of acrylamide gel in pores of silica-alumina thin membrane

Acrylamide of primary monomer and N,N'-methylenebisacrylamide as crosslinking agent were copolymerized in the pores of silica-alumina thin membrane. In polymerizing, ammonium persulfate and N,N,N',N'-tetramethylethylenediamine were used as polymerization initiator and accelerator, respectively. In preparing the aqueous solutions of these agent, degassed and distilled water was used. In Table 1 was shown the concentrations of these agents, which was determined empirically through the pretests on the gelation time, the swelling degree of gel, the hardness of gel and so on. The polymerization procedures are as follows;

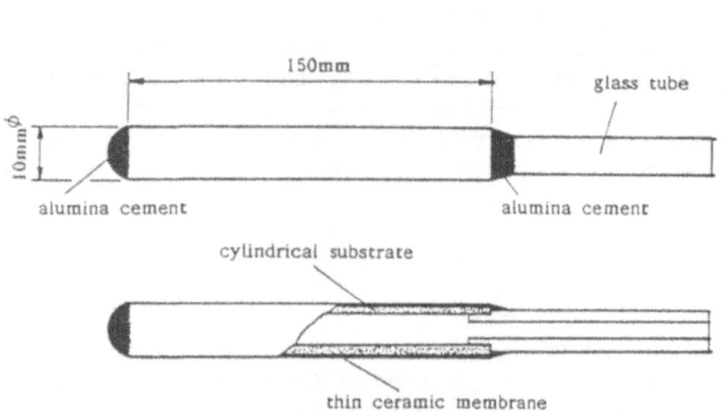

Fig.1 Schematic figure of membrane module

Table 1 Composition of acrylamide gel

acrylamide	1180 mol/m^3-gel
N,N'-methylenebisacrylamide	40.9 mol/m^3-gel
N,N,N',N'-tetramethylethylenediamine	8.9 mol/m^3-gel
ammonium persulfate	53.0 mol/m^3-solution

The pores of silica-alumina thin membrane, the gas in which was previously removed by boiling the membrane module in distilled water, were filled with ammonium persulfate solution by immersing the membrane module into the solution. After the solution of outside of membrane module was removed, the outer surface of membrane was washed with distilled water, and then the membrane module was quickly soaked into the mixture of acrylamide, N,N'-methylenebisacrylamide and N,N,N',N'-tetramethylethylenediamine solutions at 70°C. These operations were done under nitrogen circumstance. The gelation was observed immediately at the surface of membrane, and within several minutes the gelation extended all over the solution. About two hours later, the membrane module was taken out, and the gel attached on the surface of membrane was scraped off. And then the membrane module was washed in boiling water for about two hours.

These preparation procedures were repeated until the leakage of organic solvent molecules through the membrane, which was measured by the pervapouration method mentioned later, became satisfactorily small. In this work, satisfactory result was obtained by repeating more than two times as mentioned later.

EXPERIMENTAL APPARATUS AND PROCEDURE

Separation experiments were performed by the pervapouration method. Figure 2 shows the setup of experimental apparatus which comprises a

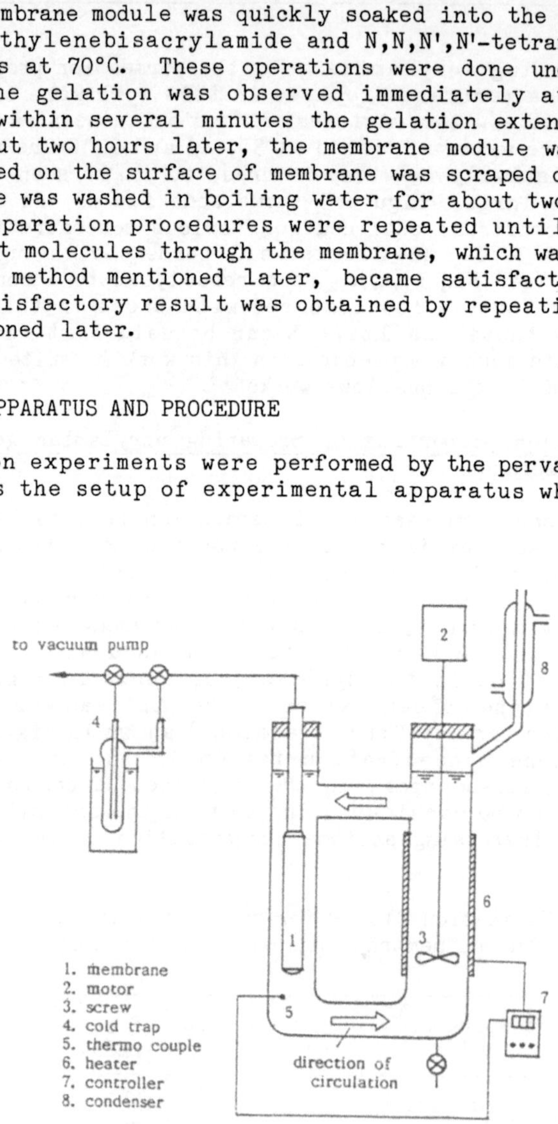

Fig.2 Experimental setup for pervapouration

membrane module, a temperature control system, a vacuum system and a cold trap. Organic solvent/water mixture in the outside of the membrane (upstream) was circulated at a linear speed above 1.0m/s by a screw, and its temperature was controlled at the specified temperature. After the inside of the membrane module (downstream) had been evacuated, the trap was cooled by liquid nitrogen to collect the permeated vapours. The flux was estimated by weighing the trapped permeate. The solvent concentrations in the upstream and in the downstream were determined by a gas chromatograph. The organic solvent/water mixtures used for experiments are methanol/water, ethanol/water, 1-propanol/water, acetone/water, tetrahydrofuran (or THF)/water and 1,4-dioxane/water mixtures.

RESULTS AND DISCUSSION

Permeation properties through silica-alumina thin membrane to support acrylamide gel

Before preparing acrylamide gel, the permeation properties through the silica-alumina membrane prepared in this work were examined by the separation of acetone/water mixtures of various concentrations. Permeation fluxes of pure water measured at 30, 40 and 50°C are shown in Table 2. Despite of relatively low temperatures, it can be seen that the fluxes of water are quite large. Figure 3 shows some fluxes and acetone concentrations in the downstream obtained for various concentrations of acetone in the upstream at 30°C. Acetone is well permeable through the membrane as well as water (see Fig.3 (a)). As a result, acetone concentrations in the downstream are not so low compared with those in the upstream (see Fig.3 (b)). From these results, it can be said that the pore size of silica-alumina thin membrane prepared in this work is quite large compared with that prepared in the previous works[2,3].

Change of permeation properties by preparing acrylamide gel in pores of silica-alumina membrane

In Fig.4 is shown an example of separation results by the composite membrane prepared acrylamide gel in the pores of silica-alumina membrane for acetone/water mixtures of various concentrations at 50°C. By the first preparation of gel, the permeation flux for pure water stream decreased 223mol/m^{2}h from 353mol/m^{2}h which was measured before preparing gel (see Table 2). The flux of water decreased with the increase of acetone concentration in the upstream, as shown in Fig.4 (a). On the other hand, as acetone concentration in the upstream was increased, the leakage of acetone (or the flux of acetone) shown in Fig.4 (b) increased in the low acetone concentration region in the upstream, and then decreased, and increased again in the high acetone concentration region. This phenomenon can be considered due to that the network of acrylamide gel shrinks with increasing acetone concentration in the upstream, which

Table 2 Permeation fluxes for pure water through a silica-alumina membrane prepared in this work

temperature [°C]	flux of water [mol/m^{2}h]
30	149
40	229
50	343

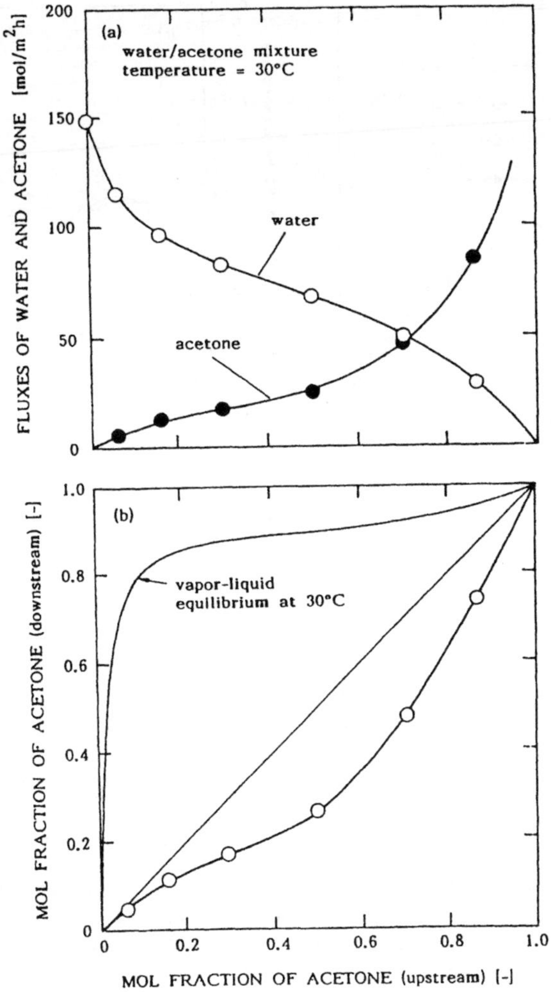

Fig.3 Observed result for separation of acetone/water mixture by a thin silica-alumina membrane.
(a) Fluxes of water and acetone.
(b) Mol fraction of acetone in the downstream.

is discussed in detail later. Acetone concentration in the downstream shown in Fig.4 (c) was rather high, and separation factor, α, defined by Eq.(1) was not high, less than 200.

$$\alpha = x/(1-x)\cdot(1-y)/y \qquad (1)$$

where, x and y are the mol fractions of acetone in the upstream and in the downstream, respectively.

To make the leakage of acetone decrease, the preparation procedures of acrylamide gel were repeated. The fluxes of water slightly decreased in the low acetone concentration region in the upstream with the repetition times. However, in the high acetone concentration region, the decrease of water flux scarcely observed (see Fig.4 (a)). On the other hand, the leakage of acetone remarkably decreased by repeating the preparation of gel more than two times, and it became satisfactorily small. And the

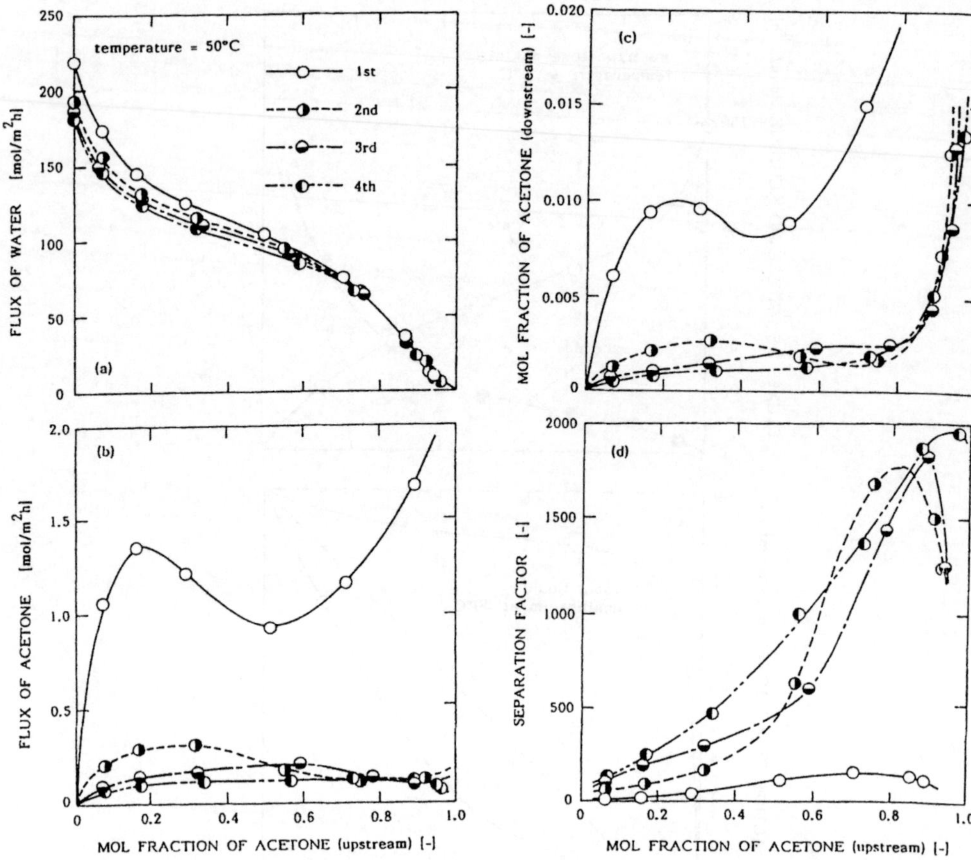

Fig.4 Observed results for separation of acetone/water mixture by a composite membrane. Keys show the number of repeating times of preparation procedures of gel.
(a) Flux of water.
(b) Leakage or flux of acetone.
(c) Mol fraction of acetone in the downstream.
(d) Separation factor, $\alpha = x/(1-x)\cdot(1-y)/y$,
where, x and y are the mol fractions of acetone in the upstream and in the downstream, respectively.

peculiar change of leakage with the increase of acetone concentration in the upstream mentioned above was similarly observed (see Fig.4 (b)). After the 4th preparation, acetone concentration in the downstream became less than 0.5mol% at 90mol% in the upstream, and the separation factor reached about 2000 (see Figs.4 (c) and (d)). From these results, it can be said that the separation factor of this kind of membrane can be made extremely high by repeating the preparation of gel without remarkably decreasing the flux of water.

Separation results for various kinds of organic solvent/water mixtures

In Fig.5 is shown some separation results for organic solvent/water mixtures at 50°C. These results were obtained with the improved composite membrane, which was prepared by repeating the preparation procedures of gel 6 times using a silica-alumina membrane having larger flux of water

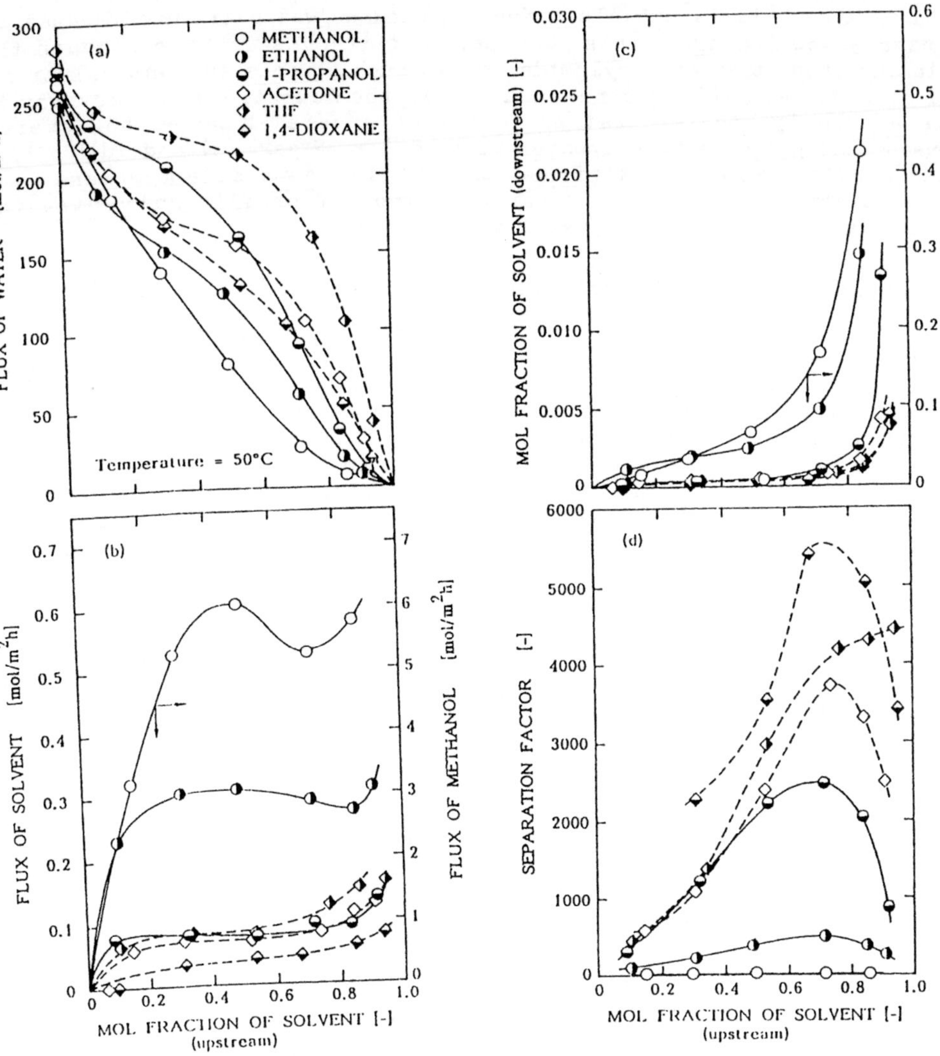

Fig.5 Some separation results of organic solvent/water mixtures by an improved composite membrane.
(a) Flux of water.
(b) Leakage of organic solvent.
(c) Mol fraction of organic solvent in the downstream.
(d) Separation factor.

than that shown in Fig.3. The flux of water is quite large for each mixture (see Fig.5 (a)). But, it depends on the component contained in the mixture, which is considered closely related to the affinity between water and solvent molecules, i.e., the osmotic pressure[4]. Furthermore, if examined in detail, the permeation flux for pure water increases by repeating separation experiments which were done in the order of ethanol/water, acetone/water, methanol/water, 1,4-dioxane/water, 1-propanol/water and THF/water mixtures. This phenomenon is considered owing to the hydrolysis of amide groups in the gel, which is discussed later. On the other hand, the leakage of solvent is extremely small except for methanol (see Fig.5 (b)). The peculiar change of leakage with the increase of solvent concentration in the upstream mentioned above can be seen for each mixture, though it is not clear for 1,4-dioxane/water mixture because of extremely low leakage. This phenomenon is discussed later through the shrinking properties of acrylamide gel in such aqueous solvent solutions. As a result, the solvent concentration in the downstream is very low to give extremely high separation factor of several hundred to several thousand except for methanol/water mixture (see Figs.5 (c) and (d)). From these results, it can be said that this kind of composite membrane have high separation performance for various kinds of organic solvent/water mixtures except for methanol/water mixture.

Shrinking properties of acrylamide gel in aqueous solvent solutions

The separation properties of organic solvent/water mixtures by the composite membrane shown above may be closely related to the shrinking properties of acrylamide gel in aqueous solvent solutions. From this point of view, acrylamide gel of rod-shape composed of the same composition as that used for the preparation of gel membrane (see Table 1) was synthesized, and the shrinking degree and the solvent concentration in the gel were examined in methanol/water, ethanol/water, acetone/water and THF/water solutions. The synthesis procedure of rod-shape gel and the measurement of shrinking degree are similar to those of many publications[5,6], and the solvent concentration in the gel was estimated from the changes of weight of gel and the solvent concentration in the solution surrounding the gel.

In Figs.6 (a) and (b) are shown the dimensionless gel volume and the solvent concentration in the gel, respectively. The gel volume decreases

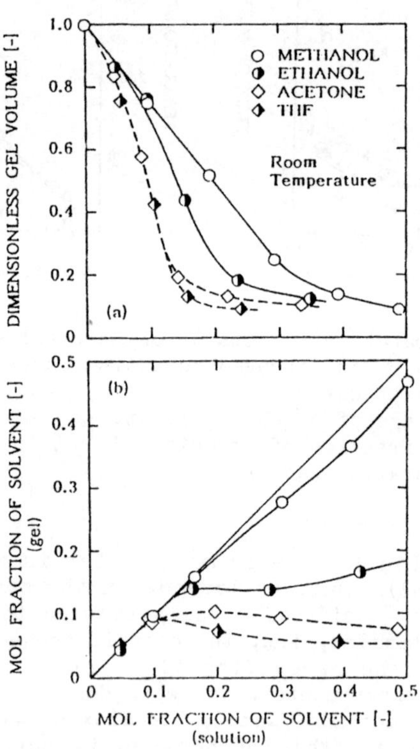

Fig.6 Shrinking properties of acrylamide gel in aqueous organic solvent solutions.
(a) Dimensionless gel volume.
(b) Mol fraction of solvent in the gel.

sharply with increasing solvent concentration in the solution for each mixture. But, even if the solvent concentrations in the solutions are the same, the gel shrinks smaller in the order of methanol, ethanol, acetone and THF solutions (see Fig.6 (a)). The order corresponds to that of molecular size. The solvent concentrations in the gel are also lower in the same order (see Fig.6 (b)). In methanol solution, methanol concentration in the gel increases monotonously with increasing methanol concentration in the solution, though it is slightly lower compared with that in the solution. On the other hands, in other solvent solutions, solvent concentrations in the gel increase at the first, and then remaine almost unchanged or decrease gradually in the experimental range. From these results, it can be said that relatively large size molecules such as acetone or THF are hard to come into the gel according to the shrinking of network, and that relatively small size molecules such as methanol are easy to come into the gel with water molecules.

Comparing these results and the separation results, it can be said that the separation performance of this kind of membrane depends on the shrinking degree of gel and the solvent molecular size. The permeation of relatively large size molecules such as 1,4-dioxane or THF through the membrane is blocked easily by the shrunk network of gel. Therefore, the separation factors for such solvent/water mixtures are extremely high. But, for relatively small size molecules, it is difficult to block the permeation through the membrane, even if the network of gel membrane shrinks. Therefore, the separation factor for methanol/water mixture is lower compared with that for other mixtures. Furthermore, the peculiar behavior of solvent leakage with the increase of solvent concentration in the upstream of membrane mentioned above is considered due to that the network of gel membrane shrinks with the increase of solvent concentration in the upstream.

However, the solvent concentration in the downstream of the composite membrane shown in Fig.5 (c) seems to be extremely low compared with that in the shrunk gel of rod-shape examined here. This result cannot be explained only by the shrinking properties of acrylamide gel. It is probably owing to the preparation method of gel membrane employed in this work. In preparing gel membrane, the preparation procedures were repeated until the leakage of solvent became satisfactorily small as mentioned above. In such preparation method, the network formed later will be tangled in the previously formed network to make the size of network smaller enough to block the permeation of solvent molecules. It is considered that the preparation method of gel membrane employed in this work also contributes to the high separation performance as well as the shrinking properties of acrylamide gel.

Change of separation performance of composite membrane with time

The permeation flux for pure water became large by repeating the separation experiments, which were done in the order of ethanol/water, acetone/water, methanol/water, 1,4-dioxane/water, 1-propanol/water and THF/water mixtures as mentioned above (see Fig.5 (a)). To examine this phenomenon in detail, the separation experiments for the same kind of mixture were repeated after 1 week. After the first experiment was finished, the membrane module was washed and immersed in distilled water until the second experiment was begun. In Fig.7 is shown an example of the results obtained for acetone/water mixture. The flux of water increases while the leakage of acetone decreases (see Fig.7 (a)). As a result, the separation factor increases (see Fig.7 (b)). The similar results were obtained also for other solvent/water mixtures. In general, it is well known that the $-CONH_2$ groups are easily hydrolyzed to the $-COOH$ groups. Therefore, it can be considered that the $-CONH_2$ groups in the original gel were partially hydrolized to the $-COOH$ groups which are probably more hydrophilic than the $-CONH_2$ group. Although the separation

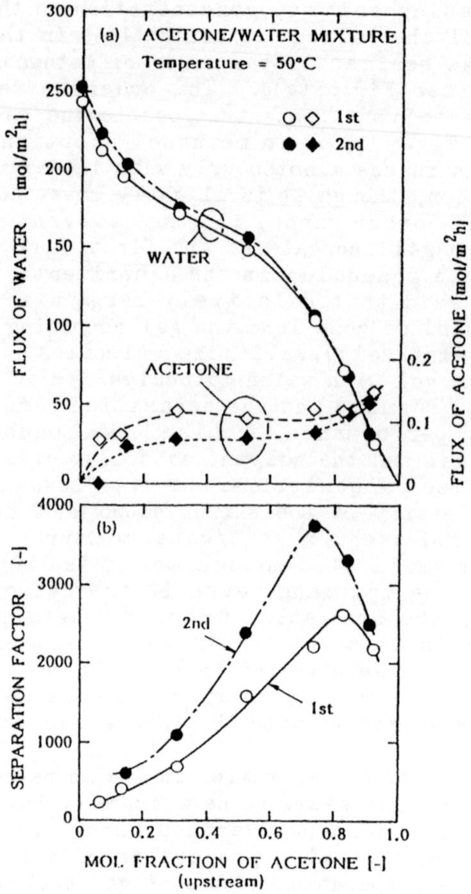

Fig.7 An example of changes of separation performance with time. The 2nd experiment was done after one week of the 1st experiment.
(a) Fluxes of water and acetone.
(b) Separation factor.

performance changed like that with time, it did not decraese during experimental period, for about three months. It is considered that the gel is prepared stably in the pores of silica-alumina membrane.

CONCLUSION

Acrylamide gel membrane prepared in the pores of silica-alumina thin membrane have fairly high separation performance for various kinds of organic solvent/water mixtures. The separation factor reached several hundred to several thousand except for methanol/water solution. Such high separation factor is considered due to that the permeation of organic solvent molecules is blocked by the shrunk network of gel. But, for relatively small size molecules such as methanol, it is difficult to block the permeation even if the network shrinks. The preparation method of gel membrane employed in this work and the properties of silica-alumina thin membrane such as pore size and structures seem to give large contribution to membrane characteristics, which are to be investigated further.

Reference

1. M.Yoshikawa, T.Yukoshi, K.Sanui and N.Ogata, J. Polym. Sci., Part A, 24: 1585 (1988)
2. M.Asaeda and Luong D. Du, Kagaku Kogaku, 48: 682 (1984)
3. M.Asaeda and Luong D. Du, J. Chem. Eng. Japan, 19: 72 (1986)
4. S.Sakohara, S.Kitao, M.Ishizaki and M.Asaeda, Proc. 1st Int. Conf. Inorganic Membrane (1989)
5. T.Tanaka, Scientific American, 244: 110 (1981)
6. Y.Nakano, H.Naruoka and M.Murase, J. Chem. Eng. Japan, 19: 274 (1986)

PERMEATION MECHANISM FOR A THERMO-SENSITIVE SWITCHING-FUNCTIONAL COMPOSITE MEMBRANE OF POROUS GLASS AND HYDROGEL

Mikio Konno, Tomoya Tsuji and Shozaburo Saito

Department of Chemistry and Engineering
Tohoku University
Sendai 980, Japan

ABSTRACT

An attempt was made to prepare a thermo-sensitive membrane with switching ability. The membrane was made of porous glass and N-isopropylacrylamide gel, whose volume-phase transition is of a thermo-shrinking type. Permeation experiments have demonstrated that the composite membrane functioned as an ultrafiltration membrane, of which the cut-off molecular weight and solution permeation rate dramatically changed on the volume-phase transition. To make clear the permeation mechanism of the membrane, additional experiments were carried out, in which the diffusion rates of ^{3}H-labeled water through the membrane were measured at various temperatures. Experimental results enabled the pore radii in the gel to be estimated. It is shown that the rejection characteristics of the membrane are predictable from the pore radii estimated.

INTRODUCTION

Volume-phase transition is a peculiar phenomenon observed with hydrogels (Tanaka, 1981). At the transition, the volume of a hydrogel will change discontinuously in response to continuous changes in surrounding conditions such as temperature (Hirokawa and Tanaka, 1984), solvent composition (Tanaka et al.,1980), pH (Tanaka et al., 1980), electrolyte concentration (Ohmine and Tanaka, 1982) and electric field (Tanaka et al., 1982). Since the volume-phase transition brings about dramatic changes in physical properties of the gels, this phenomenon is expected to be applied to the creation of new types of materials with switching abilities.

One application is the synthesis of switching-functional membranes. However, the mechanical weakness of hydrogels must be overcome for such an application. This can be attained by making a composite membrane of a gel and other materials. Recently the authors (Otake et al., 1988) have proposed a new preparation method for the composite membrane, which should be applicable to various combinations of hydrogels and porous inorganic substances. In the present work, an attempt has been made to prepare a thermo-sensitive membrane consisting of porous glass and N-isopropylacrylamide gel whose volume-phase transition was induced by the change of temperature (Hirokawa et al., 1984). The membrane prepared was subjected to permeation experiments at various temperatures, and also to diffusion experiments which were carried out to make clear the permeation mechanism of the membrane. From the experimental results, we proposed a permeation model for the composite membrane.

1. EXPERIMENTAL

1.1 Preparation of composite membrane

Porous glass plate (Asahi Glass Co. Ltd., average pore size=6000A) was used as a support, on which crosslinked poly(N-isopropylacrylamide) was chemically fixed according to the following procedure.

At first, the surface of the porous glass was modified by the following reaction in order to introduce double bonds:

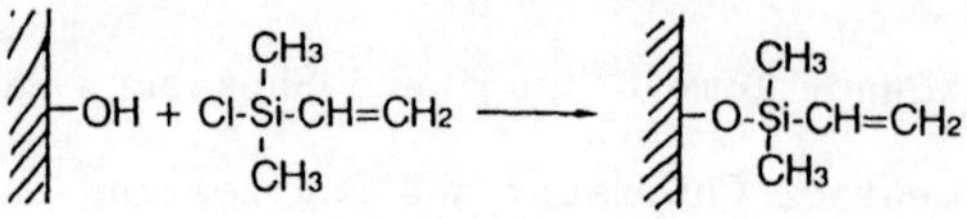

The surface modification was performed in a benzene solution of the surface modification reagent, dimethylvinylchlorosilane, at a concentration of 4.2wt.% and 25°C for 4 hours. Then the porous glass plate was soaked in an aqueous solution, which contained reactants, N-isopropylacrylamide as a monomer, N,N'-methylenebisacrylamide as a crosslinking reagent and $K_2S_2O_8$ as an initiator. To start radical polymerization, an accelerator reagent N,N,N',N'-tetramethylethylenediamine was added to the aqueous solution up to a concentration of 3.33×10^{-4}vol.%. The polymerization was performed for 40 minutes at 0°C. The initial concentration of the monomer was 10wt.%, that of the initiator was 0.023wt.%, and the mole ratio of the monomer to the crosslinking reagent was 82. During the polymerization, the gel formed was chemically fixed on the surface of the porous glass by reaction with the vinyl groups introduced by the surface modification.

Figure 1 presents scanning electron micrographs of the composite membrane. From the cross-sectional view of the membrane, the gel layer was found to be formed on the surface of the porous glass. The thickness of the layer was about 3-5 μm. No pinholes were observed in the top view of the membrane.

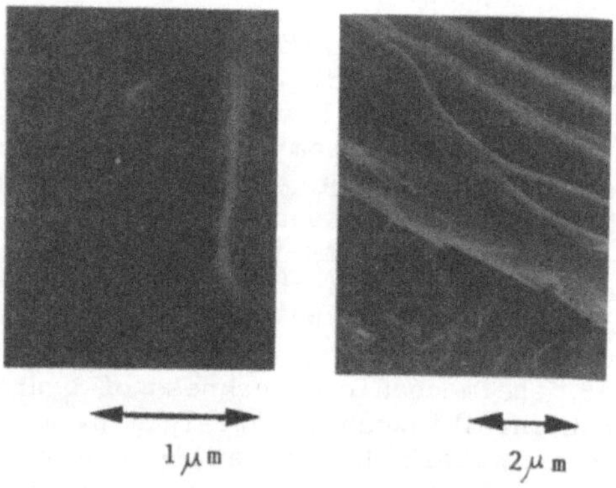

Fig.1. Scanning electron micrograph of a N-isopropylacrylamide gel and porous glass composite membrane

1.2 Swelling experiments

Prior to permeation experiments, the swelling behavior of the gel was observed to check its transition temperature. Since it was difficult to observe the gel supported on the porous glass, a bulk gel of the same composition as that in the membrane was prepared.

The gel was rod-shaped, 10mm long and 1.65mm in diameter. Equilibrium swelling volume of the gel was measured in water to which a small amount of 2-propanol was added in order to prevent the breeding of microorganisms in the gel (Chmelir et al, 1980).

Figure 2 shows equilibrium swelling ratio V/V_0, where V is the equilibrium volume of the gel and V_0 is its initial volume. The swelling curve clearly shows the occurrence of the volume-phase transition at 33°C.

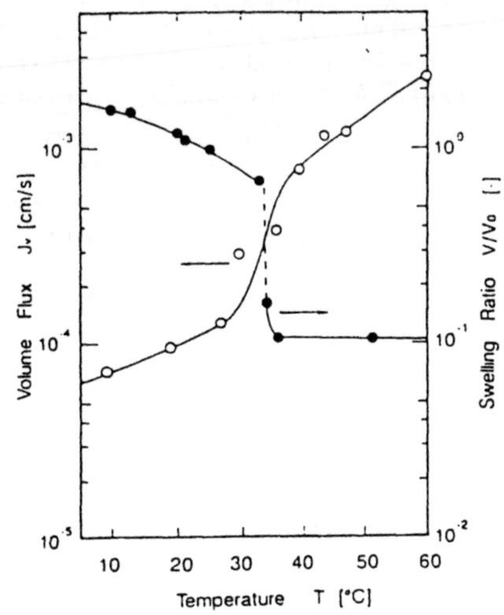

Fig.2. Swelling ratios of bulk gel and volume fluxes across composite membrane at various temperatures

1.3 Permeation experiments

In the permeation experiments, the volume fluxes of solvent and the rejections of polymer solutes were measured at various temperatures around the transition temperature. The solvent used was 3wt.% 2-propanol aqueous solution, and the solutes used were polyethylene glycol(PEG) in six different molecular weights, which are listed in Table 1.

The membrane was placed in a permeation cell equipped with a stirrer, as shown in Fig.3. Effective membrane area in the cell was $3.14cm^2$, cell volume 500 cm^3(diameter=53mm, height=218.4mm), and the diameter of the stirrer 43mm. All experiments were carried out at a solute concentration of 0.001wt.% and an impeller speed of 500 rpm in a temperature range of 5-60°C. Pressure difference between the feed and permeate solutions was set at $10Kg/cm^2$, and the pressure of the permeate solution was maintained at atmospheric pressure. The solute concentration in the permeate solution was measured by liquid chromatography, and the observed rejection R_{obs} defined by Eq.(1) was calculated:

$$R_{obs}=1-C_p/C_f \qquad (1)$$

Table 1. Molecular weight and molecular distribution of PEG

Solute	M_n	M_w/M_n
PEG 400	554	1.07
PEG 1000	985	1.03
PEG 2000	1520	1.07
PEG 7500	4524	1.07
PEG 20000	16259	1.36
PEG 45000	19263	1.36
PEG 85000	57717	1.18

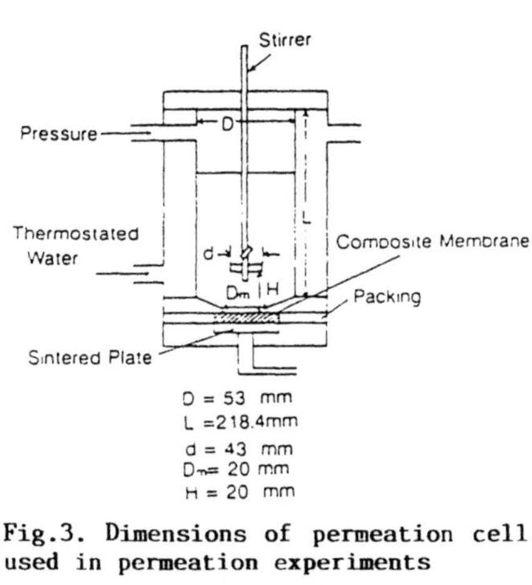

Fig.3. Dimensions of permeation cell used in permeation experiments

where C_p is the solute concentration in the permeate solution and C_f that in the feed solution.

Figure 2 shows the volume fluxes through the composite membrane together with the swelling ratio of a bulk gel. The volume flux changed significantly around the transition temperature of the bulk gel.

Figure 4 shows the rejections of PEG measured with the composite membrane at two different temperatures, 25 and 45°C. As can be seen from Fig. 2, the former temperature corresponds to the swollen state of the gel, and the latter to the shrunken state. Fig. 4 shows that the composite membrane is usable in ultrafiltration and that its rejection behavior drastically changes according to the swollen state of the gel.

In a previous work (Otake et al., 1988; Tsuji et al., 1989), the authors reported the permeation characteristics of a composite membrane of porous glass and acrylamide/acrylic acid random copolymer gel. In the swollen state of the gel, the membrane revealed low volume flux and low cut-off molecular weight of polymer solutes. Qualitatively, this permeation behavior is similar to that shown in Figs. 2 and 4.

1.4 Diffusion experiments

The results of the permeation experiments suggest that the gel supported on the porous glass has micro-pores with sizes comparable to those of the polymer solutes. Possibly the number and size of the pores might thus change greatly according to the swollen state of the gel. To obtain other information on this problem, diffusion experiments were performed using a radio active tracer, whose diffusion rates across the membrane were measured at different temperatures. The radio active tracer used was ^{3}H-labeled water (HTO), whose specific activity was 5mCi/cm^3. In the experiments the membrane was placed in a diffusion cell which had a chamber on each side of membrane. Effective membrane area in the cell was 5.73 cm^2 and the volume of each chamber 30 cm^3. The aqueous solution of HTO was put into one of the chamber and, at same time, the same amount of unlabeled water was added to the opposite chamber. Then the time-variation of ^{3}H concentration was detected, and used for the analysis with a model which will be described in the next section.

2. A PERMEATION MODEL FOR THE COMPOSITE MEMBRANE

To facilitate analysis for permeation and diffusion experiments, we now propose a permeation model for the composite membrane. In the

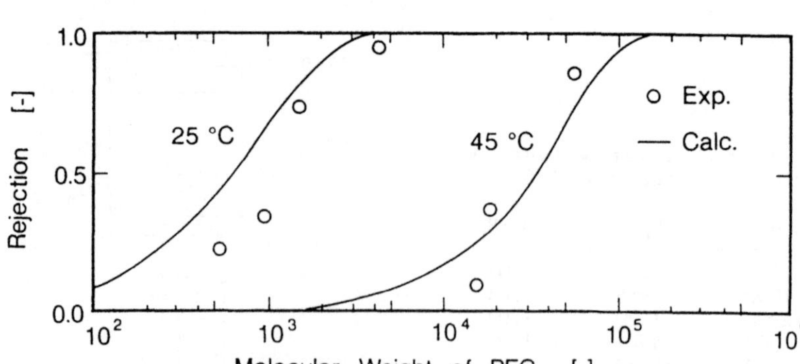

Fig.4. Experimental and calculated results of rejection

permeation model, the following assumptions were made for the structures of the composite membrane:

1) A gel layer with a uniform thickness of l_g spreads over the surface of the porous glass plate with a thickness of l_s.

2) The tracer, HTO, diffuses along the pores in both the gel and the support, but does not diffuse through regions other than the pores.

3) Concentration profile in the composite membrane can be approximated by a combination of linear distributions, as illustrated in Fig.5, since characteristic diffusion times defined as D_{HTO}/l_g^2 and D_{HTO}/l_s^2 are very small, where D_{HTO} is the diffusion coefficient of the solute.

4) The pores in the gel have an uniform size with a radius r_p.
Therefore, the following diffusion equation is obtained:

$$V_c \frac{dC_{Tf}}{dt} = -D_{HTO}A\frac{S_g}{l_g}(C_{Tf}-C_{Ti}) = -D_{HTO}A\frac{S_s}{l_s}(C_{Ti}-C_{Tp})$$

$$= -V_c\frac{dC_{Tp}}{dt} \tag{2}$$

where C_{Tf} and C_{Tp} are the ^3H concentrations in the solutions on the upstream and downstream sides respectively, A is effective membrane area, V_c the volume of the solutions, C_{Ti} the ^3H concentration at the boundary between the gel and the porous glass, S_g the ratio of total cross-sectional pore area to the effective membrane area in the gel and S_s is that in the porous glass. From Eq.(2) the following equation is derived

$$\ln\left\{1-2\frac{C_{Tp}}{C_{Tf,0}}\right\} = -\frac{2D_{HTO}A}{V_c}\left(\frac{l_g}{S_g}+\frac{l_s}{S_s}\right)^{-1} t \tag{3}$$

where $C_{Tf,0}$ is the initial value of C_{Tf}. According to this equation, the results of the diffusion experiments at 25 and 45°C were plotted in semi-logarithmic coordinates against diffusion time in Fig. 6. As expected, the experimental results are expressed by a linear relation. As a consequence, the values of $2D_{HTO}A/V_c(l_g/S_g+l_s/S_s)^{-1}$ were determined from the results of the diffusion experiments.

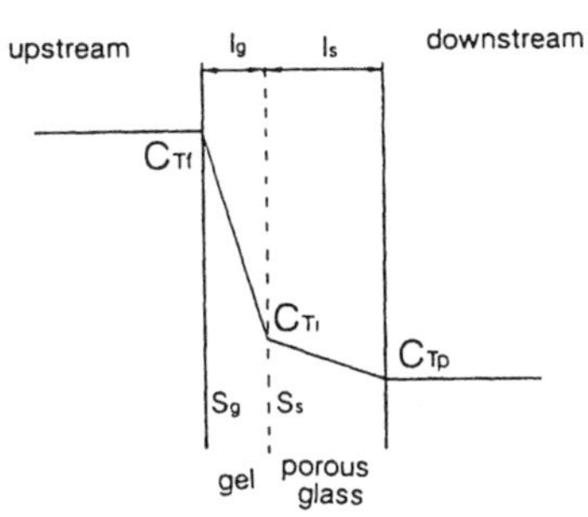

Fig.5. Profile of concentration distribution of ^{3}H-labeled water during diffusion.

In diffusion across the porous glass with no gel supported, the term of l_g/S_g in Eq.(3) is neglected and, therefore, the value of l_s/S_s in Eq.(3) can be determined. Fig. 6 presents some examples of diffusion experiments conducted solely with the porous glass. The experimental results indicate that the time-variation of the concentration accords with the expectations of Eq.(3). Thus the values of l_s/S_s were determined.

The combination of the diffusion experiments with the composite membrane and the porous glass made possible the determination of the values of $2D_{HTO}A/V_c(l_g/S_g)^{-1}$. From these values we can calculate the values of l_g/S_g by using the following correlation equation, which was obtained for the self-diffusion coefficients of HTO (Wang et al., 1953).

$$D_{HTO}=2.31x10^{-5}exp(1.26x10^{-2}/T) \quad (4)$$

Figure 7 presents the plots of S_g/l_g against temperature. The value of S_g/l_g dropped suddenly on the collapse of the gel, resulting in a sudden decrease in the diffusion rate. The value of S_g/l_g can be related to the pore structure in the gel by the following relation (Wendt et al., 1979):

$$S_g/l_g=n\pi r_p^2/l_g \quad (5)$$

where r_p is pore radius and n the number density of the pores.

Other information on the pore structure can be obtained from the solution permeation induced by pressure difference. In the permeation

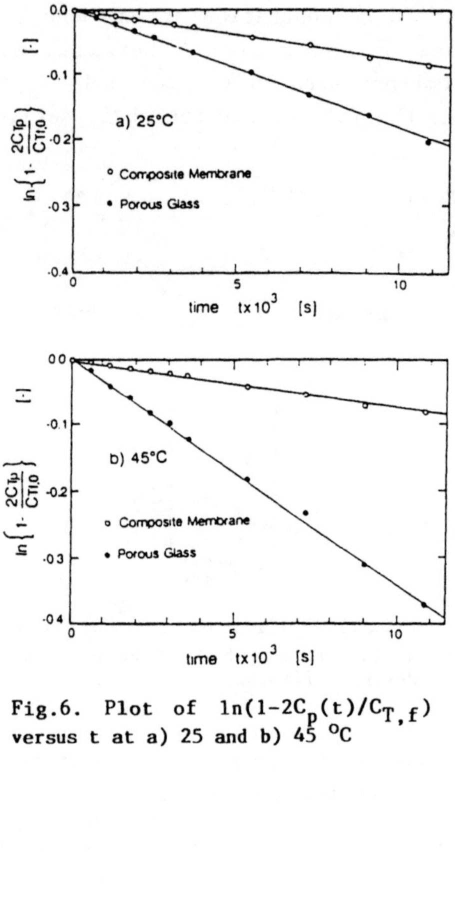

Fig.6. Plot of $\ln(1-2C_p(t)/C_{T,f})$ versus t at a) 25 and b) 45 °C

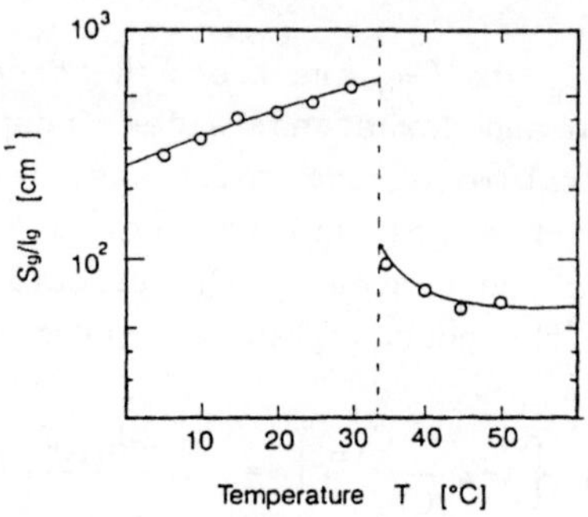

Fig.7. Temperature dependence of S_g/l_g

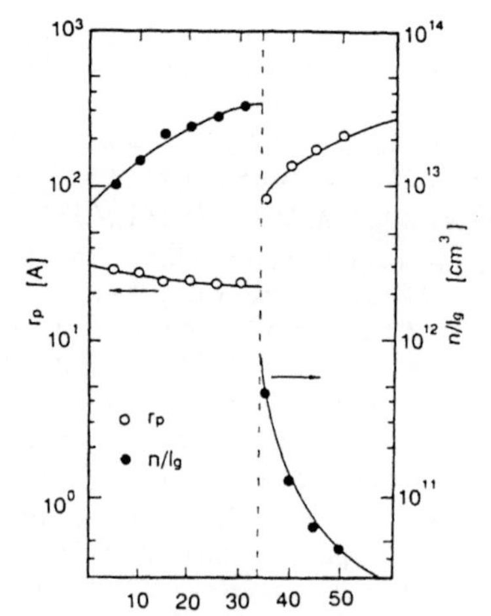

Fig.8. Temperature dependence of r_p and n/l_g of the gel

experiments it was found that the resistance of the porous glass to the permeation can be neglected in comparison with that of the gel supported. Therefore the following equation can be assumed for the solution permeation rate under the viscous flow conditions.

$$J_v = \frac{n \pi r_p^4}{8 \eta l_g} \Delta P \tag{6}$$

where η is the viscosity of permeation solution. The strong dependence of J_v on r_p in the equation suggests the reason why the gel supported on the porous glass acted as the main resistance to the solution permeation across the composite membrane. From the experimental values of volume flux, the values of nr_p^4/l_g were determined on the basis of Eq.(6). Since the values of nr_p^2/l_g were already determined in the diffusion experiments, the values of r_p and n/l_g were calculated, and are presented in Fig. 8. As can be seen, the values of r_p and n/l_g changed discontinuously at the temperature of the volume-phase transition of the gel. Since l_g is thought to decrease on the collapse of the gel, Fig. 8 indicates that the number density of the pores decreased by more than one order on the collapse of the gel. On the other hand, Fig. 8 shows that the pore radius r_p increases with temperature around the transition temperature. Qualitatively the increase in r_p explains the change in the rejection shown in Figure 4. To examine the validity of the pore radius determined, the rejection of the composite membrane was calculated.

On the basis of nonequilibrium thermodynamics the true rejection R_{true} is expressed by the following equation (Spiegler and Kedem, 1966; Verniory et al., 1973):

$$R_{true} = \sigma(1-F)/(1-\sigma F) \tag{7}$$

where $F = \exp\{-J_v(1-\sigma)/P_s\}$ (8)

and σ is a refraction coefficient, and P_s the solute permeability. The following relations are given by a pore model (Verniory et al., 1973; Wendt et al., 1979)

$$\sigma = 1 - g(q)\{2(1-q)^2 - (1-q)^4\} \tag{9}$$

$$P_s = D_s f(q)(1-q)^2 n \pi r_p^2 / l_g \tag{10}$$

$$J_v = (n \pi r_p^4 / 8 \eta l_g)(\Delta P - \sigma \Delta \pi) \tag{11}$$

where $f(q) = (1-2.1q+2.1q^3-1.75q^5+0.76q^6)/(1-0.76q^5)$ (12)

$$g(q) = (1-0.67q^2-0.2q^5)/(1-0.76q^5) \tag{13}$$

$$q = r_s/r_p \tag{14}$$

and r_s is the radius of a solute and $\Delta\pi$ the osmotic pressure difference. According to the Flory-Fox equation, r_s is given by

$$[\eta] = 3.1 \times 10^{33} r_s^{3/2}/M_w \tag{15}$$

where $[\eta]$ is the limiting viscosity, which is calculated from the following correlation equations determined for Bailey and Nishi's experimental data:

$$[\eta] = 7.96 \times 10^{-4} M_w^{0.62} \quad \text{(for } 25^{\circ}\text{C)} \tag{16}$$

$$[\eta]=6.90\times10^{-5}M_w^{0.87} \quad \text{(for 45}^{o}\text{C)} \tag{17}$$

The value of D_s is calculated from the following equation:

$$D_s=kT/6\pi\eta r_s \tag{18}$$

where k is the Boltzmann constant. The osmotic pressure difference in Eq.(11) is calculated from the van't Hoff equation:

$$\Delta\pi=(C_f-C_p)RT \tag{19}$$

The calculated results of R_{true} are presented in Fig.4. This figure shows that, when temperature changes from 25 to 45°C, the true rejection curve shifts by the same extent in molecular weight as the curve of R_{obs}. This strongly suggests that the change of the pore radius with temperature mainly controls the rejection characteristics of the composite membrane.

4. CONCLUSION

A thermo-sensitive membrane with switching ability was synthesized, which was composed of porous glass and N-isopropylacrylamide gel. The switching ability of the membrane was demonstrated in permeation experiments. To make clear the permeation mechanism of the membrane, additional experiments were also conducted, in which the diffusion rates of HTO across the membrane were measured. In the permeation experiments the membrane was found to function as an ultrafiltration membrane. The cut-off molecular weight of the membrane increased by more than one order on the collapse of the gel, accompanied by a sudden increase in the permeation rate. On the other hand, in the diffusion experiments, it was shown that the diffusion rate decreased on the collapse of the gel.

These experimental results were interpreted by a model which assumes changes in the number and radius of pores in the gel according to the degree to which it is swollen. The values of the pores were determined from the permeation rates and the diffusion rates of HTO. The values determined showed that the pore radius increased on the collapse of the gel, while the number of pores decreased. The determined values of the pore radii were used to calculate the rejection characteristics of the membrane. The model succeeded in explaining the differences in the permeation characteristics observed in the different swollen states of the gel.

NOMENCLATURE

A	=effective membrane area	$[cm^2]$
C_f	=solute concentration in feed solution	$[mol/cm^3]$
C_p	=solute concentration in permeate solution	$[mol/cm^3]$
C_{Tf}	=^{3}H concentration on upstream side	$[Bq/cm^3]$
$C_{Tf,0}$	=initial value of C_{Tf}	$[Bq/cm^3]$
C_{Tp}	=^{3}H concentration on downstream side	$[Bq/cm^3]$
D_{HTO}	=self-diffusion coefficient of ^{3}H-labeled water	$[cm^2/s]$
D_s	=diffusion coefficient of solute	$[cm^2/s]$
J_v	=volume flux	$[cm^3/cm^2\cdot s]$
k	=Boltzmann's constant	$[J/K]$
l_g	=thickness of gel layer	$[cm]$
l_s	=thickness of porous glass	$[cm]$
M_w	=molecular weight of solute	$[-]$

n	=pore density	$[cm^{-2}]$
P_s	=solute permeability	$[mol/cm^2 \cdot Pa]$
ΔP	=pressure difference	[Pa]
R	=gas constant	[J/mol K]
R_{obs}	=observed rejection	[-]
R_{true}	=true rejection	[-]
r_p	=pore radius	[cm]
r_s	=solute radius	[cm]
S_g	=ratio of total cross-sectional pore area to effective membrane area in gel	[-]
S_s	=ratio of total cross-sectional pore area to effective membrane area in porous glass	[-]
T	=absolute temperature	[K]
t	=diffusion time	[s]
V	=volume of gel	$[cm^3]$
V_0	=initial volume of gel	$[cm^3]$
V_c	=cell volume in diffusion experiment	$[cm^3]$
η	=solution viscosity	[Pa·s]
$[\eta]$	=limiting viscosity	$[g/100cm^3]$
$\Delta\pi$	=osmotic pressure difference	[Pa]

REFERENCE CITATION

Chmelir, M., Kunschner, A. and Barthell, E., 1980, Water soluble acrylamide polymer. 2. Aging and viscous flow of aqueous solution of polyacrylamde and hydrolyzed polyacrylamide, Antewalt. Makromol. Chem., **89**:145

Hirokawa, Y. and Tanaka, T., 1984, Volume phase transition in a nonionic gel, J. Chem. Phys., **81**:6379

Ohmine, I. and Tanaka, T., 1982, Salt effect on the phase transition of ionic gels, : J. Chem. Phys., **77**:5735

Otake, K., Tsuji, T., Konno, M. and Saito, S., 1988, Permeation of a new hydrogel and porous glass composite membrane, J. Chem. Eng. Japan, **21**:443

Spiegler, K. S. and Kedem, O., 1966, Thermodynamics of hyperfiltration (reverse osmosis):criteria for efficient membranes, Desalination, **1**:311(1966)

Tanaka, T., 1981, Gel, Science, **244**:1241

Tanaka, T., Fillmore, D. J., Sun, S. T., Nishio, I., Swislow, G. and Shah, A., 1980, Phase transition in ionic gels, Phys. Rev. Lett., **45**:1636

Tanaka, T., Nishio, I., Sun, S. T. and Nishio, S., 1982, Colaspes of gels in electric field, Science, **218**:467

Tsuji, T., Otake, K., Konno, M. and Saito, S., 1989, Permeation characteristics of porous glass and hydrogel composite membrane, submitted to J. Appl. Polym. Sci.

Verniory, A., Dubois, R., Decoodt, P., Gassee, J. P. and Lambert, P. P., 1973, Measurement of the permeability of biological membranes, J. Gen. Physiol., **62**:489

Wang, J. H., Robinson C. V. and Edelman, I. S., 1953, Self-diffusion and structure of liquid water. 3. Measurement of the self-diffusion of liquid water with H^2, H^3 and O^{18} as tracers, J. Am. Chem. Soc., **76**: 446

Wendt, R. P., Klein, E., Bresler, E. H., Holland, F. F., Serino, R. M. and Villa, H., 1979, Sieving propaties of hemodialysis membrane, J. Memb. Sci., 5:23

GEL DEWATERING PROCESS FOR BIOLOGICAL SLURRY

Hajime Unno

Dept. of Bioengineering
Tokyo Institute of Technology
Tokyo 152, Japan

Xia Huang

Dept. of Environmental Engineering
Tsinghua University
Beijin, China

Takashi Akehata

Dept. of Environmental Chemistry and Engineering
Tokyo Institute of Technology
Yokohama 227, Japan

Okihiko Hirasa

Research Institute for Polymers and Textiles
Tsukuba 305, Japan

INTRODUCTION

Biological slurries, such as excess activated sludge from wastewater treatment plants, organic waste slurry from alcohol production process, etc., generally contain tremendous amount of water. Most of these slurries have to be concentrated/dewatered before further processing so as to make the relevant process economically feasible. However, they are composed of highly hydrophilic organic solids and difficult to be dewatered. Filtration is one of the conventional dewatering methods, but it carries some drawbacks for the efficient treatment of large amount of highly hydrophilic slurry. This article is proposing a novel process which utilizes temperature-sensitive hydrogel as filtration media.

BACKGROUND OF THE PROCESS CONCEPT[1]

The principle of conventional dewatering apparatuses such as belt-press, vacuum filter, etc., is based on utilization of filtration phenomenon. The conventional filtration process is inefficient in the respect that 1) it is difficult to increase the filtration area in proportion to the equipment volume, 2) filter cake formed on filter medium produces a large resistance to filtration, 3) water to be separated from bulk of slurry must travel a long distance, L, to filter medium, and 4) exchange of slurry in the vicinity to filter medium with the slurry in the bulk is scarcely expected to increase the filtration rate.

To overcome the drawbacks mentioned above, an idea of dispersing the filtration surface throughout the slurry to be processed is considered. A concept for explaining the idea is shown in Figure 1. A medium offering the area will be a reversiblly swelling and shrinking hydrogel particle. The new process, named as ' gel dewatering process', is expected to improve each item mentioned above as follows: 1) the filtration area can be increased in proportion to the material to be processed by increasing the amount of gel particles, 2) the structure of filter cake can be destroyed and reconstructed by enlarging the size of water-absorbent gel particle in the cource of water absorption, 3) the moving distance, l , of water to the filtration surface is decreased by dispersing them throughout the slurry, and 4) the filtration rate can be increased by stirring the slurry to transport it from the bulk to the vicinity of the filtration surface.

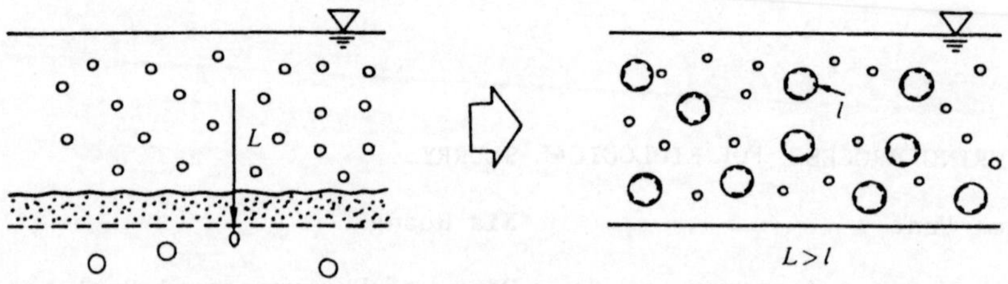

Figure 1. Idea of 'gel dewatering process'

CHARACTERISTICS OF BIOLOGICAL SLURRY

Biological slurries containing microbial cells, biological metabolites, medium for culture, etc., are generally highly hydrophilic and difficult to be dewatered. Figure 2 shows filtration curves observed by a conventional constant-pressure filtration apparatus, the parameter being the applied pressure. These organic slurries are generally of highly compressible nature and therefore the pressurized filtration, greater than 6.2×10^3 Pa in case of the figure, is not efficient. This result suggests a possibility of utilizing gel particles of highly water absorbing nature and of low/moderate swelling pressure.

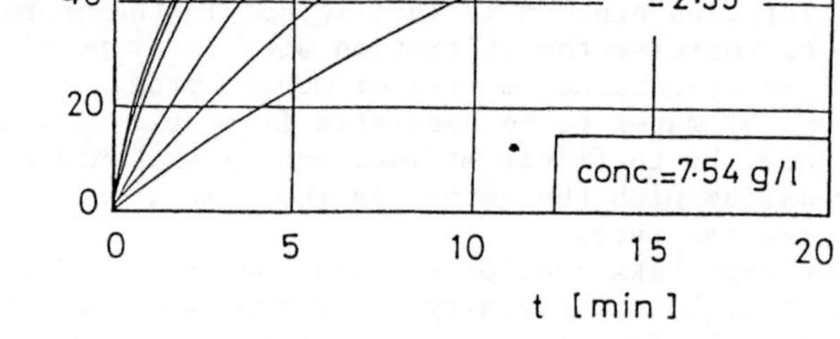

Figure 2. Filtration curve under constant-pressure filtration

Most biological slurries contain many kinds of ions. As is well known, the kinetic behavior of ionic gels is extremely affected by ions. Therefore, gels to be applied to the above proposed process should be insensitive to ions as far as possible.

CHARACTERISTICS OF A WATER-ABSORBING GEL, POLY(VINYLMETHYLETHER) GEL

One of the candidates for gels being applied to the process is a poly (vinylmethylether) gel (abbreviated as PVMEG). PVMEG is a nonionic gel and, swells and shrinks reversibly in response to the temperature due to hydration and dehydration reactions, respectively at low and high temperatures as shown in equation(1).

$$\underset{\text{(hydrated gel)}}{-(CH_2-\underset{\underset{\underset{HOH}{\vdots}}{OCH_3}}{\underset{|}{CH}})_n-} \underset{\text{cool}}{\overset{\text{heat}}{\rightleftharpoons}} \underset{\text{(dehydrated gel)}}{-(CH_2-\underset{OCH_3}{\underset{|}{CH}})_n-} + H_2O \qquad (1)$$

The transition heat required for the hydration or dehydration reaction is about 87.3 J/(g-dry gel) and the transition temperature is 310K. These characteristic values imply that it is possible for the present process to utilize waste heat of low temperature, which is usually discarded.

PVMEG, prepared by irradiating γ-ray to poly(vinylmethylether) aqueous solution, carries such properties as 1) little effect of ions, 2) rapid swelling and shrinking, 3) completely reversible change of swelling and shrinking, 4) small energy amounts necessary for the change. The rapid response to the temperature change is secured by making the gel as small as possible since the time constant, T, for swelling or shrinking process is roughly proportional to the square of gel size, d, as shown in equation(2), if the heat transfer rate is not the rate-determining process.

$$T \propto d^2 / D \qquad (2)$$

where D is the diffusion coefficient of gel matrix.
However, too small size is inconvenient for the handring of gels. Therefore, the structure of gel should be of relatively large size, but being an assembly of the smaller gel unit. One of those structures is of a macro-porous structure. Macro-porous PVMEG can be easily prepared by controling the condition of γ-ray irradiation.

Preparation of PVMEG[2]

A PVMEG was prepared by 1.09 Mrad/h for 9.3 hours irradiation of γ-ray from ^{60}Co to a mixture of 30 wt% poly(vinylmethylether) aqueous solution with 15% ferric oxide powder. Ferric oxide powder increased about 3 times the intra-gel heat transfer rate to a gel with no ferric oxide powder. Ferric oxide powder has also strengthened the gel structure.

Behavior of PVMEG[3,4]

A response characteristic of a macro-porous PVMEG particle is shown in Figure 3 with photographs of the corresponding gel particles. The gel particle responded very quickly and reversibly to the ambient temperature. The transition temperature of the PVMEG was 310K as shown in Figure 4.

Eventhough PVMEG is nonionic, its response to temperature is not completely insensitive to ionic environment. Figure 5 shows some examples of the effects of ions, i.e., salts on the volume change according to the temperature. The ions tested are mostly those considered to be contained within biological slurries. The results are summalized as follows. 1) The gel's behavior depended solely on the anions contained in the solution. 2) The change of equilibrium volume of gel in salt solution was explained by the valency and the radius of the anion group. 3) The transition temperature was independent of the kind and concentrations of ions.

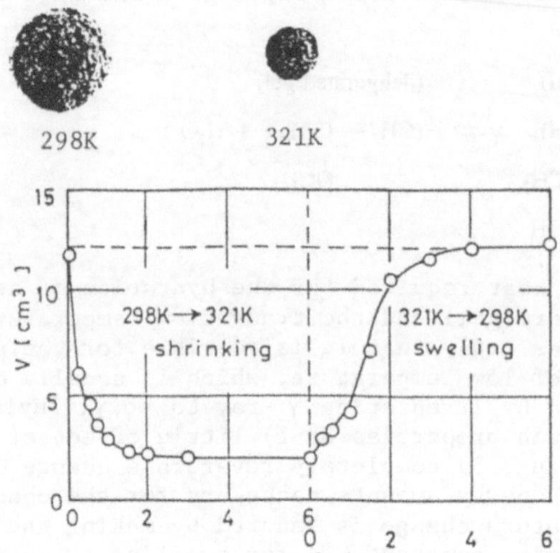

Figure 3. Reversible volume change of a PVMEG between two different temperatures, including photographs of the gel corresponding swollen and shrunken states

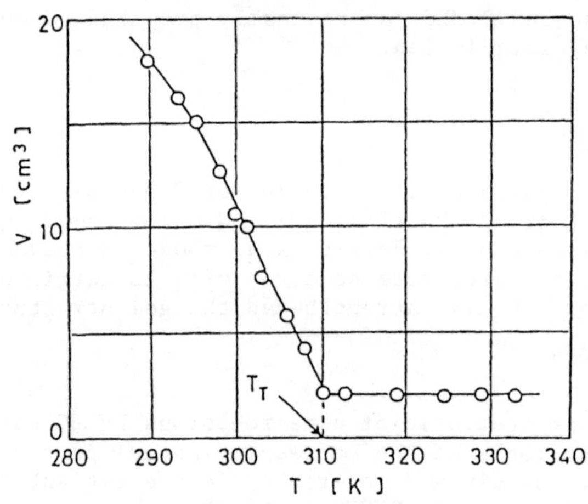

Figure 4. Equilibrium relationship between the volume V of a PVMEG and temperature T

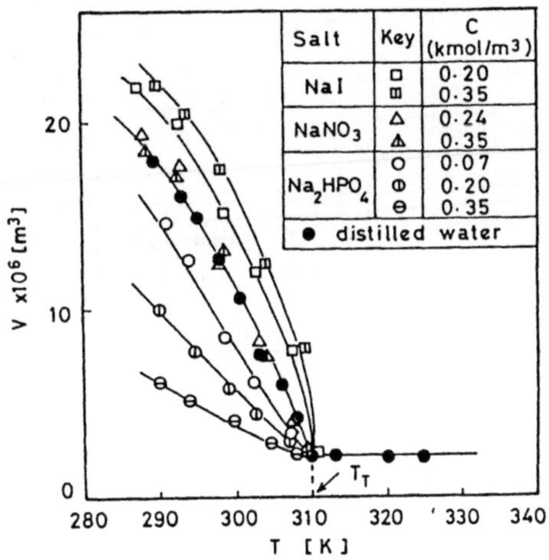

Figure 5. Equilibrium volume-temperature curve of a PVMEG in salt solution

Swelling pressure[5]

Figure 6 shows an example of swelling pressure of a PVMEG measured by the apparatus shown in the figure. The swelling pressure was a strong function of the porosity of the gel. The relationship between the porosity and the swelling pressure was affected a little by ions as shown in the figure. The effect of ions on the swelling pressure was correlated well by the ratio of the valency to the radius of the anions[4].

Dewatering excess activated sludge[6]

Figure 7 shows water absorbing behaviors of a PVMEG in slurries of several different solid concentrations, where the measurements were made by the apparatus shown in the figure. The water absorption rate, i.e., the swelling rate decreased with increase of the solid concentration. This is primarily due to the increase in the resistance-to-flow of slurry with increase of solid concentration. For the process application the decrease of water absorption rate may become a serious problem. However, since the decrease is not so great as is seen in the figure, this drawback can be easily overcome by using a number of gel particles of appropriate size.

Figure 8 shows a result of a series of batch concentration of activated sludge by using one gel particle, the size being 3.12 cm in diameter at 293.5K. The experimental procedure was as follows: 1) the gel was shrunken by warming up to 321K, 2) the shrunken gel was put into a slurry in a beaker, where the slurry temperature was kept 298K, 3) the gel was kept to swell by absorbing water from the surrounding slurry for 10 minutes, 4) the swollen gel was took out to measure the size for evaluating the absorbed water. Then the procedures from 1) through 4) were repeated using the same gel and the same slurry.

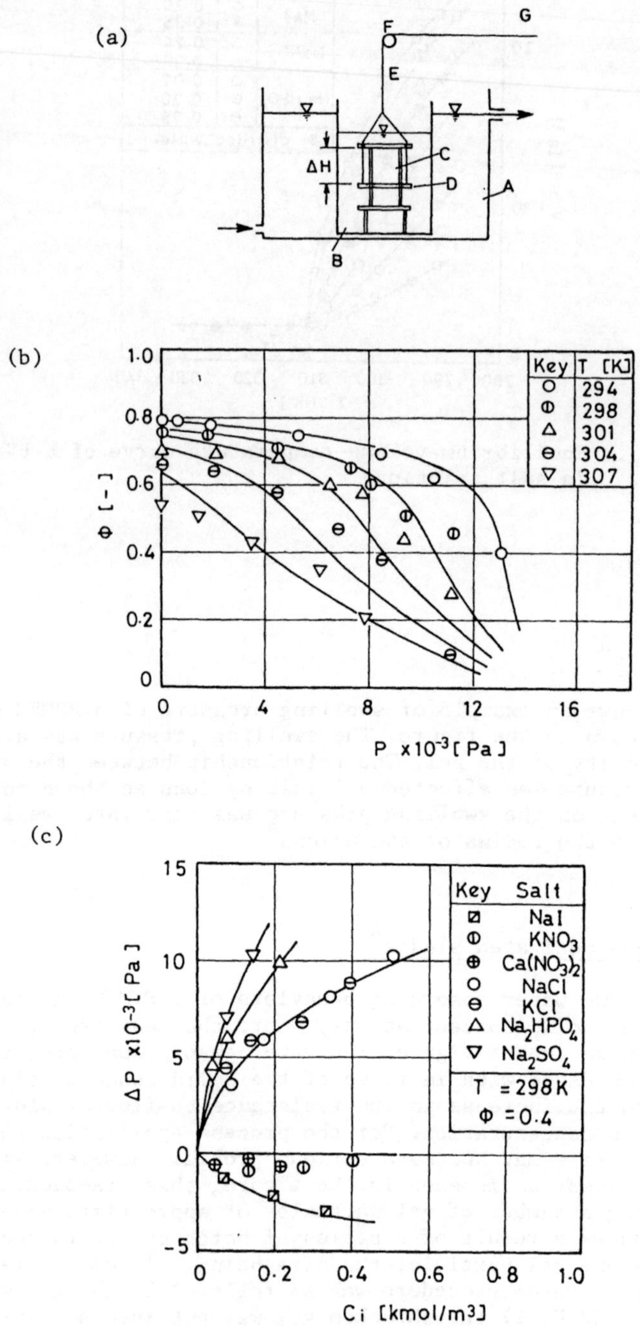

Figure 6. Effect of salts on the swelling pressure of PVMEG
(a) Measurement system
(b) Relationship between gel porosity and swelling pressure at different temperature
(c) The difference of swelling pressure between that in distilled water and that in salt solution at 298K

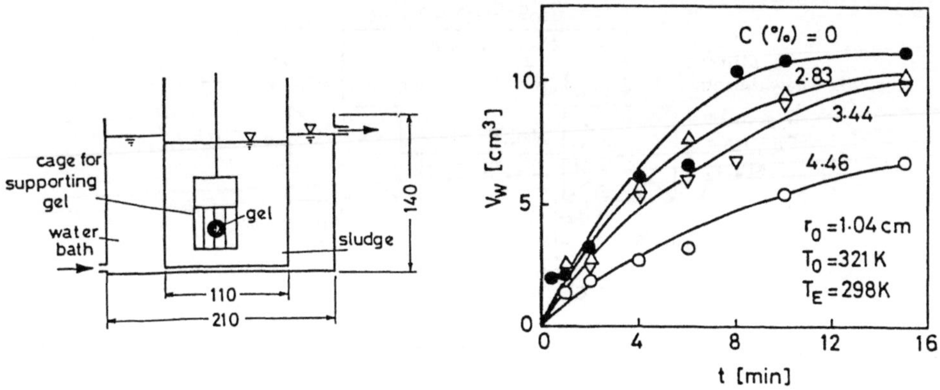

Figure 7. Time change of water amount absorbed by a gel in excess activated sludge with experimental set-up

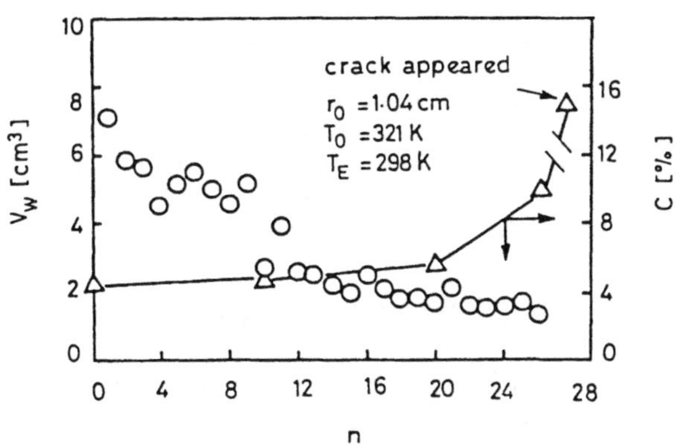

Figure 8. Changes in water amount absorbed by a gel and sludge concentration with time

The amount of absorbed water decreased with the repeated operation since the solid concentration increased with the repetition. The slurry was finally concentrated upto the solid concentration of 16%. Some cracks were observed finally on the slurry surface, which shows the loss of slurry fluidity. Figure 8 suggests that, as suggested in the section of process concept, utilizing an appropriate amount of gel particles makes it feasible to construct a dewatering system by using PVMEG particles.

PROCESS EVALUATION[7,8]

Concept of 'Gel Dewatering Process'

A conceptual flow sheet of 'gel dewatering process' is shown in figure 9. The process consists of a slurry-dewatering reactor and a gel-regeneration reactor. The gel particles are recycled for reuse.

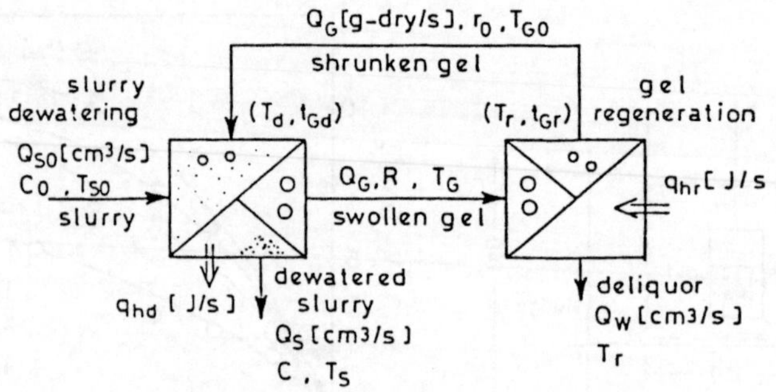

Figure 9. A basic concept of 'gel dewatering process' by use of PVMRG

Gels shrunken by raising the temperature are fed and mixed with the slurry in the slurry-dewatering reactor. The slurry is dewatered to a solid concentration C and the gels increase their volume by absorbing water during the dewatering period. The swollen gels separated from the dewatered slurry are transported to the gel-regeneration reactor and sent back to the slurry-dewatering reactor again after being shrunken.

Process Model

To evaluate the process feasibility a model analysis of gel dewatering process was made. The water-absorbing and regeneration processes of the gel were analyzed by considering the material balance of both the gel matrix and the water content in the gel (equations (6) and (7)), and the heat balance in and around the gel (equation (3)) with the boundary conditions expressing the filtration on the gel's surface (equations (8)-(12)). The mathematical models are summarized as follows.

$$\frac{\partial}{\partial t}[\{\rho_w C_{PW}\varepsilon + \rho_P C_{PP}(1-\varepsilon)\}Tr^2] = \frac{\partial}{\partial r}\left[\{K_w\varepsilon + K_P(1-\varepsilon)\}r^2\frac{\partial T}{\partial r}\right] - \frac{\partial}{\partial r}[\{U_w\rho_w C_{PW}\varepsilon + U_P\rho_P C_{PP}(1-\varepsilon)\}Tr^2] + (1-\varepsilon)\rho P r^2 Q_T \qquad (3)$$

$$\int_0^{R(t)} 4\pi r^2\varepsilon(r)\,dr = M(\text{const}) \qquad (4)$$

where

$$\varepsilon = f(T) \qquad (5)$$

$$U_P = -\int_0^r 4\pi r^2\,\partial(1-\varepsilon)/\partial t\,dr/\{4\pi r^2(1-\varepsilon)\} \qquad (6)$$

$$U_W = -\int_0^r 4\pi r^2\,\partial\varepsilon/\partial t\,dr/\{4\pi r^2\varepsilon\} \qquad (7)$$

Constant M is total amount of polymer matrix of one gel particle. The function of equation(5) was experimentally determined by the equilibrium relationship shown in figures 3 and 4. Up and Uw are local moving velocities for polymer matrix and water within a gel at radial position r, respectively.

$$\begin{cases} T|_{t=0} = T_{G0} & (8)\\ R(t)|_{t=0} = r_0. & (9)\end{cases}$$

$$\begin{cases} \partial T/\partial r|_{r=0} = 0 & (10)\\ \partial T/\partial r\{K_w\varepsilon + K_P(1-\varepsilon)\}|_{r=R(t)} = h(T|R - T_E) & (11)\\ U_{W\varepsilon}|_{r=R(t)} = P_f A(1-mC)/\{(V_f + V_{f0})\mu\rho\alpha C\} & (12)\end{cases}$$

The right-hand side of equation(12) is Ruth's filtration equation, which describes the filtration process occurred on the gel surface. For this analysis specific resistance to filtration was expressed as a function of filtration pressure for each slurry as shown in figure 2. The effect of mixing condition of slurry with gels on the dewatering reactor performance was also took into consideration.

Table 1. Dewatering capacity of various dewatering processes

Item	Vacuum filter	Pressure filter	Centrifuge	Belt press	Gel dewatering
Slurry	Raw Sludge				Excess sludge
Slurry conc.	3–4%				3%
Dewatering aid	Inorganic coagulant		Polymer flocculant		PVMEG
Addition amount (kg/kg-ds)	$Ca(OH)_2$ 0.6–0.2 $FeCl_3$ 0.1		0.015–0.005	0.01–0.004	7–8
Dewatering velocity	10–15 (kg/m^2h)	3–5 kg/m^2h	—	100–170 (kg/mh)	—
Recovery rate of solid (%)	81	100	92	100	83
Water content in cake (%)	72–78	60–65	70–80	60–76	85

Process Feasibility[7]

By using the model various kinds of process simulations were performed. Table 1 shows a comparison of dewatering capacity of various dewatering processes in the light of economical feasibility of 'gel dewatering process'. The process needs polymer gel as a dewatering aid and its necessary amount is estimated 7 - 8 kg-dry gel/kg-solid from the model analysis when the condition of mixing of slurry with gels in a dewatering reactor is assumed piston-like flow. The slurry is dewatered to about 15%, along with solid recovery 83%, which was partly due to adhesion of small amount of slurry to the gel's surface. To be comparable in the efficiency to a belt-press, which is considered at present a most efficient process for treating biological slurry, the gels used in 'gel dewatering process' should be used repeatedly more than 2000 times. The estimation indicates that gel must have enough strength against mechanical collision/shearing for many times.

By comaring the present process with conventional dewatering processes, it has been clarified that 'gel dewatering process' is simple in both operation and system structure, its set-up area is smaller and running cost is lower under the condition that gel has enough strength to be used repeatedly for sufficiently long time.

CONCLUDING REMARKS

For the efficient dewatering of mocrobe-rich slurry, which is difficult to be dewatered, a new process named 'gel dewatering process' with the use of poly(vinylmethylether) gel (PVMEG) was proposed. By a dewatering experiment of excess activated sludge using one spherical PVMEG and through some engineering calculations of the dewatering process, the process was promising.

In the application of the dewateing process there remained some problems to be solved. Some improvements in the gel's properties such as mechanical strength for thouthands times of repeated use, large water-absorbability, etc. Additional engineering investigations about handling of gel and the design of dewatering equipment are also necessitated.

LITERATURE CITED

1. H.Unno, Functional structure of hydrogel and its engineering application, Huntai to Kougyou(J.for Powder Industry). 20 (3):31 (1988)
2. O.Hirasa, T.Akehata, H.Unno and X.Huang, Koukai Tokkyo Kouhou (A) (Japan Patent). S62-250940 (1987)
3. X.Huang, H.Unno, T.Akehata and O.Hirasa, Analysis of kinetic behavior of temperature-sensitive water-absorbing hydrogel, J. Chem.Engng.Japan. 20:123 (1987)
4. X.Huang, H.Unno, T.Akehata and O.Hirasa, Effect of salt solution on swelling or shrinking behavior of poly(vinylmethylether) gel (PVMEG), J.Chem.Engng.Japan. 21:10 (1988)
5. X.Huang, H.Unno, T.Akehata and O.Hirasa, Swelling pressure of poly(vinylmethylether) gel (PVMEG) in swelling process, J.Chem. Engng.Japan. 21:651 (1988)
6. X.Huang, H.Unno, T.Akehata and O.Hirasa, Applicability of temperature-sensitive water-absorbing hydrogel (polyvinylmethylether gel) to dewatering of sludges, Kagaku Kougaku Ronbunshu (Japanese J.Chem.Engng.). 13:518 (1988)
7. X.Huang, H.Unno, T.Akehata and O.Hirasa, Engineering application of temperature-sensitive hydrogel to sludge dewatering, Mizushori Gijutsu (Water Treatment Journal). 29:277 (1988)
8. X.Huang, H.Unno, T.Akehata and O.Hirasa, Dewatering of biological slurry by using water-absorbent polymer gel, Biotechn. and Bioengng. 34:102 (1989)

SENSORS AND ACTUATORS

A POLYMER COMPOSITE SHOWING ELECTROCONTRACTILE RESPONSE

Piero Chiarelli, Kayo Umezawa and Danilo De Rossi

Centro "E.Piaggio" University of Pisa, Pisa, Italy

CNR Institute of Clinical Physiology, Pisa, Italy

INTRODUCTION

In recent years renewed interest in polymer mechanochemistry (1) has provoked several studies aimed at overcoming the intrinsic limitations of previous systems in respect to their practical use as soft linear actuators for orthotics, prosthetics and advanced robotics (2). Electrolytically activated (3,4) and electrokinetically driven (5,6) mechanochemical (MC) systems have been studied to eventually provide a more convenient method of excitation, while other attemps have focused on preparing faster (7) or stronger(8) polymers. Despite these efforts, the availability of a material capable to reversibly contract and relax under electric stimuli with contraction times and forces comparable to skeletal muscles is still an elusive goal. It is the authors' opinion that new materials and methods are definitely needed to accomplish a distributed "delivery" of electrochemical stimuli of controllable and suitable nature to trigger and sustain the mechanical action.

An intimate contact between the contractile component and the electromechanical energy source would provide shorter diffusion-reaction time constant and, in case of the kinetics being rate limited by these phenomena, a much enhanced speed of response of the actuator. In addition, the activating electrochemical reaction should not generate by-products negatively affecting the device life-time and the contractile component integrity.

We report here the preparation and characterization of a composite made of a polyelectrolyte gel and a π-conjugated electron conducting polymer. Samples are prepared in thin strip form by a two step process: thermal crosslinking of a polyacrylic acid (PAA)-polyvinyl alcohol (PVA) mixture, followed by gas state polymerization of polypyrrole (PPy), catalyzed by ferric chloride, to obtain an interpenetrated network.

Polymer Gels, Edited by D. DeRossi *et al.*
Plenum Press, New York, 1991

The composite shows both MC and electromechanochemical (EMC) response and, under appropriate electric driving conditions, contractility is observed with no gas evolution at the polymeric electrode.

EXPERIMENTAL

A thermally crosslinked PVA (average mol.weight 10,000)-PAA (average mol. weight 250,000) hydrogel was prepared following the method of Kuhn and Hargitay (9), curing at 130 °C for 40 minutes. The gel samples were cast in thin sheets from which strips were cut, the dimensions of which were typically 5.0 cm long, 0.5 cm large and 0.015 cm thick. The samples were then immersed for 24 hours in an aqueous ferric chloride (40% by weight) solution brought to pH=0.5 by adding hydrochloric acid.

The gas state polymerization of PPy was performed following the method suggested by Ojio and Miyata (10), by putting a pyrex flask, containing the samples in a deaerated, saturated atmosphere of water and pyrrole vapours, into a cryostat at a temperature ranging from-15°C to-5°C for a time period ranging from a minimum of 2 hours to a maximum of 120 hours. Samples were then repeatedly washed in distilled water. The presence of residual iron in ionic form within the gel after the washing procedure was ascertained by chemical analysis on the ashes of control samples.

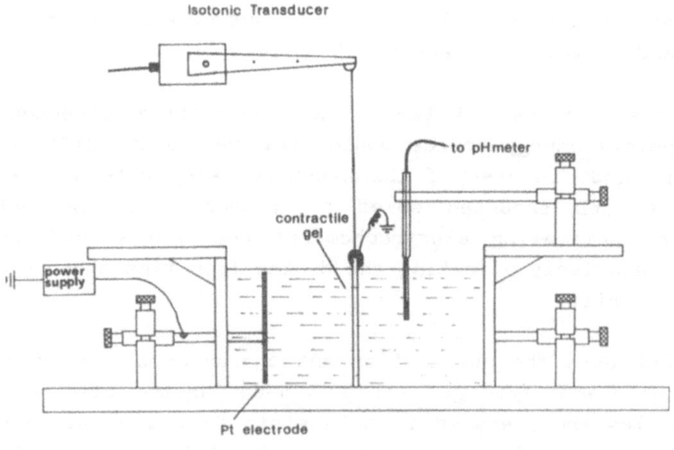

Figure 1. Schematic drawing of the experimental cell used to perform isotonic experiments under chemical or electrochemical excitation.

The conductivity of the samples was measured by a four-point technique using vacuum deposited indium electrodes, after partial dehydration achieved by keeping the (PAA-PVA)-PPy strips at 20°C in air (R.H. 50%) for 24 hours and subsequently in a oven at 35°C for 30 minutes.

MC and EMC experiments were performed on various samples prepared at -5°C and -10°C using the apparatus shown in Fig. 1, under isotonic conditions.In the course of the MC experiments, a strip was clamped at the bottom of a perspex cell, filled with stirred 0.05 M NaCl solution, and fixed at its upper end to the shaft of an isotonic assembly composed of a rotatory displacement transducer and a variable counterload. Step changes in the pH of the solution from 2 to 12 were obtained by repeated, alternate additions of HCl and NaOH, also resulting in a progressive change in the salt concentration of the solution equal to 0.02 M per cycle.

The samples used in the EMC experiments (exposed to pyrrole and water vapors for 120 hours at -10°C) were carefully washed in distilled water, their upper end was put in electrical contact with a platinum wire by means of a silver-loaded glue (the contact was then covered with a thin layer of epoxy to provide mechanical stregth and electrical insulation from the external environment), and then re-equilibrated with a 0.05 M NaCl solution. A flat platinum electrode was used as counterelectrode in the course of the EMC experiments, in which steps of dc potential difference were applied and the sample length was recorded as a function of time. The electrolytic solution was magnetically stirred and bubbled with nitrogen during the experiments. The electric current

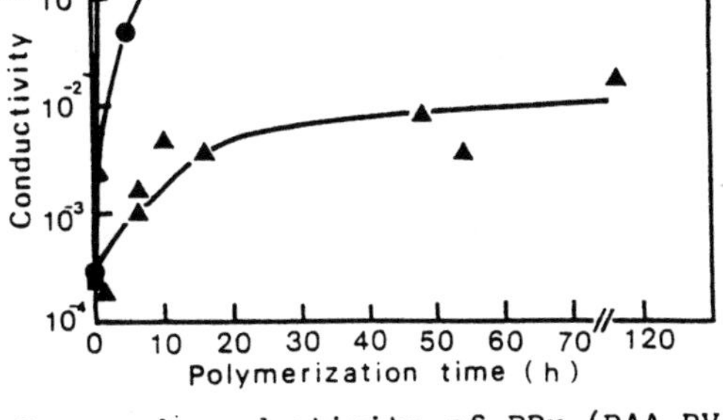

Figure 2. Measured conductivity of PPy-(PAA-PVA) samples as a function of gas polymerization time. The data refer to samples prepared at different temperatures: (▲)-15°C, (●)-10°C, (■)-5°C.

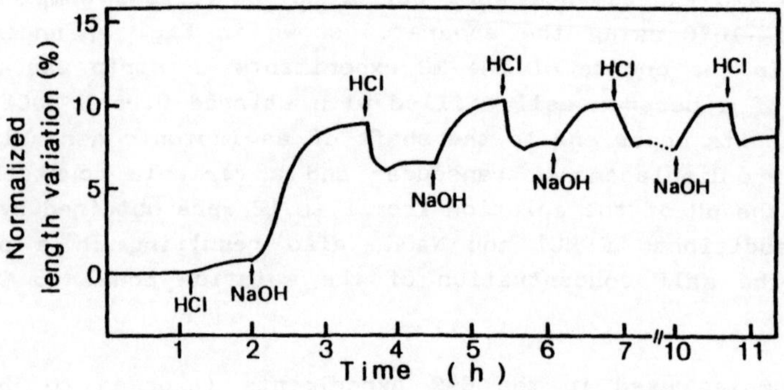

Figure 3a. Mechanochemical response.
Typical fractional length change of a PPy-(PAA-PVA) sample with residual $FeCl_3$ under repeated additions of acid and base.

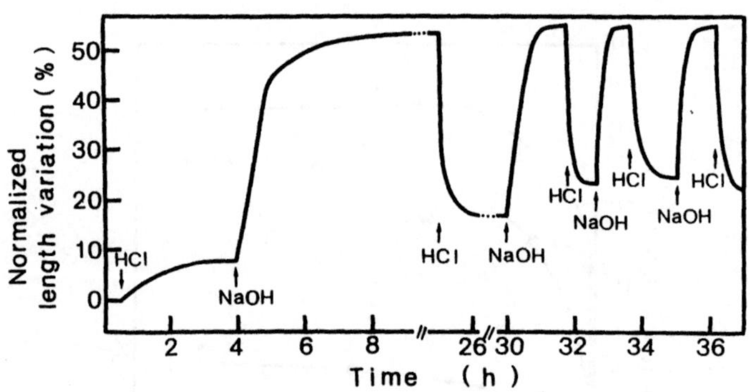

Figure 3b. Mechanochemical response.
Response of a (PAA-PVA) control sample with residual FeCl which does not contain PPy.

flowing in the circuit and the pH of the electrolytic solution were also recorded with time.

RESULTS

Fig. 2 shows the dependence of the conductivity of samples prepared at different polymerization temperatures on polymerization time. Samples prepared at -15°C were found typically quite inhomogenenous and the level of conductivity ranging from 10^{-4} to 10^{-3} S/cm, of the same order of magnitude of control samples which did not contain PPy. Samples prepared at -10°C and -5°C showed a steep increase in conductivity in the first 10 hours of polymerization, reaching saturated conductivity values of the order of 10^{0} and 10^{1} S/cm respectively.

The MC response under isotonic conditions (counterload 4.35 grams) of a sample (initial lenght 1.4 cm, thickness 0.014 cm) loaded with PPy at -10°C for 120 hours is reported in Fig. 3a. In Figure 3b the MC response(counterload 4.35 grams) of a control sample (initial length 1.37 cm, thickness 0.022 cm) of PAA-PVA which contains FeCl but no PPy, is shown.

In both cases a large sample elongation was observed in the course of the first MC cycle; moreover in the case of samples containing PPy (fig. 3a) their length was observed to increase slightly with the number of MC cycles, becoming stable at about 1.1 times the initial length; fractional length change of the sample decreased from 8.8% to 2.5% during these cycles.

Fig. 4 shows the recording of the length changes of the sample (initial lenght 3.05cm; thickness 0.016 cm) in response to the stepwise application of dc potential differences between the composite and the counterelectrode while the concomitant electric currents flowing into the cell are shown in fig. 5. Macroscopic length changes have been observed only when the potential difference was of the order of 2 volts or higher (see fig. 4).

Gas bubbling was not observed at the polymeric electrode, irrespective of its polarity.

The demonstrated capacity of PPy to absorb large quantities of doping agents (11) and to incorporate electrolytic gaseous products (12) is thought to be retained by the composite.

The pH of the solution surrounding the sample varied as expected (11) from an initial value of 5.1 to 6.5, when potential difference of +2 volts was applied, and to 4.3 when the potential difference was -2 volts (we assume by convention that a positive potential difference is applied when the couterelectrode has an electric potential higher than that of the polymeric sample).

In this respect it has been observed that, during EMC experiments, the sample contracts when the pH of the external solution increases, while it swells when the bulk solution becomes more acid.

The behavior only apparently contradicts the classical MC observations (1) since the effects of elecrodiffusion and chemical reactions inside the polyelectrolyte-PPy composite create a local ionic environment which is very different from the solution bulk.

DISCUSSION

By referring to the results illustrated in fig. 2 it can be observed that the samples with the highest specific conductivity are obtained when the gas state polymerization of PPy is performed at -5°C; the level of specific conductivity saturates at approximately 24 hours exposure time. The order of magnitude of the specific conductivity in this case reaches a value of 32 S cm^{-1} which is four times higher of the sample prepared by Ojio and Miyata (10) using PVA as host polymeric matrix, when polymerization time was also 24 hours and $FeCl_3$ concentration was 40% by weight.

The samples we prepared were very stable in air and in NaCl acqueous solution. Although remarkably high, the value of the attained specific conductivity is still insufficient to avoid the electric current flow to be largely shunted by the saline solution in which the sample is immersed.

In addition, the current carrying capability of the sample is still too low, considerably rate limiting the EMC Kinetics.

The considerably lower extent of the MC fractional length changes (fig.3a, 3b) can be ascribed to the presence of the stiff PPy mesh entangled within the polyelectrolyte gel network.

The analysis of the electrochemical stimulated mechanical response of the composite after few cycles of preconditioning must take in account the electrolytic reactions occuring at the PPy electrode in the course of EMC cycles:

$PPy+OH^- ---- (PPy^+ OH^-)+e^-$ 1)

$PPy+Cl^- ---- (PPy^+ Cl^-)+e^-$ 2)

and the electrolysis of water.

The effect of reaction 1) and 2) is to lower the pH and the ionic strength of the solution around PPy (12).

The effect of pH is much more relevant (13) that the antagonist effect induced by ionic strength changes, causing the sample to shrink.

Inverting the electrode polarity causes the reactions 1) and 2) to proceed in the reverse direction with a resulting increase of ionic strength and pH close to the PPy interface in the interstitial gel solution. Accordingly the sample elongates.

Looking at fig. 4 it can be noted how a substantial EMC response is obtained only when the potential difference applied between the (PAA-PVA)-PPy electrode and the platinum counterelectrode is of the order of 2 volts or higher.

The observation that a substantial EMC response is only obtained when the voltage is of the order of 2 volts or higher can be supported by the possible occurrence of two different reactions, i.e. PPy oxidation-reduction and water electrolysis.The activation potentials of these reactions appear to be close and, at the present stage of our work, no conclusive evidence can be given to support the occurrence of one excluding the other.

In facts, considering that the internal resistance of the cell we used is of the order of 500 Ω and the electric current of the order of 1 mA, it can be calculated that the point of maximum rate of change of PPy redox reaction is at a potential of + 1.92 (using our convention) in respect to a Pt counterelectrode at normal conditions (14) while the electrolysis discharge potential is 2.02 volts.

Looking again at fig. 4 it can be also observed how, after repeated EMC cycles, the sample rest length increases, while the electric current time course (fig. 5) after a voltage step application show a progressive peak decay.

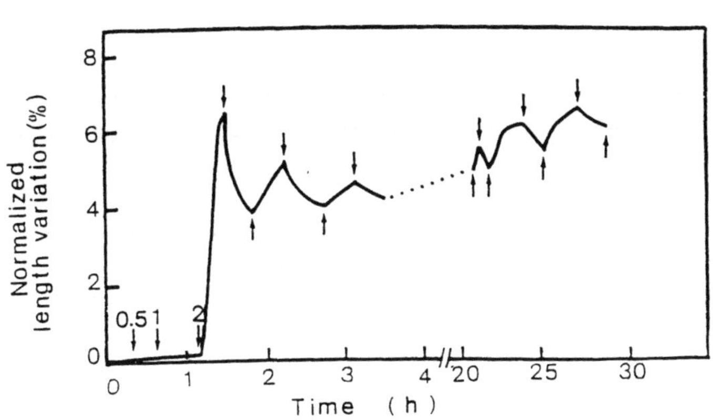

Figure 4. Electromechanochemical response. The isotonic length changes are reported as a function of time, following application of various potential differences (0.5, 1, 2V) between the sample and the Pt counterelectrode. The arrows indicate application of 2 volts of alternate polarity.

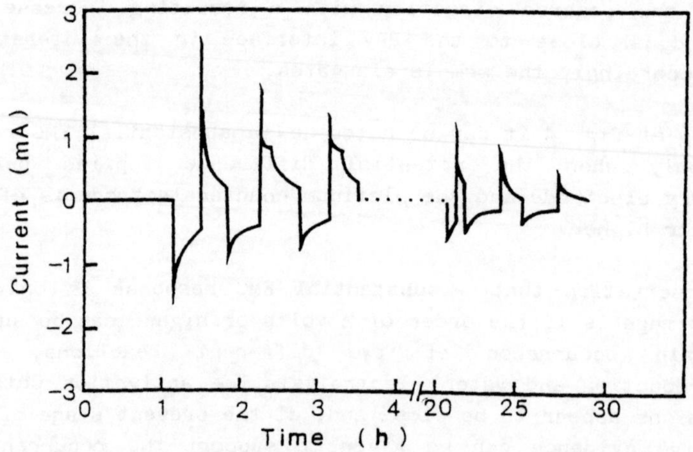

Figure 5: Electromechanochemical response. The resulting electric current flow vs time following the voltage steps application as in fig. 4.

The progressive decay of the amplitude of the current peaks is conceivably due to a decrease in conductivity caused by the repeated swelling and shrinking cycles which tend to break the PPy connectivity. An additional effect may also be ascribed to "chemical fatigue" at the nitrogen of the pyrrole (15).

CONCLUSIONS

The polymer composite we have prepared shows both MC and EMC response.

EMC response is not accompanied, when properly driven, by gas evolution at the working polymeric electrode, avoiding problems related to its pratical usability.

The concept of blending electron conducting polymers and MC polyelectrolytes appears to the authors to offer substantial possibilities in improving the state of the art in electrically driven soft actuators.

However, considerable problems exist, mostly related to the poor mechanical properties and still insufficient electronic conductivity.

Work is in progress also using different polymers possessing a better match of elastic compliances and higher specific conductivity.

REFERENCES

1. A. Katchalsky and A. Oplatka, "Mechanochemical conversion" in Principles of receptor physiology (W.R. Lownstein, Ed.), pp. 1-17, Springer Verlag, New York (1971).
2. Y. Osada, "Conversion of chemical into mechanical energy by synthetic polymers(Chemomechanical systems)", Adv. Polym. Sci., 82, 1-46 (1987).
3. R.P. Hamlen, C.E. Kent and S.N. Shafer, "Electrolytically activated contractile polymers", Nature, 206, 1149-1150 (1965).
4. D. De Rossi, P. Chiarelli, G. Buzzigoli, C. Domenici, L. Lazzeri, "Contractile behavior of electrically activated mechanochemical polymer actuators", Trans. Am. Soc. Artif. Inter. Organs, XXXII, 157-162 (1986).
5. A.J. Grodzinsky and J.R. Melcher, "Electrochemical trasduction with charged polyelectrolyte membranes", IEEE Trans. Biomed. Eng., BME-23, 421-433 (1976).
6. Y. Osada and M. Hasebe, "Electrically activated devices using polyelectrolyte gels", Chemistry Letters, pp. 1285-1288 (1985).
7. M. Suzuki, "Responsive mechanochemical actuator materials by PVA hydrogel", Proc. JUPAC CHEMRAWN VI, p. IB11, May 17-22, Tokyo (1987).
8. Y. Itoh, T. Matsumura, S. Umemoto, N. Okui and T. Sakai, "Contraction/elongation mechanism of acrylonitrile gel fibers", Polymer Preprints-Japan (English Edition), 36 (Nos. 5-10), p. E184 (1987).
9. W. Kuhn and B. Hargitay, "Muskelahniche Arbeitleisstung Kunstlicher Hochpolymer Stoffe", Z. Elektrokem., 55, 490-502 (1951).
10. T. Ojio and S. Miyata, "Highly transparent and conducting polypyrrole-poly(vinyl alcohol) composite films prepared by gas polymerization", Polymer Journal, 18, 95-98 (19886).
11. M. Salmon, A.F. Diaz, A.J. Logan, M. Kroumbi and J. Bargon, "Chemical modification of conducting polypyrrole films", Mol. Cryst. Liq. Cryst., 83, 265-276 (1982).
12. M. Okano, A. Fujishima and K. Honda, "A new method of pH control by use of a polypyrrole coated electrode", Electroanal. Chem. Interfacial Electrochem., 185, 393-396 (1985).
13. Unpublished observations
14. A.F. Diaz and J.I. Castillo, "A polymer electrode with variable conductivity: polypyrrole", J.C.S. Chem. Comm., 397-398 (1980).
15. O. Iganas, R. Erlandsson, C. Nylander and I. Lundstrom, "Proton modification of conducting polypyrrole", J. Phys. Chem. Solids, 45, 427-432 (1984).

CONVERSION OF CHEMICAL INTO MECHANICAL ENERGY BY SYNSTHETIC POLYMER GELS (Chemomechanical System)

R. Kishi, M. Hara, K. Sawahata, and Y. Osada
Department of Chemistry, Ibaraki University
2-1-1 Bunkyou, Mito, Ibaraki 310, Japan

1. INTRODUCTION

System which undergoes shape change and develop contractile force responsing to outside stimulus is called "Chemomechanical (or Mechanochemical) System"[1] and refers to thermodynamic systems capable of transforming chemical energy directly into mechanical work or conversely of transforming mechanical into chemical potential energy[2]. The isothermal conversion of chemical energy into mechanical work underlies the motility of all living organisms and can easily be seen, for instance, in muscle, flagella and ciliary movement. All these biological systems are characterized by an extremely high efficiency of energy conversion. The high efficiency of the biological systems is largely due to direct conversion of chemical energy without unnecessary intermediate passes producing heat.

Chemomechanical systems are the only artificial systems at present which can convert chemical energy directly into mechanical work. They may have potential uses, for example, under water and in space where any power supply is limited or difficult to obtain. This type of compounds can also be considered as "metamorphic-active" polymer, meaning that the chemomechanical reaction may be used not only to generate mechanical energy on a macroscopic level but also to transform informations as a "signal" or "receptor", whereby semimicroscopic deformation or strain play an essential role for displaying functions such as those of a switch or a sensor.

This kind of "semimicroscopic" chemomechanical work can be seen in many biological systems. Enzymes change their conformation and accumulate strain when they form an enzyme-substrate complex ("activated state") and partially liberate

Polymer Gels, Edited by D. DeRossi *et al.*
Plenum Press, New York, 1991

as chemical energy and utilize to promote the enzyme reaction. Hemoglobin also changes the conformational state if one of the four globulin molecules binds with an oxygen molecule. This induces a submicroscopic displacement of the globulin chain which is transferred to the adjacent globulin as a "signal" and provides for displaying an "allosteric effect". Swelling of the squid giant axon during the rising phase of the action potential was demonstrated by T. Tasaki[3]. Authors say that the observed simultaneity of the peak of the mechanical response on the peak of the electrical response suggests that there is a common physico-chemical process,i.e., swelling brought about by sorption of water is invariably associated with a large increase in conductance and this enhance the intramembrane ion mobilities and alters the ion selectivities.
Thus, chemomechanical process can be considered as a common phenomenon dominating the dynamics and functions of biological systems.

The chemomechanical systems using synthetic polymer gels may be classified according to the principle and type of reaction as follows:

1)hydrogen-ion transfer (pH muscle),
2)ion exchange or chelation,
3)redox reaction (redox muscle),
4)steric isomerization,
5)phase transition or order-disorder transition,
6)polymer-polymer association or aggregation,
7)electrokinetic processes,

During the last decade a number of chemomechanical systems have been investigated. They include polyelectrolyte fibers and membranes which expand and contract upon changing their solubility and or degree of ionization; non-changed polymer fibers which collapse upon changing temperature, polymer membranes which associate cooperatively with "complementary" polymers. Chemomechanical systems contracting reversibly under electric stimulus and photo-irradiations have also been developed.[4]

Some of the chemomechanical gels are able to generate tensile force up to several Kg/cm^2 within seconds, which is in the same order of that of muscle. Investigation of stimuli-sensitive drug delivery systems (DDS)[5] and enzymes immobilized in the thermally reversible hydrogels[6] on the base of shape change of the gel are also under progress.
The chemomechanical system induced by electrokinetic process (electro-activated chemomechanical system) is specially intereting from the points of the control of system and high efficiency of energy conversion.

In this paper, principle, behavior and mechanism of the electro-activated chemomechanical system is described. some kinds of electro-activated chemomechanical devices are also introduced.

2. PRINCIPLE AND BEHAVIOR

(A) Contractile behaviors

We found that any polyelectrolyte gels undergo shape changes by applying DC current and the velocity of shape change is proportional to the charge density in the gel. This maybe the first model of an electrically activated artificial muscle working in aerobic and aqueous medium system because the gel contracts and delates reversibly by an electric stimulus under isothermal condition.

The electrical control makes use of crosslinked polyelectrolyte gels. The system is quite symple, for instance, a water-swollen polymer gel is inserted between a pair of electrodes which is connected to a DC source[4].
When the electricity was turned on, the polymer gel starts to shrink releasing water droplet one after another. As shown in Fig.1, a water swollen poly(2-acrylamido-2-methyl-1-propanesulfonic acid,AMPS) gel (12mm wide, 19mm long, 30mm high, 5.6 g weight, 1g dry poly(APMS) adsorbed 2250g water), reduced in weight as large as 70% of its original weight due to loss of water when 6.3 V/cm electric field or 0.7 mA/cm^2 DC current was imposed for 20 min. The minimum value of electric field to induce contraction of the gel was 1.6 V/cm and the rate of water release increased with increasing electric current.

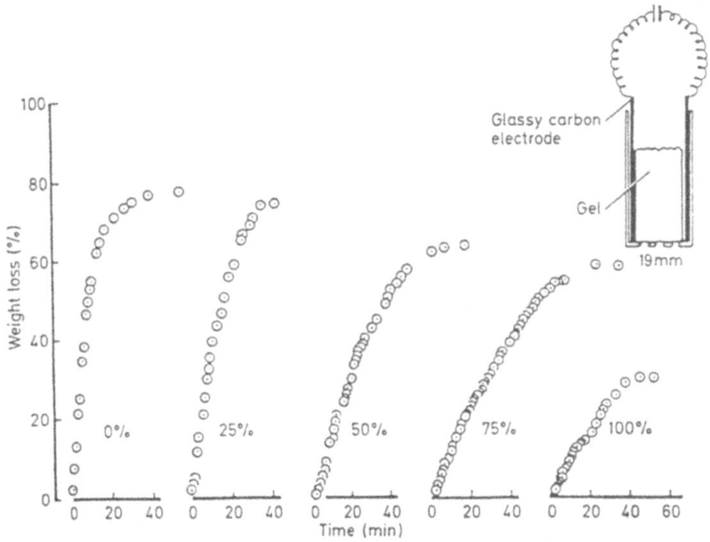

Fig.1 Apparatus for the electro-activated chemomechanical system and time profile of weight change of the gel equilibrated in water-ethanol mixtures. Gel: poly(2-acrylamido-2-methyl-1- propanesulfonic acid), electric field: 6.3V/cm, Degree of swelling of the gel before the experi ment was 2250 times and 1000 times in ethanol,
Numbers denote volume% of ethanol.

In order to clarify the local and spacial shape change of the gel, several marks were made on the surface of the hydrogel using sewing needles with a distance of 10 mm intervals each. The gel near anode made an extensive contraction and allowed us to push and displace the electrode continuously in order to keep it contact with the gel. On the contrary, the gel near cathode did not contract, but even expanded with time and water droplet preferentially dripped out from the gel near cathode. The quantitative response of the poly(AMPS-2-hydroxyethylmethacrylate (HEMA) copolymer, poly(AMPS-co-HEMA)gel, to the electric field with time is shown in Fig.2. It is seen that the application of the electric field to the gel causes an extensive but asymmetrical contraction and forms the shape which is reminiscent of the bottle.

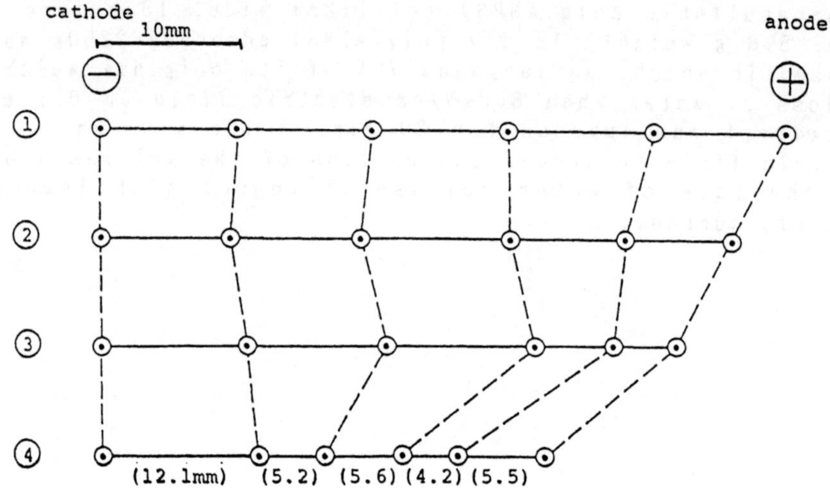

Fig.2 Time profiles of shape change of poly(AMPS- co-HEMA) gel induced by electric field(30V). 1, no electric field: 2, after 30min: 3, after 90min: 4, after 210 min.

We previously reported that a threshold potential is required to cause an extensive contraction by electric field[7]. When Pt is used as an electrode, the threshold potential is ca. 2 V which is almost same as the value of overpotential of water electrolysis. Below this value the gel practically shows no contraction. However, above this value the rate of contraction increased in proportion to the current. The rate of volume change was proportional to the amount of current and the contraction could be explained in terms of ion transport of counter-microions of particles in the electric field. These facts strongly suggest that the contraction is occurred by electrokinetic reaction.

(B) Mechanism

As reported in the earlier papers, we have several experimental facts supporting an interpretation that the chemomechanical behaviors observed are essentially electrochemical phenomenon. They are 1) absolute absence of contraction for neutral (non-charged) hydrogel. 2) swelling near the anode and contraction near the cathode when the gel is negatively charged and the reversal observation for the positively charged polyelectrolyte gel. 3) parallel relation of the rate of contraction to the amount of electrical current. Thus, electrokinetic process involving the transport of water and electric charge should be accounted for.

The most striking support, however, for the presumed electrokinetic nature is direct observation of migration of water (electroosmosis) and of the charged ions (electrophoresis) towards the electrode bearing a normal charge opposite in sign to the net charge borne by the gel. The observed chemomechanical behaviors of the hydrogels led us to consider the possibility that the external electric field underwent the direct interaction with the internal electric field in the gel, particularly the coupling of electroosmosis of water and electrophoresis of charged molecules.

Experiment on electroosmosis was carried out with two measuring vertical capillaries and the apparatus used is shown in Fig.3. A disk-shaped poly(AMPS-co-HEMA) gel, 80 mm in diameter and 20 mm thick was placed in the Petri dish and immersed partly in water. A pair of glass capillaries with 2 mm of inner diameter was allowed to stand in parallel on the top of the gel with a distance of 10 mm. Water was then poured to both capillaries and a pair of Pt wire electrodes was placed in each of them. Distance of the electrodes was controlled by changing the position of the electrode in water. DC voltage (30 V) was applied and the position of liquid levels in the measuring capillaries were determined with time using a cathetometer. When DC current was applied through the electrodes, a water level of cathode quickly elevated and that of anode lowered.

Fig.3 Schematic representation of the apparatus for electroosmotic transport of water in the polyelectrolyte gel.
1,platinum wire electrode
2,capillaries
3, polyelectrolyte gel
4, water 5, petri dish.

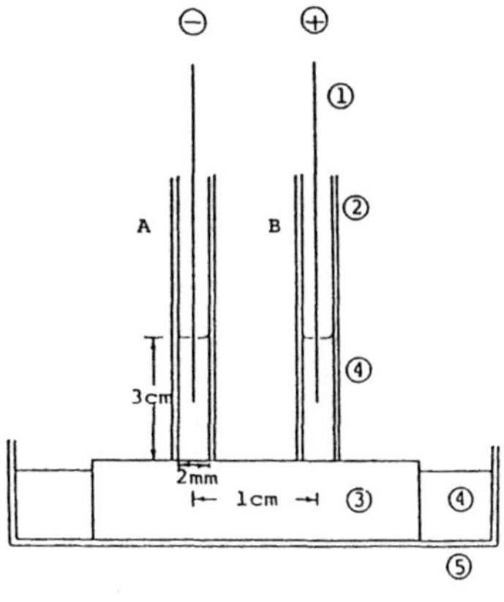

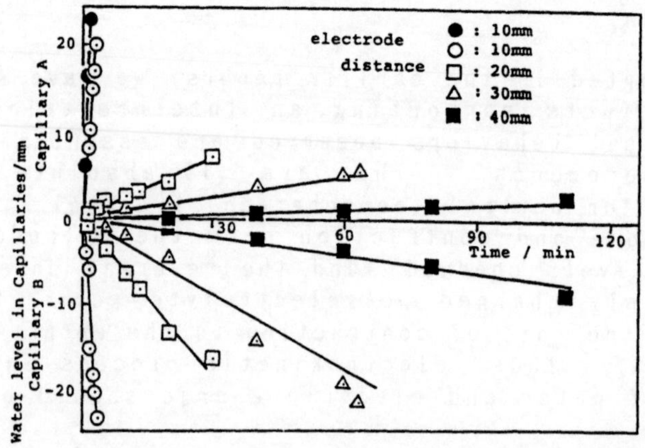

Fig.4 Time dependence of water level in the measuring capillaries when DC voltage (30V) was applied. Electrode distance and current observed:
(○) 10mm, 4mA: (●) 10mm (no water was poured in the capillaries), 4mA: (□) 20mm, 0.1mA: (△) 30mm, 0.03mA: (■) 40mm, 0.01mA. Sample: PAMPS gel.

The velocity of change in water level varied with change in the distance between the electrodes and the value was almost proportional to the current. The direction of flow was always in the direction of counter ion movement (Fig.4). When an electrode was placed on the surface of the gel (in this case the electrode distance is 10 mm) and when no water was poured into the capillary, elevation of water level of cathode was also observed and the velocity was almost same as the case when water was poured. Thus, water undergoes electro-osmotic migration from anode to cathode in the crosslinked anionic polymer network.

The movement of ions in an electric field is called electrophoresis and this also occurs in our case. Electrophoretic movement of poly(methacrylic acid)(PMAA) solution (concentration; 0.12 mol l^{-1}) is easily demonstrated by applying electric field to the charged macromolecular solution, for example, 30 V for 15 min to 1 wt% (0.12 mol l^{-1}) aqueous solution of PMAA and by measuring the amount of adsorbed PMAA on each of electrode. By this experiment it was found that PMAA adsorbed on anode was ca. 3.3 times larger than on cathode and 2 times larger than on the electrode without electric field. Thus, it could be assumed that mildly-crosslinked and highly-swollen polyelectrolyte gel used in this experiment undergoes electrophoretic displacement. An extensive adhension and sticking to the electrode bearing charge opposite in sign to that of gel was also observed when the electric field is applied for a certain period of experiment.

The physico-chemical property of a gel is associated with the strength of interactions between macromolecular network and the liquid. The internal force acting on the polyelectrolytic gel is expressed in terms of osmotic pressure, rubber elasticity of polymer network, affinity of solvent with polymer and electrostatic repulsions of macroions. If the gel is perturbed from one state to other, it rearranges the structure to reach an thermodynamically favorable state[8]. Thus, a rearrangement of the network of alginic acid to give oriented ordered-structure[9] due to an electrophoretic migration of macromolecular network was found.

An qualitative interpretation of the phenomena observed is that the application of an electric field causes a pumping of largely mobile ions and partly macro-network ions and surrounding hydrated water to the opposite directions until they reach the electrode. In general, the velocity of migration and the velocity of gel contraction are governed both by the quantity of mobile ions q_n and the gradient of electric field ΔE which is the driving force of a pumping of these ions.

The force produced in the gel F is proportional to the sproduct of q_n, and ΔE, i.e., $F = \Delta E \times q_n$.

If we ignore the effect of elastic forces due to covalently crosslinked network and if the gel are free to move as an aggregate of particles in the electric field, a slightly crosslinked and highly swollen polymer gel eventually attains a steady state velocity U as in solution. This value is apparently proportional to the velocity of contraction of the gel υ under certain period of time.

$$\upsilon \times U = \Delta E \times q_n / r = \Delta E \times I/\mu \qquad (1)$$

Where r is a hydrodynamic frictional coefficient which depends on the geometry of the gel, thickness of double layer, viscosity of medium and others. If the average mobility of the medium in the gel is μ, the rate of contraction may be proportional to the product of the gradient of electric field and current I.

In order to prove this simple model, a contractile experiment was carried out under various salt concentrations. The change in salt concentration was dominated by NaCl, and water-swollen PMAA gels were equilibrated immersing in 1000 ml of NaCl solution with various concentrations. These gels, 50 mm long, 10 mm in diameter and equilibrated in different NaCl concentrations were allowed to interpose between a pair of Pt wire electrodes and DC voltage 10 V was applied through the electrodes. Potential gradient was calculated by measuring the voltage at every 5 mm intervals. As a result, it was found that the electric field decreases almost in proportional to the distance from anode except near the both electrodes where rapid potential drops were observed. Assuming that gradient of electric field in the central part of the gel is constant, one can calculate the potential gradient. It was found that the gradient of the potential drop decreases with increase in salt concentration in the gel, while the electric current increased. Thus, the product of "E and I at each interval in the gel has a maximum value and the normalized results of $\Delta E \times I$ for the different gels are shown in Fig.5 (dotted line). The shape of this curve indicates that the rate of contraction increases rapidly with increase in salt concentration, which via maximum value at the concentration near 0.08 mol l^{-1} decreases with further increase in salt concentration.

Experiments of electro-activated contraction using PAMPS gel and PMAA gel equilibrated in various salt concentrations were carried out and the results are shown in Fig.5.[9] As presumed from the model, the rate of contraction increased significantly with increase in salt concentration, which via a maximum value at the concentration near 0.1 M, decreased rapidly with further increase of salt concentration.

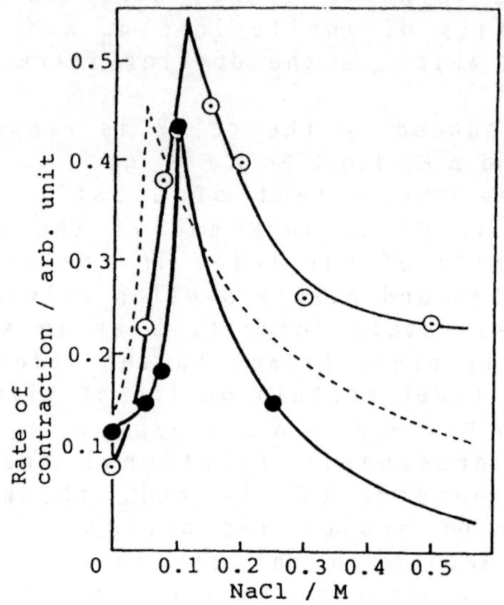

Fig.5 Dependence of the contraction of PVA-PAA composite membrane (○) and PAMPS (●) gel on NaCl concentration. Dotted line indicates theoretical curve. Experiments of gel contraction were carried out in the similar manner as in Fig.2 using gels, 50mm long and 10mm in diameter. 10V DC was applied.

The obtained profile is the same for two different gels and the result appears satisfactorily consistent with the curve calculated from the simple electrophoretic model described. On the other hand, if PAMPS gel is equilibrated in water-acetone mixture with different mixing ratios and contractile experiment is carried out, one can observe a gradual and monotonous decrease in the rate of contraction with increase in acetone through the entire mixing ratios. The result is explained by decrease in "E and I with increase in acetone.

(C)Contraction of Microparticle[7)]

In order to establish the system with minimized response time, microparticle of sodium salt of poly(acrylic acid) gel was synthesized and size change of particle with time under various electric fields was measured.

A microparticle was placed between two SnO_2 electrodes. 10X10mm and was mounted on a polymeric frame with spacing of 1.3 mm. A direct-current density of 0-0.3mA cm-2 was applied via SnO_2 electrodes, and the volume change of the microparticles was observed with an optical microscope. In order to eliminate the effect of change in pH of the fluid, a salt bridge was used instead of $SnO_2$the glass electrode but the some contractile behavior was obtained.

Fig.6 shows time profiles of the contraction of a particle(180 um in diameter) in the cell when a direct current ranging from 0 to 0.3 mA cm-2 was applied. Below 2.4 V (18.5V cm-1) no contraction or detectable electric current was observed. Shrinkage started when the d.c. voltage reached 3V or the electric current reached 0.02 uA. The rate of shrinkage increased with increasing electric field. Note that an increase in voltage increased not only the rate of contraction but the final size(Vmin) that the particle could achieve. For example, a microparticle 180 um in diameter completely contracted within 50 S when 6V dc was applied, and the diameter was reduced to 60um. No detectable shrinkage occurred after this, even if the electric field was applied for longer.

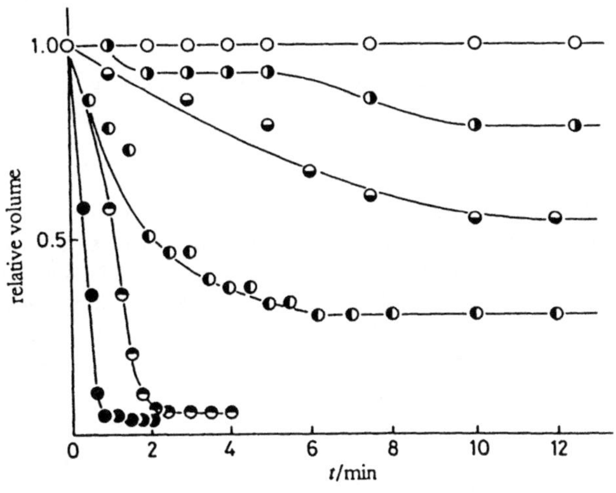

Fig.6 Time profiles of volume change of the mecroparticles. voltage: (○) 2.4V (18.5V/cm): (◑) 3.0V (21.1V/cm): (◒)3.6V (27.7V/cm): (◐) 4.2V(32.3 V/cm), (◓) 4.8V (36.9V/cm): (●) 6V (46.2 V/cm): microparticle diameter. 180um.

After the electric field turned off most microparticlesrecovered the original size and shape within 5-10 minute and showed good reproducibility.

Rate of volume change should be a function of a initial size of the particle, and the relative rate of volume change size of the particle, and the relative rate of volume change must decrease with increasing size of the particle. Experiments using particles with different sizes were carried out, and the time needed for a 96% volume change was shown in Fig.7. The experimental results are inversely proportional to the square of the particle size. From Fig.7, it was calculated that the time taken for an 96% volume change for a particle of 1 um diameter is as low as 0.23 ms under these conditions. This result coincides with Tanaka's equation [10], showing a proportionality to the square of the characteristic length of the gel.

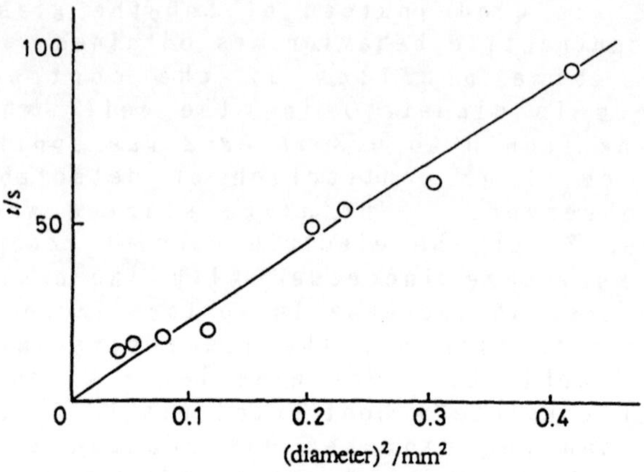

Fig.7 Dependence of the time needed for 96% volume shrinkuge on the square of the microparticle size; DC voltage, 3.5V (26.9V/cm).

3. ELECTRO-ACTIVATED CHEMOMECHANICAL DEVICES

(A) Artificial Muscle

Chemomechanical system capable of lifting a load up and down automatically and repeatedly is the simplest application of electro-shrinkable gels. The most interesting feature was that the rate of contraction of the gel increased with load. Rate of contraction initially decreased from 0.4 to 0.1 cm/h with loads ranging from 0 to 5.5 g, but beyond that the lifting rate increased with increasing load. Thus, a weight of 22 g was lifted 6 times faster than that of 5.5 g. The efficiency of the work done by the gel with load of 22g was 24 times lager than that of 5.5g (Fig.8)[11].

This anomalous lifting behavior of the gel can be explained in terms of a spontaneous ionization of ionizable groups leading to an spontaneous increased electric current when the gel is stretched (reverse chemomechanical reaction(Fig.9). It is well known that the dissociation of weak polyelectrolytes results in a more expanded conformation of a polymer chain. Conversely, the stretching of a macromolecular chain by an applied stress induces an additional ionization. This should also be observed in polyelectrolyte networks. In fact, we observed the ionization of carboxylic groups, i.e., a decrease of pH of the gel when a stress was applied as shown in Fig.10.

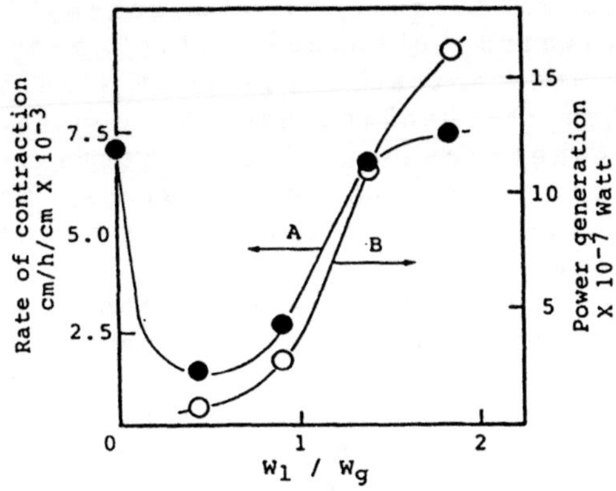

Fig.8 Rate of contraction (A) and power generated (watt)(B) vs load applied (Wl) per weight and length of PMAA gel sample(Wg). Gel: 60 mm long, 17mm in diameter, wet weight 12g, Electric field: 50V DC.

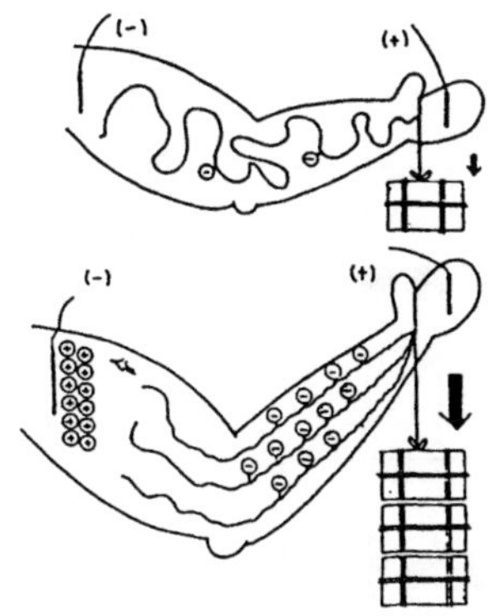

Fig.9 Schematic illustration of mechanism of anomalous contraction

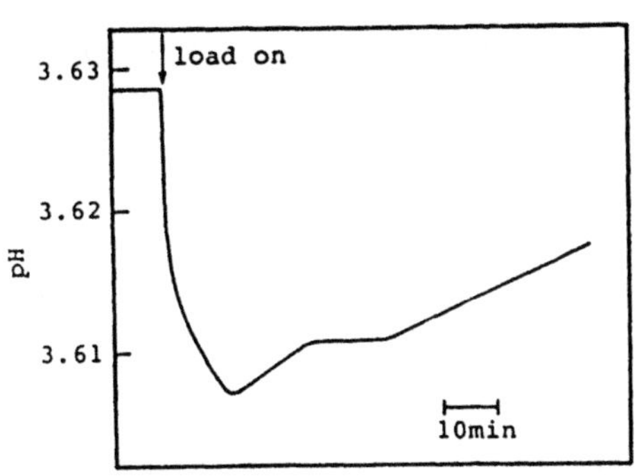

Fig.10 pH profile of the PMAA gel sample under tensile stress of 20g. Sample dimensions: length 100mm long, 17mm in diameter, wet weight 20g. pH in the gel was measured by inserting a microcom bination pH probe (MI-410) into the gel to a 5mm depth.

(B) Chemical Valve

Fig.11 shows an example of the schematic diagram for electrically-activated "chemical valve" membrane which reversibly expands and contracts the pore size by the electric stimulus.[4] When the mechanochemical contraction described is developed isometrically, i.e., keeping the membrane dimensions constant, the contractile stress appearing in the membrane should expand the pore channels though which solute and solvent permeate.

Fig.11 shows the result of a chemomechanical contraction of the PAMPS membrane on water permeation when 6.5 V of DC was alternatively imposed. It is seen that the "chemical valve" membrane can increase and decrease the water permeability many times by electric stimulus. Increase in water permeability was proportional to DC current. This makes the membrane possible to use as a permselective membrane continuously separating solute mixtures with different molecular size. This type of electrically-activated "chemical valve" membrane exhibited long term stability and could work continuously for more than one month.

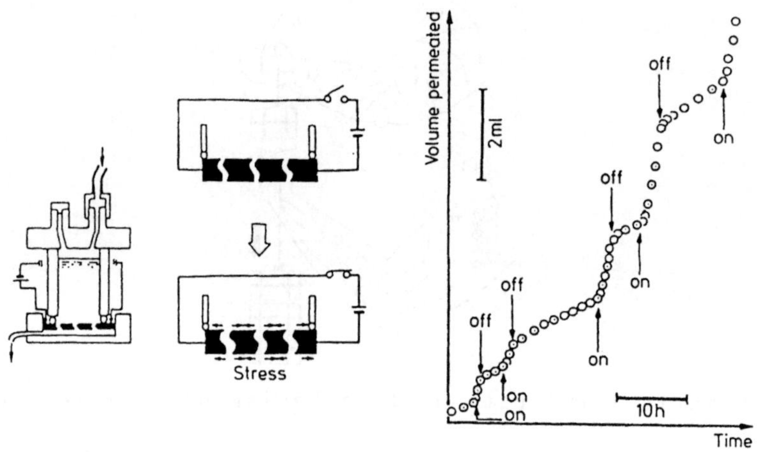

Fig.11 Apparatus for the electro-activated "chemical valve" membrane and change of the water permeability by alternative "on" and "off".

(C) Drug Delivery System (DDS)

In recent years a much attention was paid to develop drug delivery system (DDS) and utilization of various carrier membranes for an extended period of time and minimized change of delivery velocity is one of them. However, if we look the DDS from future therapentical viewpoint, we have to develop an effective DDS devices which are able to sense the needs of the body and deliver an appropriate dose of drug at appropriate moment in response to the body[5]. Chemomechanical system using polyelectrolyte gel in which contains a drug may provide one possibility to realize such kind of active DDS. When an drug molecules are entrapped within the gel which exhibits reversible shrinking and swelling as an electric stimulus is turned on and off, the drug may be switched on and off since the substrate diffusion rate should be altered by the solvent flow caused by contracting stress of the gel and by the change in the crosslinked network density (or the gel pore size). This stress will also "squeeze" out the swelling fluid containing drug as the gel shrink(Fig.12).

Fig. 12 Schematic illustration of mechanism of drug delivery system

Fig. 13 shows the changes of concentration of pilocalpine(Pil) released out of Na-PAA microparticles as a function of time.[12] It is seen that the concentration of Pil increases quickly and sharply , when DC voltage was applied in water. The rate of Pil release while "on" time was 9.8×10^{-6} mol/l.s of that of "off" time was 1.8×10^{-6} mol/l.s. Thus, the rate of release was 5.4 times larger than "off" time. The rate of release increased with increasing electric field applied. The rate of Pil release under 5.0 V was about 4 times larger than that at 3.0 V, which coincides with the ratio of the rate of contraction of microparticles[7] at 5 and 3 volts indicating that Pil was released by the contraction of gel volume. Note here that the Pil is a ionizable compound and sustains the electrophoretic migration toward cathode under electric field. In order to associate the release of bioactive compound with chemomechanical shrinkage of the swollen gel particle, glucose was entrapped into the gel and the release test was carried out. In this case no electrophoretic transport of glucose should take place and was found that glucose is initially "flashed out" quickly by applying electric field. The rate of release of glucose of the first "on" state was 45 times larger than the following "off" state.

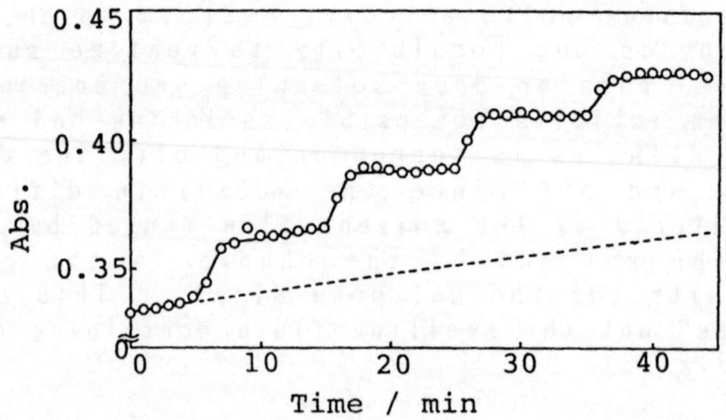

Fig. 13 Time profile of pilocarpine rerease from microparticles of sodium salt of polyacrylic acid, by electric stimulus.

(D) Bimetal

Polymeric device which functions as "bimetal" was made from gel-plastic laminated structure. The contractile force developed in the gel by an electric stimulus through a pair of platinum wire electrodes could bend the "nonshrinkable" poly(vinyl chloride) tube. The composite device consisting of crosslinked poly(methacrylic acid)(degree of swelling; 3 times) and poly(vinyl chloride) tube as illustrated in Fig. bent as large as 20 after ca. 30 min, and 43 after 70 min. It recovered the original straight form by dipping in water.

4. FUTURE ASPECTS

At present, responses obtained in these systems are rather slow, generally in the order of second to minute and they have not been applied to engineering uses such as actuators in robotics. Luck of mechanical toughness and of long-term durability are another problems to be solved. Essential improvement of low efficiency of energy conversion should also be made for practical application of the chemomechanical system. However, an attempt to solve these problems has been made using new type of hydrophilic-hydrophobic balanced gels having a liquid crystal structure[13,14].

Nevertheless, chemomechanical systems have a variety of potential applications. They may be used as self-regulating drug-delivery system, switching device, thermostat, pH-sensitive electrode and various kinds of sensors as a part of which was described. System which develop contractile force and change their dimensions may eventually lead to approaches to "biomechanics" or "molecular machines" based on water-swollen "soft" materials. From this point of view, the application to an "actuator" for robots and other machine systems has particular interest. In that case we should expect to built up a soft gel actuator possesing a senser capable of responsing automoticaly to outside environmental charges. The chemomechanical systems are also interesting as analogs to physiological muscle and biological systems exhibiting rhythmic phenomena[15-17],e.g., oscillation of nerve walls where the appearance of dissipative structures is considered to play an important role.

References

1)Mechanochemical (chemomechanical) system is well known as a model converting chemical into mechanical energy and a variety of systems have been reported. The system developed by us is considered as a prototype model of a chemomechanical (mechanochemical) system operated under electric similus. We prefer the term "chemomechanical reaction," "chemomechanical system" instead of "mechanochemical" proposed by A. Katchalsky to give more precise difinition and to avoid a confusion with other terminology.

Note that Mechanochemical reaction often refers to chemical reaction induced by mechanical stress fracture and dumpling to give radical and crosslinking formation.

[Y.Osada, Adv. Polymr Sci., 82, 1(1987)]

2)Transformation from mechanical into chemical energy may be envisaged in which the application of mechanical energy may rise the chemical potential of substance and permit its transfer from a reservoir of a lower chemical potential to a reservoir of higher potential(reverse chemomechanical system). Details are given in Y. Osada, Adv. Polym. Sci., 82,1, springer-Verlag, Heidelberg, 1987.

3)K. Iwasa and T. Tasaki, Biochem. Biophys. Res. Comm., 95, 1328(1980). I. Tasaki, K. Iwasa, Structure and Function of Excitable Gel. ed.by D.C. Chang, I. Tasaki, W.J. Adelman and H. R. Leuchtag Plenum Pub., Corp. New York, 307(1983).

4)Y. Osada, M. Hasebe, Chem. Lett., 1985, 1285.

5)Y. H. Bae, T. Okano, R. Hsu, S.W. Kim, Makromol Chem. Rapid Comm. 8, 481(1987).

6)L. C. Dong and A. Hoffman, ACS Symp. Series, 350, 236(1987).

7)Y. Osada and R. Kishi, J. Chem. Soc., Faraday Trans., I, 85, 655(1989).

8)J. Hasa, M. Ilavsky, K. Dusek, J. Polym. Sci., Polym. Phys. Ed., 13, 253(1975).

9)R. Kishi, M. Hasebe, M. Hara and Y. Osada, Polymer for advanced Technologies, 1, (1990).

10)T.Tanaka and D.J.Fillmore, J. Chem. Phys. 70,1214(1979)

11)Y. Osada, R. Kishi and M. Hasebe, J. Polym. Sci.: Part C: Polym. Lett., 25, 481(1987).

12)Y. Osada, R. Kishi, Y. Yasunaga and K. Sawahata,

13)Y. Osada and Y. Yasunaga, Koubunshi Ronbunshu, 46, 655(1989).

14)R. Kishi, M. Sisido and S. Tazuke, Polymer Preprints, Japan, 38, 3859, 3862(1989).
15)K. Umezawa and Y. Osada, Chem. Lett., 187, 1795.
16)Y. Osada, K. Umezawa and A. Yamauchi, Makromol. Chem., 189, 597(1988).
17)Y. Osada, K. Umezawa and A. Yamauchi, Bull. Chem. Soc. Jpn., 62, 3232(1989).

AMPHOTERIC POLYVINYL ALCOHOL HYDROGEL AND ELECTROHYDRODYNAMIC CONTROL METHOD FOR ARTIFICIAL MUSCLES

Makoto Suzuki

Biomechanics Division
Mechanical Engineering Laboratory
1-2 Namiki, Tsukuba, 305 Japan

INTRODUCTION

Since Kuhn and Katchalsky et al.[1] first demonstrated artificial muscle in 1950, various mechanochemical materials have been presented.[2] From the view point of mechanical engineering, great hopes are set on these materials as a new actuator, especially in the fields of robotics and medical welfare instruments, and improvements in their performance are eagerly awaited. In such fields realization of light weight, soft touch and powerful actuators is an important subject. We expected that an amphoteric polyvinyl alcohol (PVA) hydrogel might have a potential high enough to realize the mechanical active performance of skeletal muscles.[3] Because poly(vinyl alcohol) aqueous solution more than 5%w was found to become strong hydrogel by freezing and thawing by Nanbu in 1982.[4] First of our purposes is to examine how much this material can become closer to muscle in the mechanical active properties. Second as a device we have to develope a control technique for mechanical power generation by mechanochemical materials. This paper consists of two sections, first for the improvement of material properties and second for the new method of electrohydrodynamic control.

SYNTHESIS AND MECHANICAL PROPERTIES OF AMPHOTERIC PVA HYDROGEL FILMS

Synthesis of Amphoteric PVA Hydrogel

In the amphoteric PVA hydrogel polyacrylic acids and polyallylamines were fixed in the network of PVA by means of PVA-PVA intermolecular association by hydrogen bonds. It has a strength more than 1MPa in spite of high water con-tent more than 85%. A uniform aqueous solution was obtained by mixing poly-vinyl alcohol (Kurare 117H; Mw:74,800; saponification>99.6%), polyacrylic acid (-COOH, SP2; Mw: 170,000), and polyallylamine hydrochloride ($-NH_3Cl$, Nittobo; Mw: 60,000) at monomeric concentrations of 1.74 M, 0.245 M and 0.26 M, respectively. With the use of spacers, the solution was sandwiched between two sheets of optically ground heat resistant glass 10 mm thick, to form a liquid layer of constant thickness.

Amphoteric PVA hydrogel films were then obtained by subjecting the solution to repeated freezing and thawing. The temperature range was between -50 C and room temperature around 23 C, with temperature falling and rising at a rate of 3.0 degree/min just before the freezing and thawing points. This was simply realized by placing the liquid in and removing it from a refrigerator at 30-min intervals.

Polymer Gels, Edited by D. DeRossi *et al.*
Plenum Press, New York, 1991

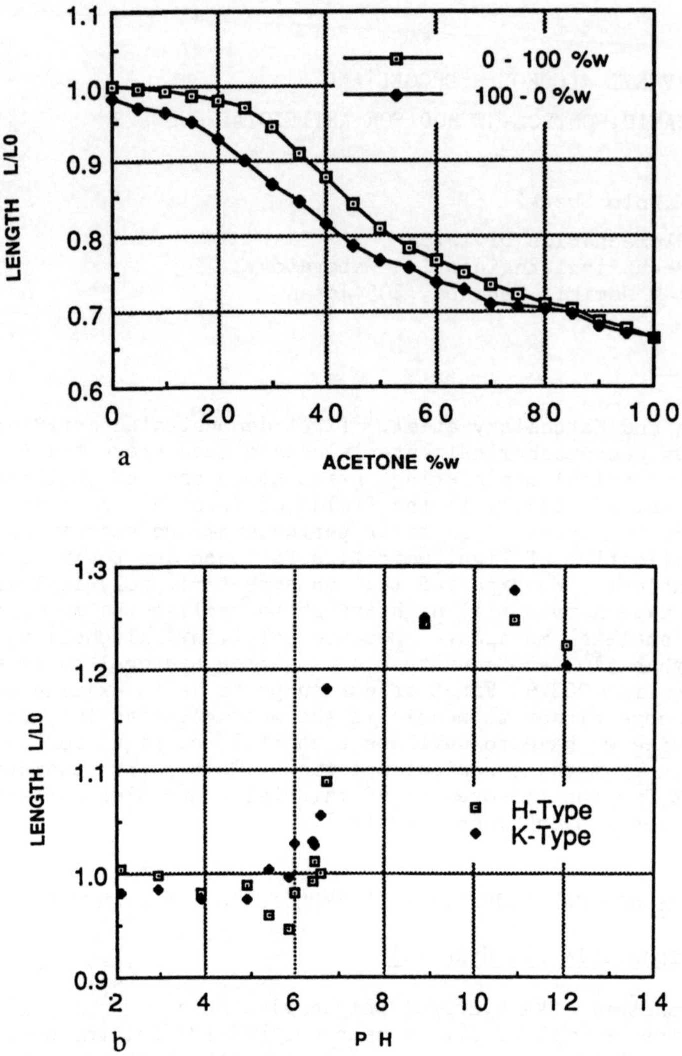

Fig.1 Equilibrium Swelling of Amphoteric PVA Hydrogel in a:water/acetone mixture, b:HCl/KOH aqueous solution. H-Type was brought from HCl 1N and K-Type from KOH 1N.

Some gel films were rendered anisotropic by stretching uniaxially. First they were frozen and thawed three times. The glass sheets of one side were then removed and the gel films stretched to 1.6 times their original length in one axial direction. They were then frozen and thawed further 10 times, and stretched again to 1.25 times their length in the same axial direction. They were finally frozen and thawed a further 10 or 20 times.

To produce finished samples, the gel films were washed three times by the following procedure: KOH of 1N, water, HCl of 1N, and water. They were lastly removed from the HCl and sufficiently washed in ultra pure water. When the spacers are set at 55 microns, the thickness of the average control gel film is about 60 microns, and for stretched gel about 35 microns.

The equilibrium swelling characteristics of these gels in a mixed solvent of acetone and water are shown in Fig.1a and in HCl/KOH aqueous solution without extra salts in Fig.1b. In the following experiments, an acetone/water solvent was used. The solvent exchange between acetone and water was more effective in increasing contraction factor and speed than that between acid and alkali or between water and ethanol.

Stress and Strain Characteristics

Control and stretched gel films were then cut into strips, and tension was applied to them in water to examine strain and stress characteristics. Fig.2 shows the results for three strips. One is of control gel, and the others of stretched gel. "Parallel" denotes that the load was applied parallel to the direction of stretching; "Perpendicular" denotes that load was applied perpendicular to the direction of stretching.

All the strips in this case had been cut from gel films frozen 23 times. It was found that when the number of freeze-thaw cycles was the same, the control and perpendicular stretched gels exhibited almost the same elasticity. In the stretched gel film the Young modulus in the parallel direction was twice larger than that in the perpendicular direction.

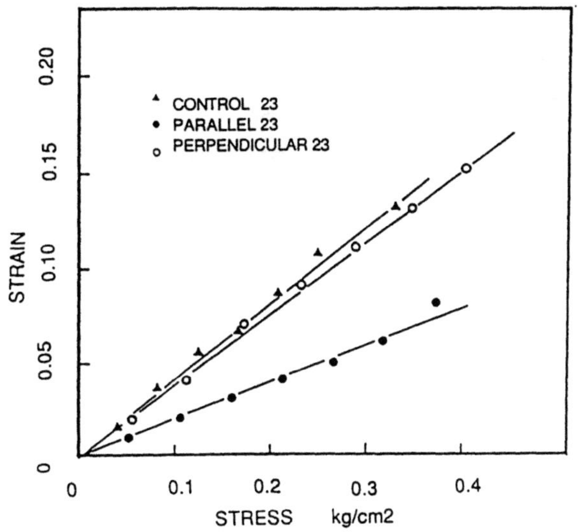

Fig.2 Stress-Strain Relations for Control (Non-stretched) Gel Film and Stretched Gel Films.

Output Characteristics of Amphoteric PVA Hydrogel Film

To examine the mechanochemical properties of the gel films, and in particular the dynamic characteristics of contraction and elongation, we used the apparatus shown in Fig.3. In the cell the gel film is soaked with water or acetone and vertical load can be applied to it. Solvent exchange is performed by draining liquid from the drain and injecting solvent into the cell by means of a pressurized syringe.

A change in the length of the gel film is converted into an electrical signal by the goniometer and recorded. The beam is extremely light and the response time of this measuring system, including liquid pouring time, is about 1 sec, but for the following experiments we were able to improve this to 0.1 sec.

Film Thickness and Response Time

Fig.4 shows the relation between contraction response time and film thickness of gel films with different water contents of original mixed solution while the polymer molar ratio was fixed. Response time here indicates the time required for 50% of total contraction by length to occur. The broken lines are proportional to squares, and the values for response time are generally seen to follow these lines.

From the results of previous experiments, it was known that the macromolecular density of a formed gel film increases with water content because the thickness of a formed gel film decreased quickly with water content of original solution increasing in the range more than 85%. Fig.4 clearly indicates that the response of a gel becomes slower as the macromolecular density increases. Raising the macromolecular density by 60% results in a response four times slower.

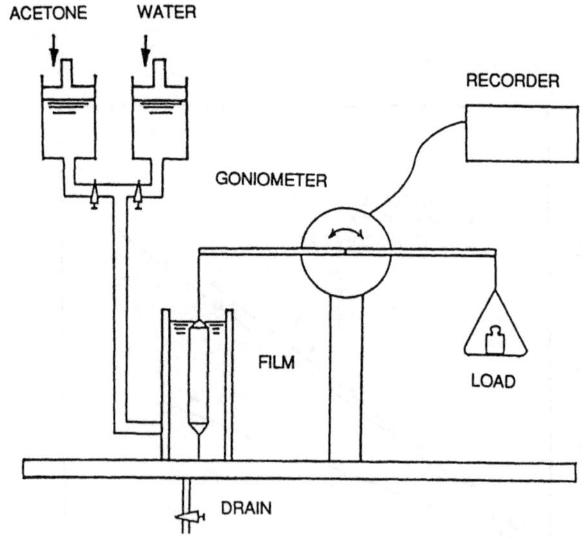

Fig.3 Apparatus to Measure Dynamic Response of Gel Film. Acetone or Water was injected to the cell with one of the pressurized syringes. When the film contracts the angle change of the bar was electrially detected.

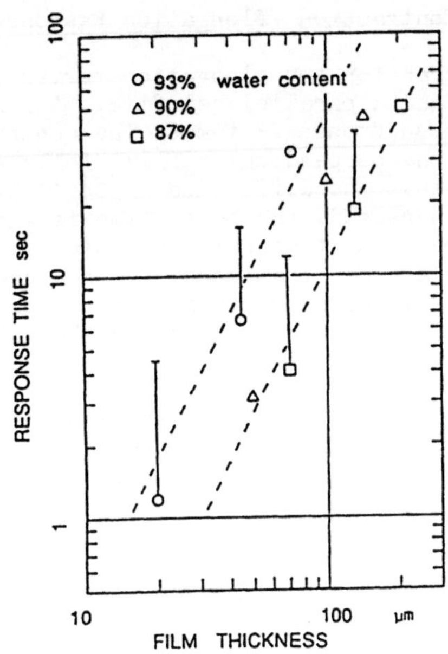

Fig.4 Response Time Versus Film Thickness As a Function of Water Content before Gelation.

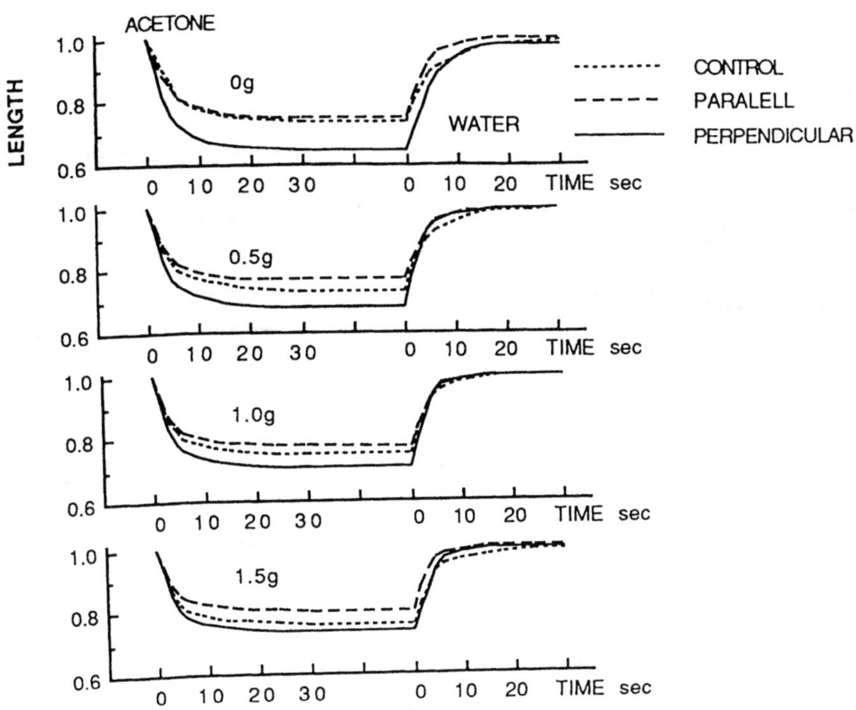

Fig.5 Curves of Contraction and Elongation. Open circle:Perpendicular,Solid circle:Parallel, Triangle:Control.

Effect of Anisotropy on Contraction /Elongation Response

Fig.5 shows the contraction and elongation curves of a control gel film, a perpendicular gel film and a parallel gel film. All measured 10 mm wide and 25 mm long, and had been frozen 23 times. The contraction factor was shown to be greatest for the perpendicular film, followed by the control gel film. It was also found that the difference in contraction factor became smaller as the load was increased. Response time is shown in Fig.6. Firstly, it can be seen that the response time for elongation is shorter than for contraction. It can also be seen that stretched gel contracts nearly twice as fast as does control gel. However, if response time is roughly proportional to the square of film thickness, it cannot be said that response time has been improved by stretching.

Another observation is that the contraction factor is greater for gels which have been frozen the most times. The largest contraction factor was obtained for a perpendicular gel which had been frozen 33 times. Right-angle stretching was shown to be effective up to a load of about 2 kg/cm^2.

Output

Output power density "G" can be expressed by the following form

$$G = L*Cr/2t$$

where Cr is the contraction factor, L is contraction tensile strength (load) and t is the contraction response time measured at 50% of total contraction length. The results are shown in Fig.7.

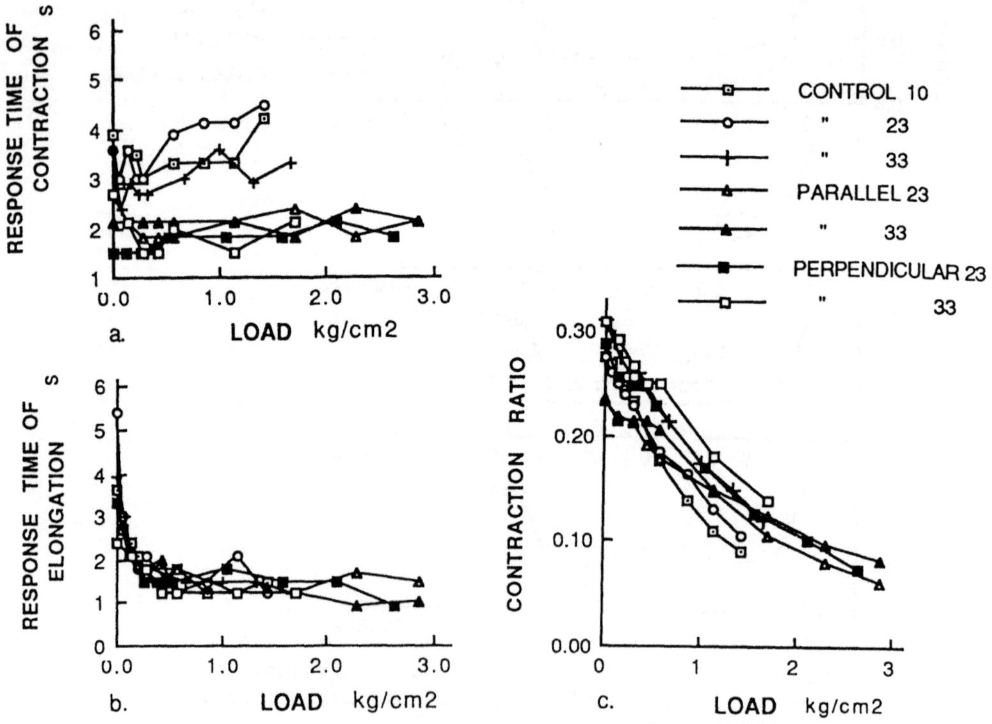

Fig.6 Effect of Anisotropy on Contraction/Elongation response.
a:contraction response time, b:elongation response time,c:contraction ratio

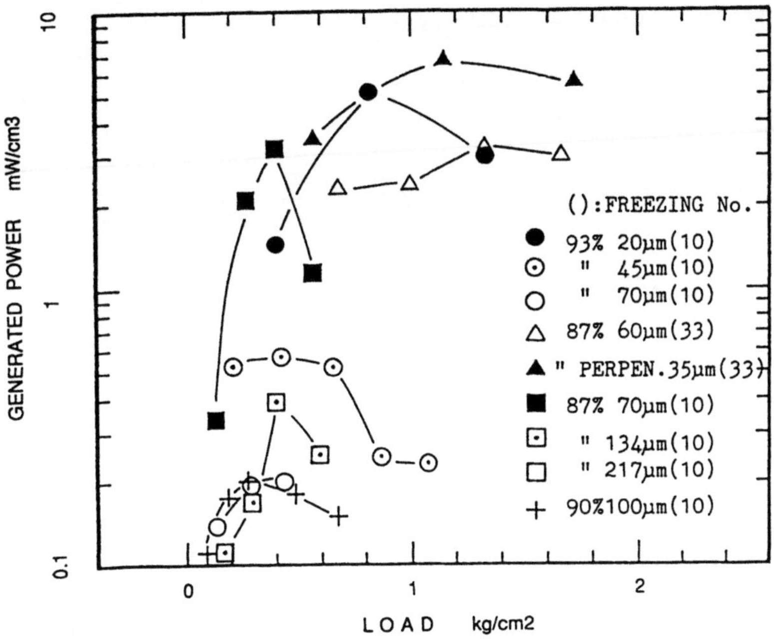

Fig.7 Mechanical Power Density of Various Gel Films. %'s indiate the water content before gelation.

Approach to High Power Density Gel Film

For a material to be suitable as a substitute for skeletal muscle, it must have an output power density around 0.1 W/cm^3. As stated previously, two possible solutions are to reduce the thickness or to make the material porous. Making it anisotropic by stretching can reduce the amount of a change in cubage against a change in the length of the gel film, thereby reducing liquidity loss and consequently improving efficiency. Based on this knowledge, we made two kinds of film, one a nonstretched thin film and the other a stretched thin film.

When producing a thin film it is important to ensure that the original mixed solution is as uniform as possible. After mixing, the viscosity of the solution gradually increases day by day. For example, a viscosity of 1,400cp immediately after mixing will, over a period 1 month, rise to 2,500cp. Overtime, the solution begins to show uniform liquidity. We found the solution most suitable for making thin gel films about 11 days after mixing.

With this solution we made one nonstretched film 10 microns thick, and one film 12 microns film thick, stretched perpendicularly to the applied load. Both were frozen 33 times. Fig.8 shows a typical response curve and response time of these two films on log-log plots with the other films. The nonstretched film took 0.12 sec to respond to a load of 1.4 kg/cm^2 (including the response time of measuring system of 0.1 sec). The stretched gel film responded in 0.3 sec to a load of 0 - 2.2 kg/cm^2. The contraction ratios were the same as in Fig.6. The output power density were close to that of skeletal muscle, the former showing an output of 0.106 W/cm^3 and the latter a maximum output of 0.044 W/cm^3.[5]

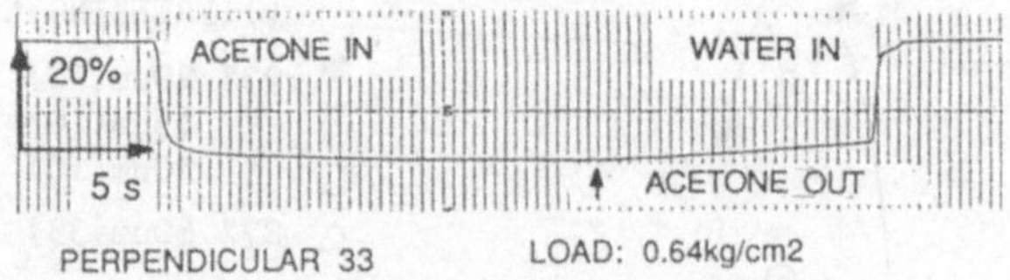

The typical curve of contraction

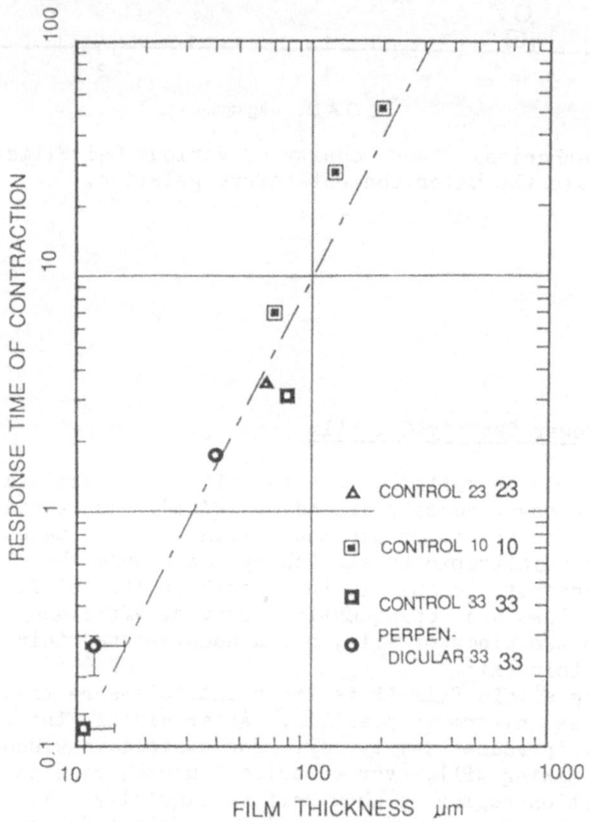

Fig.8 High Speed Contraction by Thin Films.
The upper curve shows the typical contraction.

Microscopic Observation of Hydrogel Films

The photographs of cross section of hydrogel films were obtained with cryo-SEM (scanning electron microscopy) technique by cooperation of Dr. Fujikawa.[12] First sample hydrogel films were frozen by dipping in liquid nitrogen slush and they were placed at a cupper block stage in a vacuum evaporation chamber. The temperature of the stage was controlled around -100 C to etch the cutted surface of frozen gel film. Then the surface was coated with platinum and carbon and observed with SEM. Photo 1a and 1b are the cross sectional views of non-stretched film and stretched film, respectively. Many pores can be seen together with irregular dense polymer regions which probably hold the strength but limit the response time.

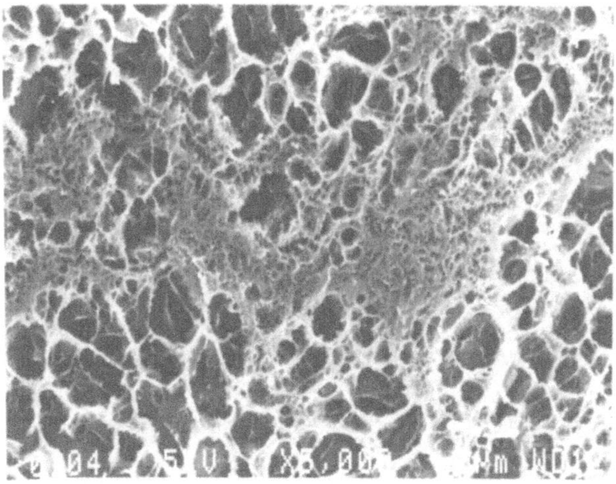

a. Cross Section of Control Gel (33 times frozen).

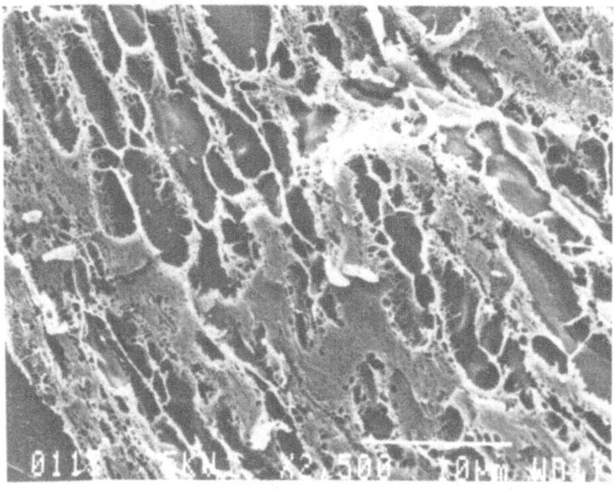

b. Cross Section of Stretched Gel (33 times Frozen).

Photo. 1 Cryo-SEM Images of Cross Section of Gel Film.

Artificial Muscle Model for Demonstration

Photo 2a and 2b shows the motion of human-arm type actuator with a bundle of hydrogel films of 60 microns thick. The muscle was 12cm long and can raise a ball of 2 grams in about 5 sec.

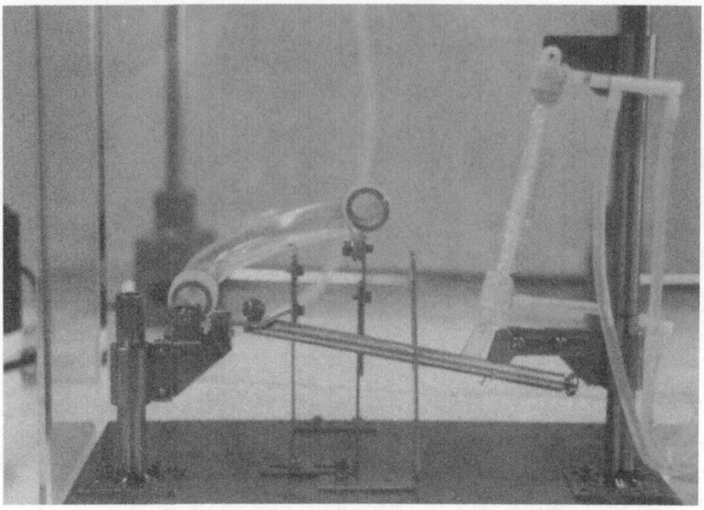

a. Relaxing in Water.

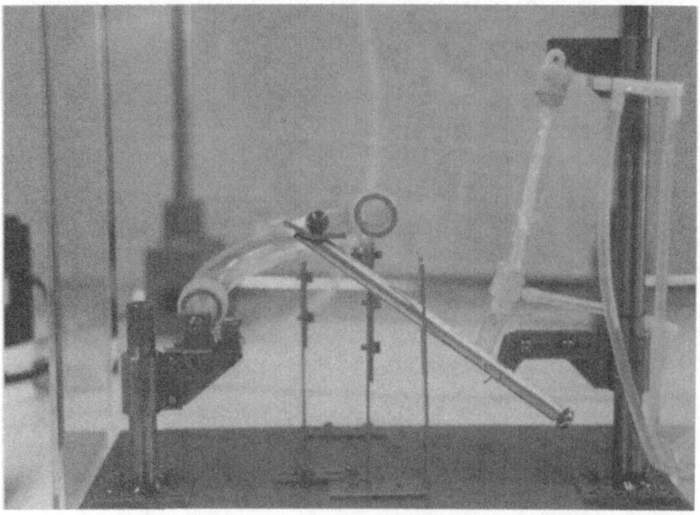

b. Shrinking in Acetone.

Photo. 2 Artificial Muscle Model by PVA Gel Film

ELECTROHYDRODYNAMIC CONTROL OF HYDROGEL CONTRACTION

Back Ground of Electrohydrodynamic Control

As shown by Osada[2], Tanaka et al.[6] and De Rossi et al.[7] hydrogels of polyelectrolytes simply placed between two electrodes subjected to a DC electric field showed remarkable change of volume. Grodzinsky[8] studied electrokinetic membrane transport where hydrodynamic transport played an important role. We have paid attention to the transport of solvent through gel by electrohydrodynamic(EHD) effect. EHD transport of solvent inside gel is considered to be effective when a sufficiently high electric field was applied on a hydrogel film.[9-10] In this study we have examined a new method to controll the mechanical properties, especially the size, of hydrogel film by means of an EHD technique.

Apparatus to Measure Electro-osmotic Flux through Hydrogel Film

The apparatus shown in Fig.9 consists of four compartments numbered I,II,III and IV from the left. In the compartments of I and IV two electrodes were placed to apply an electric field on the hydrogel film suspended between the compartments of II and III. In I and IV, 0.1N NaCl aqueous solutions were filled. The other two platinum electrodes were set closely to the hydrogel film in II and III to measure the potential difference across the hydrogel film. In II and III, NaOH or HCl solutions of concentration from 1mN to 100mN were filled and glass tubes were set to measure the pressure difference between II and III. An anionic ion exchange membrane and a cationic ion exchange membrane connected electrically I to II and III to IV,respectively. In this system when a DC electric field was applied between the power electrodes, positive for I and negative for IV, the solutions in II and III were not contaminated because of electrodialysis. The potential difference measured by the platinum electrodes due to the solution difference was at most 0.3V.

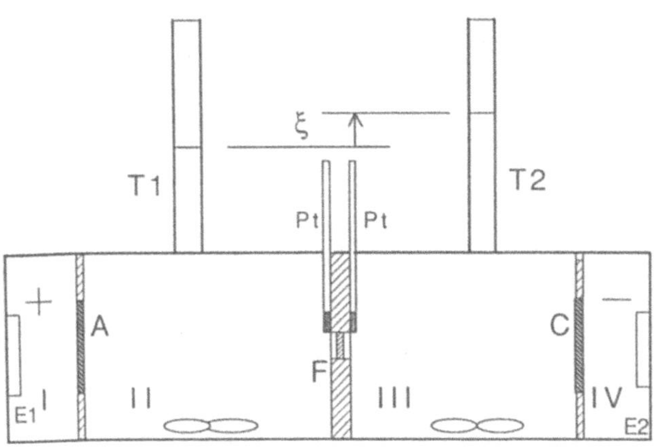

Fig.9 Apparatus to Measure Electroosmosis through Gel Film. E1,E2:electrodes, T1,T2:glass tubes to see level, F:film, A/C:anion/cation exchange membrane, Pt:electrode to measure membrane potential.

When sodium ions move through the hydrogel film co-transportation of solvent occurred. This phenomenon can be expected from ion drag pumping[8-9]. The volumes of II and III were large enough to neglect the change of concentration during the tests.

As shown in Fig.1a in the high PH region the hydrogel behaved like a polyanionic gel. It means that the hydrogel film allows transportation of cations. In the low PH region the hydrogel behaved like a polycationic gel, where anions can be transported.

Results of Co-transportation of Solvent

Fig.10 shows the I-V characteristics of the hydrogel film. "87-2" means that the original solution before freezing and thawing process had a water content of 87%. "87-2s" was made by washing "87-2" with 1N HCl and 1N NaOH solutions repeatedly. It is suspected that since polyallylamine used as polyanion was a type of hydrochloride salt the washing process replaced hydrochloride salt to water. It shows the I-V characteristics were almost linear. Fig.11 shows a good linear relation between the solution flux and the electric current through gel film. The difference of pressure head occurred between II and III increased linearly corresponding to the direction of mobile ions inside the hydrogel when the electric current was held constant.

Similar tests were done for various solution combinations as shown in Fig.12. The vertical axis represents the electrohydrodynamic mobility K_{EHD}, which is vd/V. v,d and V are the flow speed of solvent inside gel film, the film thickness and the potential difference across the film, respectively. In the vertical axis the data of the positive side were obtained by using NaOH solutions both in II and III, where the flux was from II(left) to III (right). On the other hand data of the negative side were obtained by using HCl solutions.

From the figure it was found that the solvent inside the hydrogel flew more as the concentration of NaOH or HCl became higher in the inlet side compartment. Table 1 shows electro-osmotic properties when the compartments II and III contained different type of solutions. Electric volume resistivities __ and co-transportation coefficients K^C's which were obtained from the flow speeds divided by electric current densities through the hydrogel film were presented together with the electrohydrodynamic mobilities. The electrohydrodynamic mobility represents the electric controllability of the contraction of polyelectrolyte hydrogels. Because high flux rates enable quick response of mechanochemical reaction of hydrogel. Therefore the optimum combination of solutions for II and III was found to be 30mN NaOH for II and 10mN HCl for III in this case.

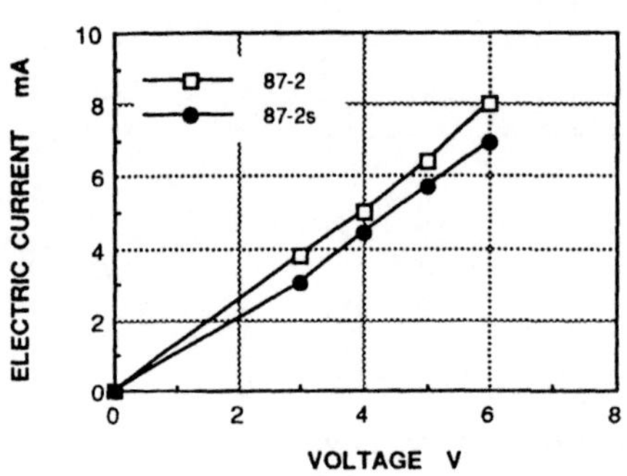

Fig.10 I-V Characteristics of Gel Film. 87:water content before gelation, 2:without HCl, 2s:with HCl.

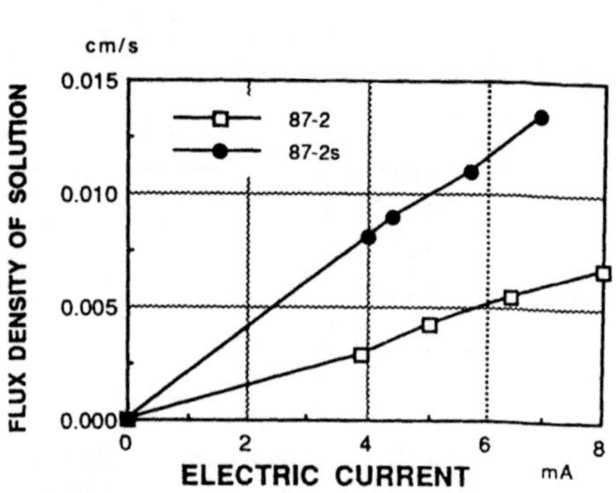

Fig.11 Flux to Current Relation 87-2:without HCl, 87-2s:with HCl.

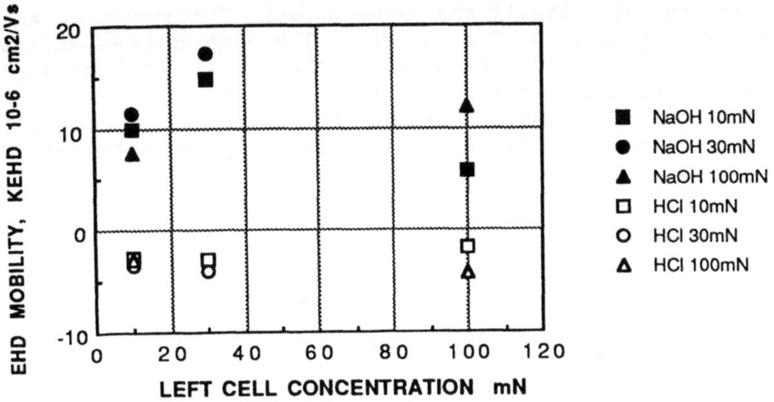

Fig.12 EHD Mobility Obtained by vd/V. Solutions of left and right cells were both NaOH or HCl.

Table 1 EHD Mobility K_{EHD}, Electric Resistivity rho, and Cotransportation coeff. K_{CT}.

NaOH \ HCl	3	10	30 mN
30 mN	+ NaOH - HCl KEHD ρ KCT - NaOH + HCl		(0.92) (0.32) 21.34 16.5 (0.02) (0.005)
10		3.63 (0.91) 17.5 42.7 0.064 (0.039)	2.79 (0.37) 9.9 15.8 0.028 (0.006)
3	4.15 1.79 14.9 48.3 0.061 0.087	11.09 2.3 7.6 30.5 0.084 0.069	KEHD 10^{-6} cm2/Vs ρ ohmcm KCT cm3/C

Electric Control of Hydrogel Contraction

Fig.13 shows the apparatus to demonstrate the electrohydrodynamic control of mechanochemical contraction of hydrogel tube. Inside the tube a solution such as 30mN NaOH solution was filled and a stainless wire electrode was set with insulating rings to prevent the tube surface from contacting the electrode. Outside the tube a solution such as 10mN HCl was filled and the other electrode was set. Fig.14 shows the curves of contraction and elongation for various combinations of solutions. The combinations of a mixed solution of acetone (90%) and 30mN NaOH solution (10%) and 30mN NaOH(100%) showed a large stroke and better controllability. The combination of $Ba(OH)_2$ solution inside and NaOH solution outside showed another possibility of electric control of mechanochemical contraction of polyelectrolytic hydrogels.[11]

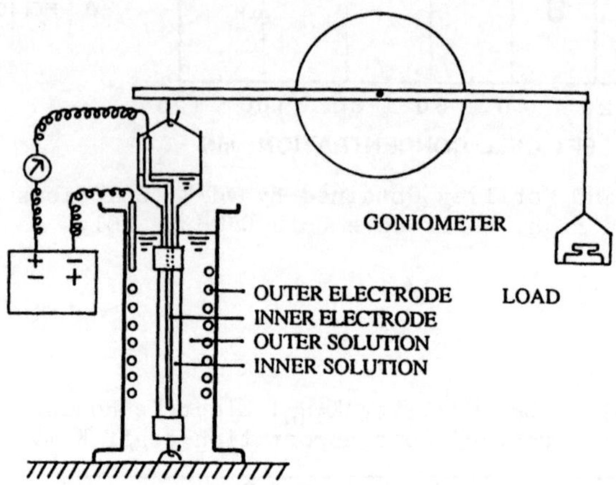

Fig.13 Apparatus to Demonstrate Electric Control of Gel Film Contraction.

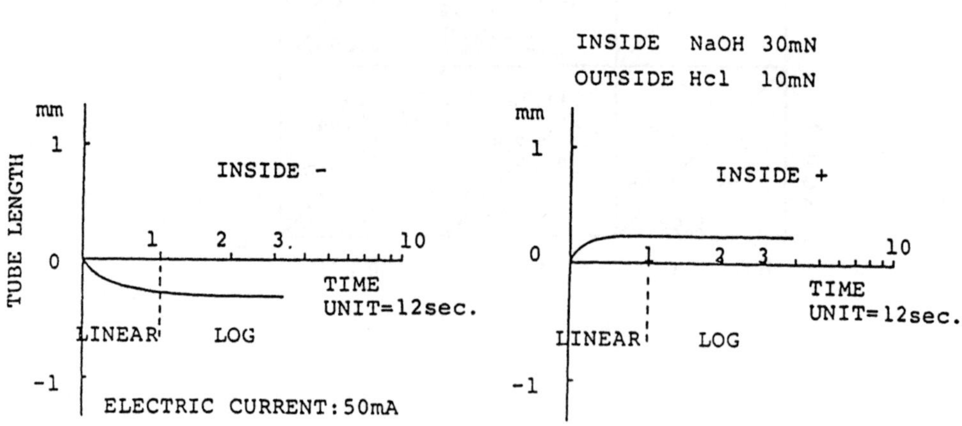

a Hcl-NaOH EXCHANGING SYSTEM

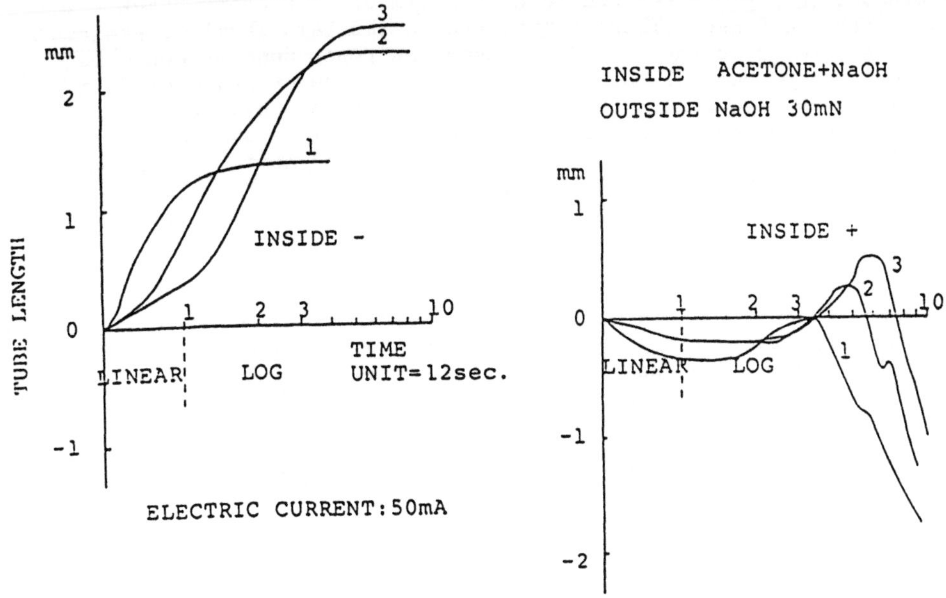

b ACETONE-NaOHaq.sol. EXCHANGING SYSTEM

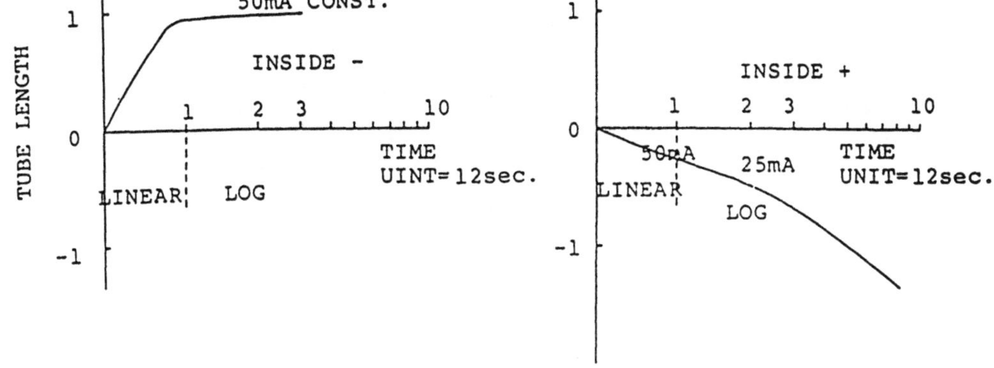

c Ba-Na EXCHANGING SYSTEM

Fig.14 Curves of Electromechanical Response.
a:HCl/NaOH system, b:NaOH 30mN/acetone+ system,and c:Ba/Na system.

CONCLUSION

In this study new methods were examined for the synthesis of high performance mechanochemical hydrogel and for the control of hydrogel motion. Uniaxially stretched (anisotropic) hydrogel films showed higher contraction ratio than non-stretched gels. The highest output power density of 0.1 W/cm^3 was obtained with a non-stretched thin gel film of 10 microns thick. It responded within 0.2 sec. These values are in the same level of skeletal muscle's. According to the cryo-SEM images the power density could be further improved by control the fine porous structure of hydrogel film.

By examining the electro-osmotic properties of the hydrogel film the electrohydrodynamic mobilities were measured for various combinations of solvent. As the result the technique of electrohydrodynamic solvent exchange was more effective than usual method simply applying electric field. It proposes a new method to realize an electrically driven actuator device.

Acknowledgement

The author would like to express his sincere thanks to Dr. S. Fujishige for his important suggestion, to Mrs.Y. Satoh, Mr. M. Matsuzawa and Mr. N. Shiba in carrying out the experiment, to Dr. T. Tateishi and the members of Biomechanics division, and also to Prof. Y. Osada, Dr. N. Okui and Dr.A. Yamauchi for fruitful discussions. The author also thanks to Dr.S. Fujikawa of Hokkaido University who kindly cooperated to take cryo-SEM photographs. This study was supported by AIST and STA funds in 1986-1989.

References

1. W. Kuhn, B. Hargitay, A. Katchalsky, and H. Eisenberg, Reversible Dilatation and Contraction by Changing the State of Ionization of High Polymer Acid Networks, Nature 165(4196): 514 (1950)
2. Y. Osada, Conversion of Chemical into Mechanical Energy by Synthetic Polymers, in:"Advances in Polymer Science82", S. Olive and G. Henrici Olive, ed., Springer-Verlag Berlin Heidelberg (1987)
3. M. Suzuki, T. Tateishi, T. Ushida, and S. Fujishige, An Artificial Muscle of Polyvinyl Alcohol Hydrogel Composites, Biorheology, 23: 274 (1986)
   M. Suzuki, Responsive Mechanochemical Actuator Materials by PVA Hydrogel, IUPAC CHEMRAWN VI, The Chemical Society of Japan,IB11 (1987).
   M. Suzuki, Expansible Macromolecular Material and Porous Macromolecular Membrane,U.S.Patent, 4717800 (Jan.5,1988),4753761 (Jun.28,1988)
4. M. Nanbu, High Water Content Rubber, Polymer Appl., 32(11): 523 (1983)
5. M. Suzuki, Amphoteric poly(vinyl alcohol) Hydrogel as a Material of Artificial Muscle, Kobunsi Ronbunshu, 46(10): 603 (1989).
6. T. Tanaka,D.Fillmore,S.T.Sun and I.Nishio, Phase Transitions in Ionic gels, Phys.Rev.Lett.,17: 1636 (1980).
7. D. De Rossi, P. Chiarelli, G. Buzzigoli, C. Domenici and L. Lazzeri, Contractile Behavior of Electrically Activated Mechanochemical Polymer Actuators, Trans.Am.Soc.Artif.Intern.Organs, 32: 157 (1986).
8. A. J. Grodzinsky and J. R. Melcher, Electromechanical Transduction with Charged Polyelectrolyte Membranes, IEEE Trans.Biomed.Eng., BME-23(6): 421 (1976)
9. W. F. Pickard, Ion Drag Pumping, J.Appl.Phys.,34: 246 (1963)
10. M. Suzuki and Y. Sawada, Nonlinear Oscillations of a Polar-Liquid Column under Unipolar-Ion Injection, J.Appl.Phys., 51: 5667 (1980)
    M. Suzuki, Propagating Transitions of Electroconvection, Phys. Rev., 31(4): 2548 (1985)
11. M. Suzuki, Mechanical Energy Generation by Hydrogel, Approaches to Higher Efficiency and higher Controlability, Proc.Int.Workshop on Intell. Materials, The Soc.of Non-traditional Technology, Tokyo, 307 (1989)
12. S. Fujikawa, Artificial Biological Membrane Ultrastructural Changes Caused by Freezing, Electron Microsc. Rev., 1: 113 (1988)

DEFORMATION BEHAVIORS OF POLYMER GELS IN ELECTRIC FIELD

Toshio Kurauchi, Tohru Shiga, Yoshiharu Hirose and Akane Okada

Toyota Central Research and Development Laboratories, Inc.
Nagakute, Aichi 480-11, Japan

INTRODUCTION

A polymer gel is a crosslinked polymer network swollen in a liquid medium. Polymer gels, solid-liquid coexistent materials, are candidate biomimetic materials. Recently, their mechanical strength has become very close to that of living muscle. In a soft structure of gels, as shown in Fig. 1, the motion of polymer networks and the diffusion of ions take place easily by an external stimulus. The large volume or shape change, induced by supplying thermal, chemical or electrical energy, is an inherent nature of swollen gels.[1-4] Therefore, polymer gels have various possibilities as advanced functional polymers.

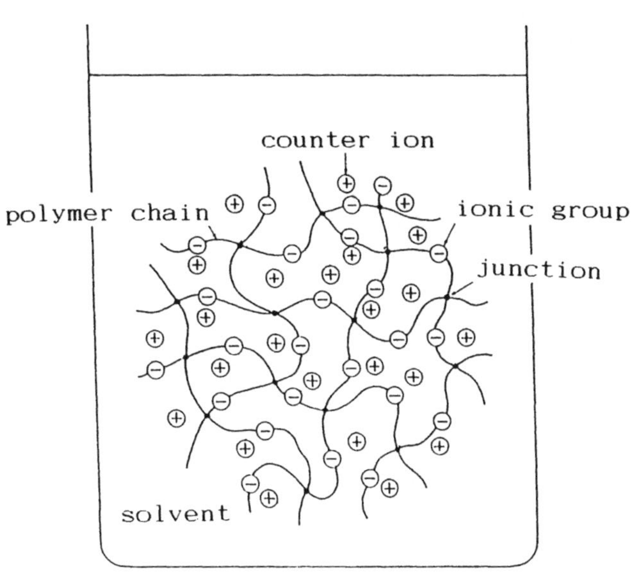

Fig.1 Cross-linked polymer network swollen in a solvent.

Polymer Gels, Edited by D. DeRossi *et al.*
Plenum Press, New York, 1991

A typical function of a gel containing ionic groups is to bend reversibly under the influence of an electric field,[5] making it useful in some actuators driven by an electric field. In this study, the deformation behaviors of poly(acrylic acid)-co-(acrylamide) gels and poly(vinyl alcohol)-poly(acrylic acid) gels under an electric field were investigated. Using the bending deformation of these ionic gels induced by an electric field, we constructed a prototype of a robot hand having soft fingers, and an artificial fish able to swim.

DEFORMATION OF POLYMER GELS

Acrylic acid-acrylamide copolymer gels (PAAm gels) swollen in aqueous electrolyte solution show three types of deformation under the influence of a d.c. electric field; shrinking, swelling and bending, as shown in Fig. 2.[5] The type of deformation depends on the fraction of sodium acrylate (AANa) in the PAAm gel, the shape of the PAAm gel and the position of the gel between the positive and negative electrodes. The PAAm gel with a low fraction of AANa shrinks on the surface facing to the positive electrode (Fig. 2(a)) and the gel with a high fraction of AANa swells on the positive electrode side (Fig. 2(b)). When a rectangular gel is placed in parallel to the electrodes, bending of the gel takes place toward the positive electrode (Fig. 2(c)) or toward the negative electrode (Fig. 2(d)), according to the fraction of AANa. The critical fraction of AANa is about 25 mol%. This bending behavior is similar to the buckling of a bimetallic strip submitted to a variation of temperature. Because we observed the bending deformation with gels having different ionic groups such as SO_3Na and NR_3Cl, the phenomenon is considered to be a general property of ionic gels. Of course, the direction of the bending is inverse, when the sign of the ionic groups is changed.

BENDING OF PAAm GELS WITH HIGH FRACTION OF AANa

PAAm gels were prepared by free radical polymerization; 0.1 mol acryic acid, 0.1 mol acrylamide, 0.001 mol methylenbisacrylamide as a cross-linking agent and 0.1 mol NaOH were dissolved in distilled water, and the solution was heated at 60°C for 3 hours.

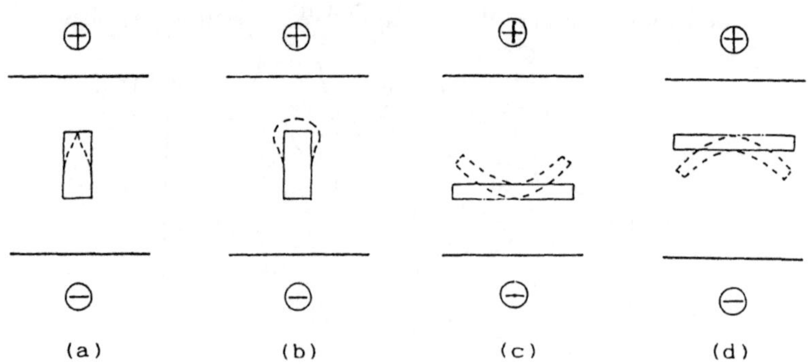

Fig.2 Deformation of PAAm gels under d.c. electric field.
(a) and (c) : for small fraction of AANa
(b) and (d) : for large fraction of AANa

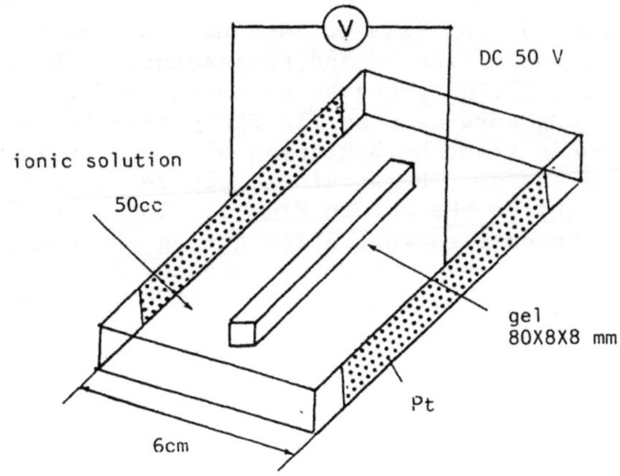

Fig.3 Apparatus of bending test.

Fig. 3 shows the diagram of the testing apparatus used in a bending test under the influence of a d.c. electric field. The specimen was 80 mm in length and its cross-section was 8×8 mm.

In Fig. 4, the weight gain of a PAAm gel is plotted as a function of the strain in bending. When the PAAm gel bends semicircularly, the strain in bending is 0.225. At that time, a weight gain of 15 % was observed, as shown in Fig. 4. From this result, it is clear that the bending is due to the swelling.

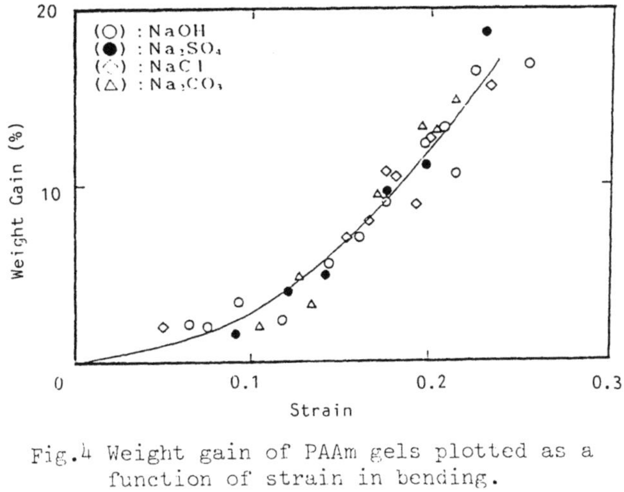

Fig.4 Weight gain of PAAm gels plotted as a function of strain in bending.

The bending speed of the PAAm gel depends on some factors including field strength, salt concentration and temperature. The bending speed is proportional to the field strength, as shown in Fig. 5, in which NaOH, NaCl, Na_2CO_3 and Na_2SO_4 were used as electrolytes. In this figure, the strain after 30 seconds from the beginning of d.c. supply of 50 V is taken as the vertical axis. The strain is plotted as a function of the salt concentration outside the gel in Fig. 6. There is a maximum in the bending speed, and the salt concentration giving the maximum depends on the valence of the ions. The temperature dependence of the bending speed was observed, and the gels could act at 5 to 50°C. The bending speed at 50°C was twice that at 5°C.

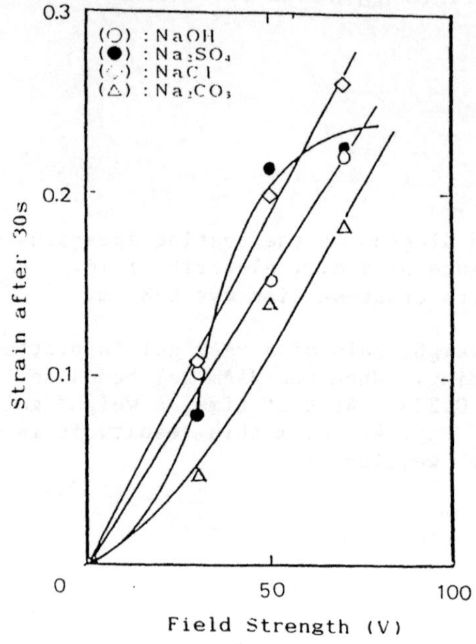

Fig.5 Bending speed plotted as a function of field strength.

By the change of polarity (+50 V, -50 V), the strain can be recovered as shown in Fig. 7. This suggests that it can be bent repeatedly by an a.c. electric field.

The number of cycles to "fatigue-rupture" depends on the kind of electrolytes in the solution. In basic electrolyte solutions such as NaOH and Na_2CO_3, the bending speed of PAAm gel was constant up to 50 cycles. In neutral salt solutions such as NaCl and Na_2SO_4, the speed slowed down gradually, as shown in Fig. 8. Some cracks occurred on the surface of the gel after 14 or 15 cycles.

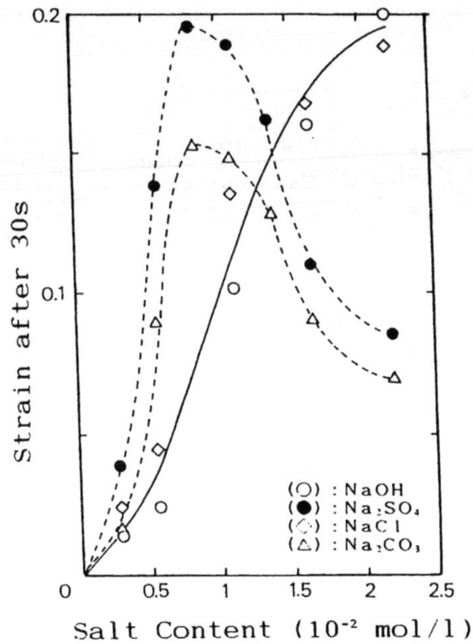

Fig.6 Bending speed plotted as a function of salt content.

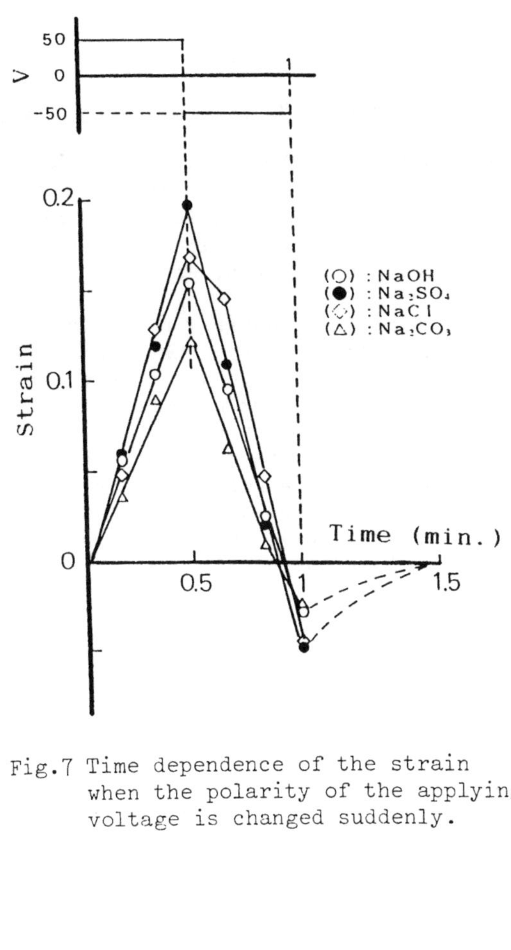

Fig.7 Time dependence of the strain when the polarity of the applying voltage is changed suddenly.

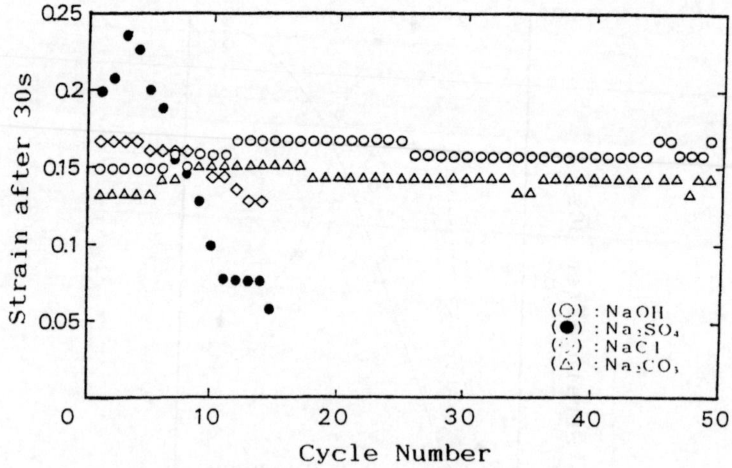

Fig.8 Fatigue rupture of PAAm gels.

BENDING MECHANISM

The volume of a gel is controlled by osmotic pressure.[6] The osmotic pressure π is given as the sum of π_1, π_2, and π_3 which correspond to the osmotic pressures due to the solubility of the solvent in the polymer chain, the rubber elasticity and the ion concentration difference between the inside and outside of the gel, respectively[7].

$$\pi = [\ \ell n\ (1 - v) + v + xv^2\ ]\ RT\ /\ V_1$$
$$+ (v^{1/3} - v/2)\ RT\nu_e\ /\ V_0 + (\Sigma C_i - \Sigma C_j)\ RT$$

where v is the volume fraction of the polymer network, x is the solubility parameter, V_0 is the volume of the polymer network under the dry condition, ν_e is the number of chains, V_1 is the molar volume of the solvent, C_i and C_j are the ion concentrations inside and outside of the gel, respectively, R is the gas constant, and T is the temperature.

At equilibrium, the osmotic pressure π of the gel is equal to that of the surrounding aqueous solution, π_0. Therefore, π_1, π_2 and π_3 have the definite values, respectively. When a d.c. electric field is applied on the gel in the aqueous solution, the counter ion and the free ion can drift to the positive or negative electrode, whereas the polyion cannot move. Consequently, π_3 varies and the value of π deviates from π_0. The swelling or the shrinking of the gel occurs until the gel reaches its new equilibrium state. Since the ions drift at the different speeds depending on their size and valencey, the osmotic pressure of the positive side is unequal to that of the negative side, and bending of the gel occurs.

PVA-PAA GEL

PAAm gels bend like a finger under an electric field, because they contain 99 % water, however, their mechanical strength is so weak that they break easily. It is necessary, therefore, to improve the mechanical properties of the polymer gels for their application as actuators.

A PVA-PAA (poly(vinyl alcohol)-poly(acric acid)) gel was prepared as follows; poly(vinyl alcohol) saponified over 95 % and poly(acric acid) were desolved in a mixed solvent of water and DMSO. The pregel solution was frozen and thawed repeatedly. The resultant gel, which turned white at this point, was immersed in NaOH aqueous solution for a long time in order to dissociate carboxyl groups.

The PVA-PAA gel obtained contains 90 % water. Polymers in this PVA-PAA gel may form an interpenetrated polymer network, as shown in Fig. 9. The X-ray diffraction profile from a dry gel is shown in Fig. 10. The dry gel without water has the same spectrum as PVA powder. Therefore, it is obviously a crystalline polymer.

In the IPN structure, it seems that PVA networks cross-linked with the fine crystallines are strong, and PAA acts as an electro-sensitive material. Fig. 11 shows the stress-strain curve of the PVA-PAA gel in a tensile test. The tensile strength and the elongation to the rupture of the PVA-PAA gel are more than 50 times as large as those of the PAAm gel and are close to those of rubber vulcanizates. The PVA-PAA gel appeared to possess high shear strength and toughness.

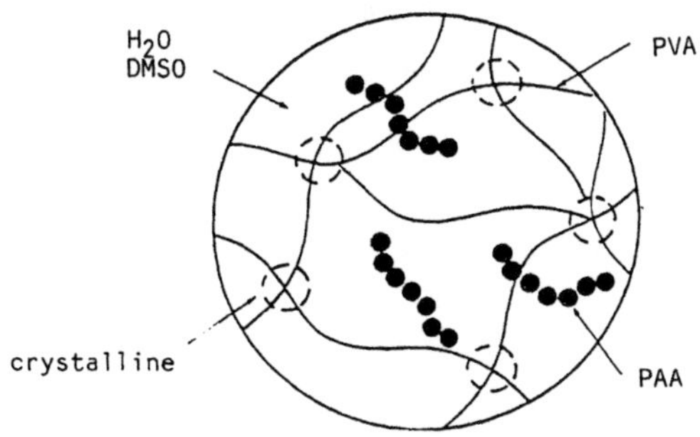

Fig.9 Conceptional picture of PVA-PAA gel.

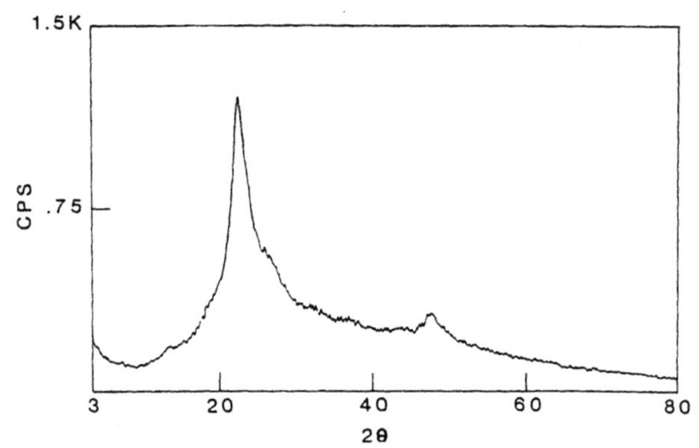

Fig.10 Diffracted X-ray from the dry PVA-PAA gel.

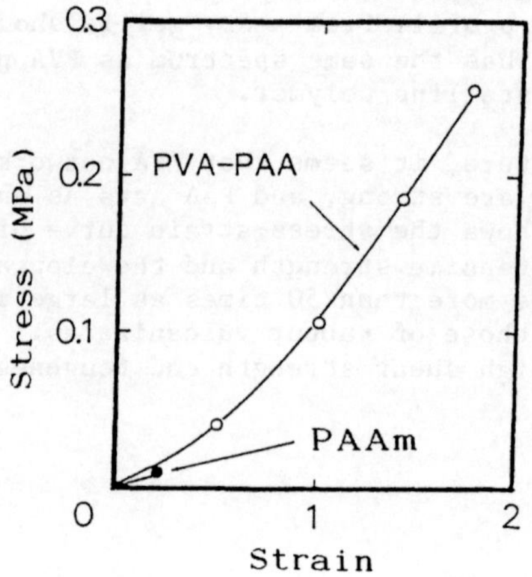

Fig.11 Tensile stress-strain curves.

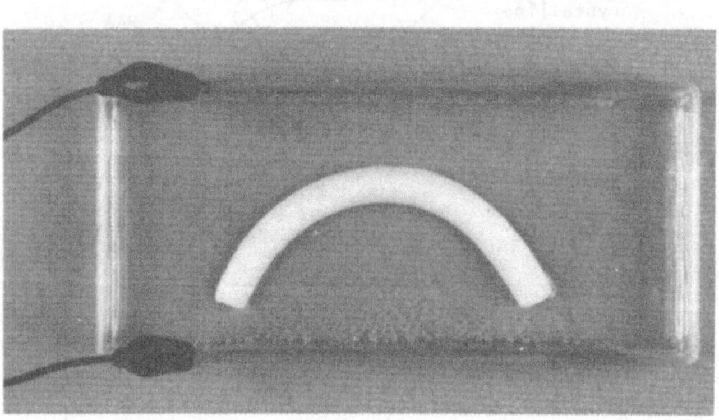

Fig.12 Bending of the PVA-PAA gel.

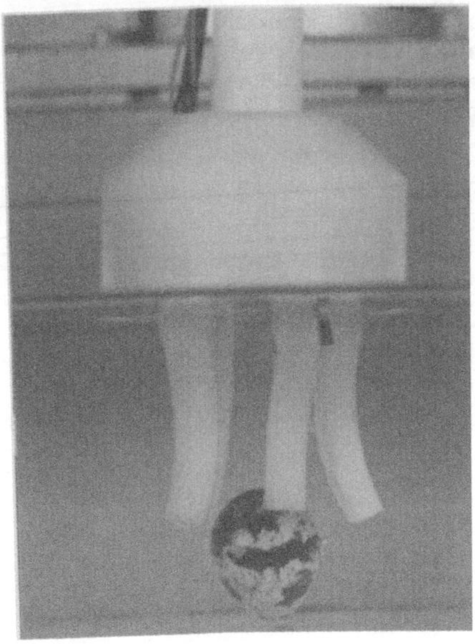

(a)

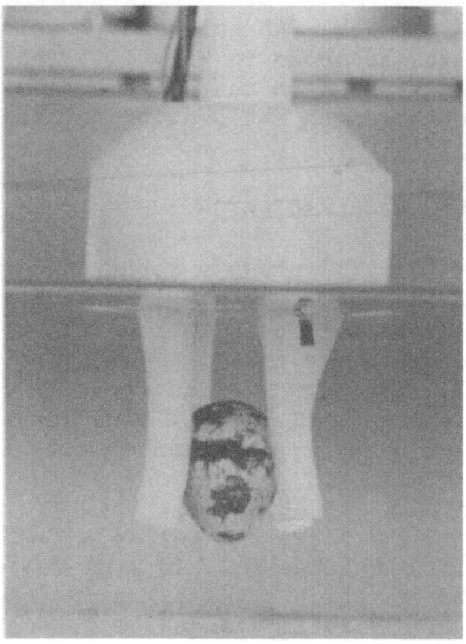

(b)

Fig.13 Robot hand made from PVA-PAA gel fingers.
(a) before catching and (b) catching an egg

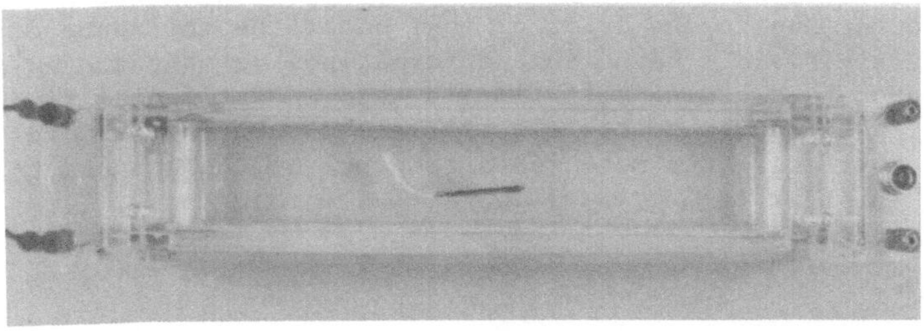

(a)

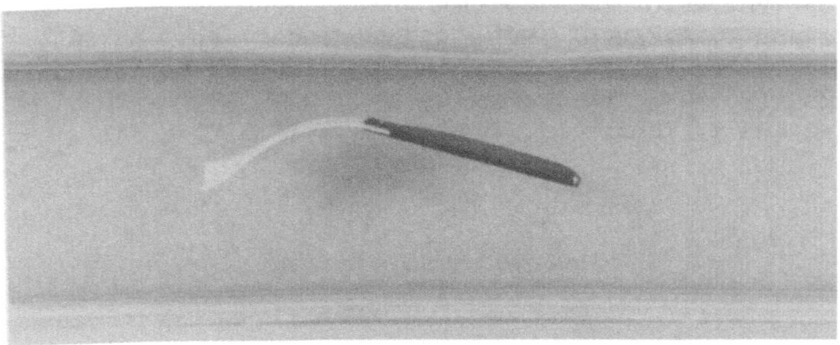

(b)

Fig.14 Artificial gel fish with PVA-PAA tail.
(a) swimming apparatus and (b) gel fish

The PVA-PAA gel bends toward the negative side under a d.c. electric field, as shown in Fig. 12. The bending speed of the PVA-PAA gel was almost equal to that of the PAAm gel. Therefore, the PVA-PAA gel is a high strength and bending deformable polymer.

POLYMER GEL ACTUATOR

Fig. 13 shows a robot hand with four gel fingers composed of a PVA-PAA gel. Each gel is a rectangular bar 60 mm in length and 6×7 mm in cross section. To bend the gel, a pair of electrodes is located near the gel. The positive electrode is fixed with a spacer apart from the gel. The negative electrode is embeded in the surface of the other side of the gel. The robot hand can safely pick up a fragile egg in a Na_2CO_3 solution by applying an electric field of 50 V.

Fig. 14 shows an artificial fish composed of a plastic plate as a float and a PVA-PAA gel as a tail (30 mm in length, 10 mm in width and 1 mm in thickness). In a Na_2CO_3 solution with an electrode set on each side, the fish swims by waving the tail as the polarity of the electric field (50 V) is changed alternately. The swimming speed is about 2 cm/s.

We demonstrated that a body can be transferred by the mechanical deformation of polymeric materials. This exhibits a new type of functionality of polymeric materials.

CONCLUSION

The mechanical response of PAAm and PVA-PAA gels upon the electric field was examined experimentally. The gels can be bent by applying a d.c. electric field across the gels. The deflection of the gel is considered to occur due to the drift of ions under the influence of the applied electric field. Using this deformation, we constructed a prototype of a robot hand having soft fingers, and an artificial fish having a soft tail, both driven by the electric field. Further studies will be carried out to realize the more rapid response of the gel and to increase the modulus in order to increase a work to the external.

REFERENCES

1. T. Tanaka, Science. 218:467 (1982).
2. Y. Osada and M. Hasebe, Chem. Lett. 1285 (1985).
3. M. Suzuki et al., Biorheology. 23:274 (1986).
4. O. Hirasa, Polymer, Jap. 35:1100 (1986).
5. T. Shiga and T. Kurauchi, Polymer Preprints, Jap. 36:2894 (1987).
6. P. J. Flory, "Principles of Polymer Chemistry," Cornell Univ. Press, Ithaca (1953).
7. I. Ohmine and T. Tanaka, J. Chem. Phys. 77:5725 (1982).

THERMORESPONSIVE POLYMER HYDROGEL

Okihiko Hirasa, Shoji Ito, Aizo Yamauchi,
Shoei Fujishige, Hisao Ichijo

Research Institute for Polymers and Textiles
1-1-4 Higashi, Tsukuba, Ibaraki 305 Japan

INTRODUCTION

In recent years, materials responding external stimuli have become of major interest and have been applied to sensor or memory media.

On the other hand, development of soft and biomimetic materials has been studying in various places. Especially, studies of mechanochemical materials responding to stimuli have been practiced very often lately.

In this paper, we describe thermoresponsive polymer hydrogels.

THERMORESPONSIVE POLYMER

Formerly there are some water-soluble polymers having negative temperature coefficient of solubility in water such as methyl cellulose[1], poly(ethylene oxide) and partially saponified poly(vinyl acetate)[2].

On atactic polymer of poly(vinyl methyl ether)(PVME), the aqueous solution shows phase transition at 38°C. The DSC diagram is shown in Figure 1. The apparent heats of phase transition are depend on concentration of solution(Fig.2). The values are 63 to 84 J/g in pure water. The transition temperature of the solution is reduced by addition of salts generally. The temperature of transition of PVME aqueous solution is reduced with concentration of NaCl linearly(Fig.3)[3].

Thermoresponsive polymers representing by N substituted acrylamide derivatives and their copolymer are synthesized and characterized by Ito and Fujishige in RIPT(Table 1 and Fig. 4)[4,5]. Additivity exists between monomer ratio and transition temperature of the copolymers[6].

THERMORESPONSIVE POYMER HYDROGEL

In case of using the polymers as thermoresponsive materials, it is necessary to form the polymers to insoluble materials in water. The formation is achieved by means of graft polymerization, copolymerization with bifunctional monomer or crosslinking by gamma ray irradiation.

On graft polymerization, Ito in RIPT studied on graft polymerization of N-isopropyl acrylyamide to cellulose. The graft polymer has the same thermal properties as poly(N-isopropyl acrylamide) homopolymer[7].

On crosslinking by gamma ray irradiation, thermoresponsive hydrogels formed from poly(vinyl methyl ether) aqueous solution are studied[8].

Polymer Gels, Edited by D. DeRossi *et al.*
Plenum Press, New York, 1991

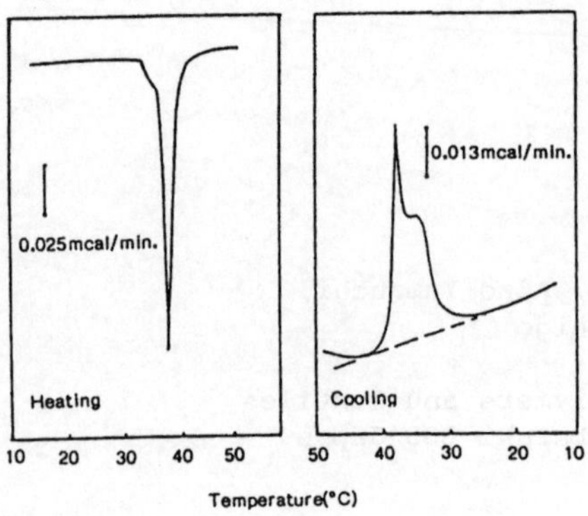

Fig.1 DSC Thermograms of PVME aqu. solution. Conc. of PVME:5wt%, Heating rate:2.0°C/min.

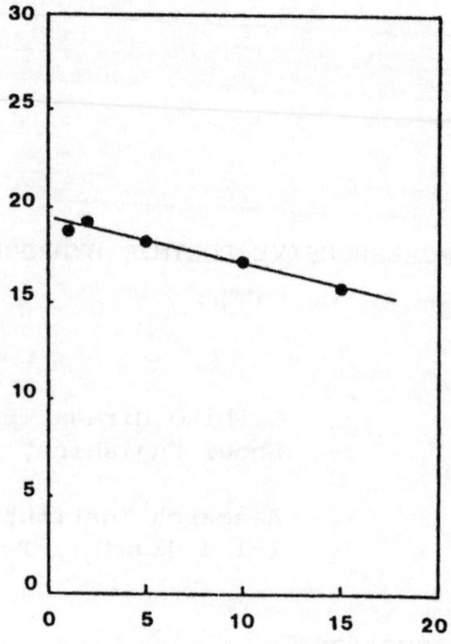

Fig.2 Relation between conc. of PVME and heat of transition

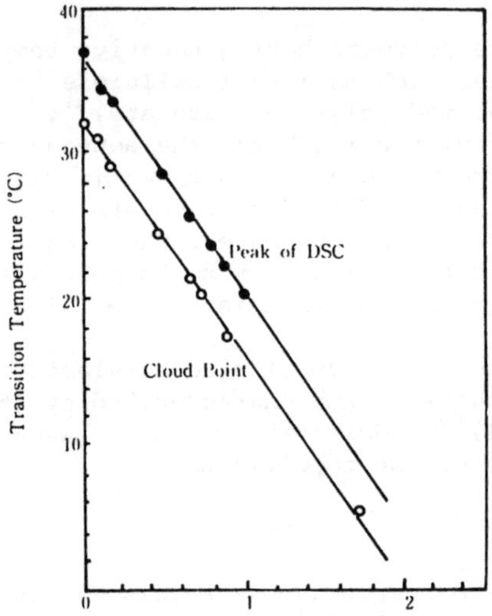

Fig.3 Relation between transition temperature and conc. of NaCl

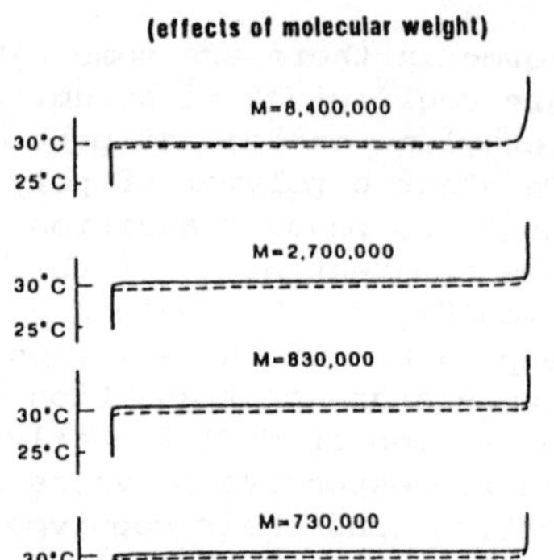

Fig.4 Phase transition of aqueous poly(NIPAM) solution

Table 1 Thermal properties of aqueous polyacrylamide derivatives

Polymer	Transition temp. (℃)	Heat of phase transition (cal/g)
Poly(*N*-ethylacrylamide)	72.0	0.1 >
Poly(*N*-ethylmethacrylamide)	50.0	0.1 >
Poly(*N*-*n*-propylacrylamide)	21.5	11.6
Poly(*N*-*n*-propylmethacrylamide)	28.0	12.9
Poly(*N*-isopropylacrylamide)	30.9	11.1
Poly(*N*-isopropylmethacrylamide)	44.0	12.0
Poly(*N*-cyclopropylacrylamide)	45.5	3.5
Poly(*N*-cyclopropylmethacrylamide)	59.0	4.2
Poly(*N*-methyl-*N*-ethylacrylamide)	56.0	5.0
Poly(*N*,*N*-diethylacrylamide)	32.0	6.3
Poly(*N*-methyl-*N*-isopropylacrylamide)	22.3	6.7
Poly(*N*-methyl-*N*-*n*-propylacrylamide)	19.8	8.3
Poly(*N*-acryloylpyrrolidine)	56.0	1.3
Poly(*N*-acryloylpiperidine)	5.5	10.0

PHASE TRANSITION OF POLYMER HYDROGEL AND STIMULI RESPONSIBILITY

In 1980, volume phase transition phenomenon of partially hydrolyzed poly(acrylamide) hydrogel is found out and clarified theoretically by Tanaka [9]. Thereafter, the phenomenon is found out on various polymergel such as gelatine, agar, DNA and some synthetic polymers generally [10,11].

The volume phase transition of gel can be expressed with functions of temperature, composition of solvent and the number of dissociated counterion per effective polymer chain [12].

Hydrogels containing ionic groups such as partially hydrolyzed poly(-acrylamide), acrylamide/acrylic acid copolymer, show volume phase transition with changing of concentration of acetone in medium. The maximum volume changes is 1000-fold[13,14]. The gels also show volume change with concentration of NaCl in mixture of acetone and water . The stimuli for volume change such as concentration of salts or solvent, pH, are generated by changing the constitution of liquid phase. The stimuli are not suitable for control in closed system. From the point of view of controlling, it is expected to control with thermal stimulus.

On thermal phase transition phenomena of hydrogel, there are many studies such as poly(N-isopropyl acrylamide)(PNIPAM) gel by Hirokawa [15], submicron gel beads of PNIPAM by Hirose [16], PNIPAM gel in pure and solvent mixture by Hirotsu[17] and copolymer gel of acrylamide/trimethyl(-N-acryloyl-3-aminopropyl) ammonium iodide by Katayama [18].

Phase diagrams of these systems appear to be useful for application of the gels to thermoresponsive materials.

At present on thermoresponsive hydrogel, many studies have been done on gel of acrylamide derivatives. On another material, vinyl ether derivatives, especially poly(vinyl methyl ether)(PVME) gel is studied in RIPT.

POLYVINYLMETHYLETHER HYDROGEL

Poly(vinyl methyl ether) is synthesized from methyl vinyl ether by means of cationic polymerization. Polymer obtained is isotactic at low polymerization temperature and insoluble in water. But on polymerization at room temperature, the polymer obtained is atactic and soluble both in cold water and organic solvent. The temperature coefficient of solubility in water is negative. The aqueous solution of atactic PVME shows phase transition at 38°C. The polymer is insoluble in hot water.

PVME molecules in aqueous solution are crosslinked by irradiation of electron beam or gamma ray. The hydrogels are formed with the dosages of 50 to 60kGy.

The gel shows phase transition by heating in the same manner as the aqueous solution. In solution, hydrated PVME molecules cause dehydration by heating above transition temperature and precipitate by aggregation.

The crosslinked hydrogel swells and shrinks in response to temperature change reversibly. Temperature dependence of degree of equilibrium swelling of the gels are shown in Figure 5. In application of the thermo-responsive hydrogel, the response velocity of swelling and shrinking becomes a serious problem of most importance. It is proved by Tanaka [19] that response time of swelling or shrinking of spherical gel is proportional to the square of the final radius of the gel. It is essential to get faster responsive gel how to prepare the gel having small diameter.

As an example, thermal response speed of spongy gel made by means of phase separation on crosslinkings is faster than that of homogeneous gel prepared at temperature under transition point (Fig.6)[20]. As in the case, it is possible to prepare the gel having various structure by controlling the phase separation processes (Fig.7). As another method for getting faster responsive gel, mixed aqueous solution of PVME and sodium alginate is spun with spinning bath of calcium chloride aqueous solution. And the fiber spun is irradiated with gamma ray subsequently. Then fibrous PVME hydrogel can be prepared. The fibrous hydrogel shows spongy structure as shown in Figure 8 and reversible isotropic swelling and shrinking in response to temperature changes. Diameter of the fibrous gel is 400 micronmeter at 20°C and 200 micronmeter at 40°C.

The response time of the gel is faster than one second.

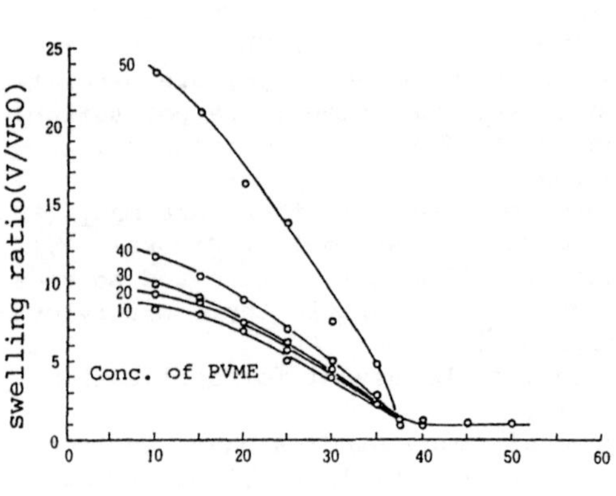

Fig.5 Swelling behavior of PVME gels
Doses of irradiated gamma ray:100kGy

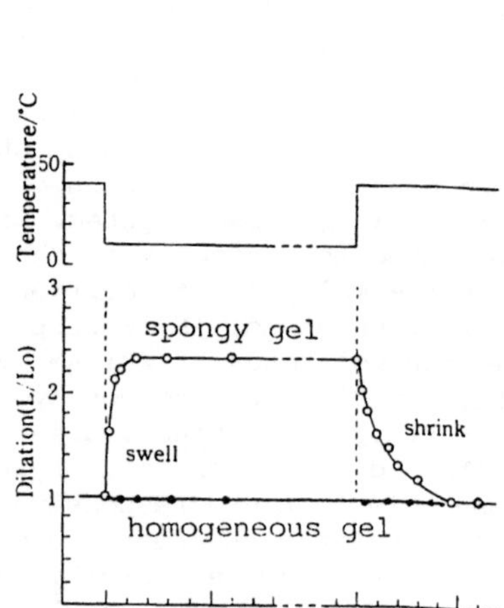

Fig.6 Temperature response of PVME gel

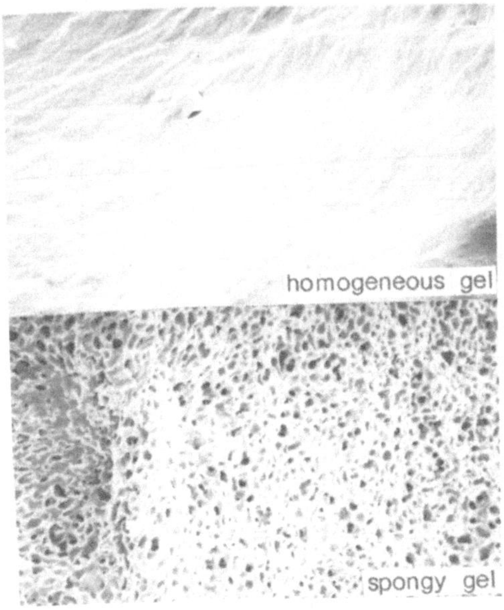

500micron

Fig.7 SEM photographs of PVME gels prepared by gamma ray irradiation

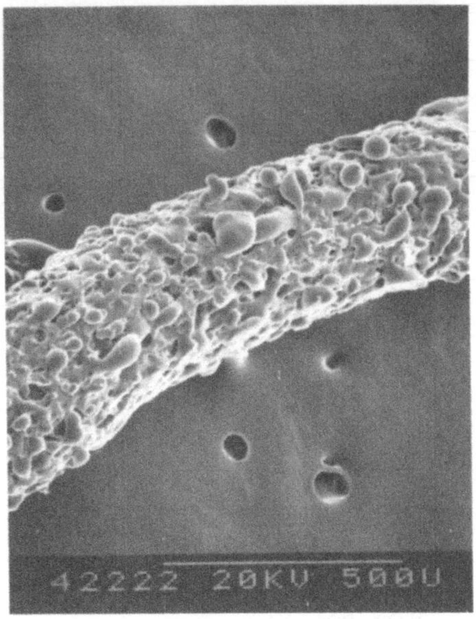

Fig.8 SEM Photograph of fibrous PVME gel

On mechanical properties of the fibrous PVME gel, the stress-strain curves are shown in Figure 9. Values of mechanical properties of the gels are shown in Table 2. By occurrence of phase transition at 38°C, the value of maximum elongation is 2.6 times and the value of Young's modulus is 5 times as much as the value of non-phase transition state respectively. The values are increased by phase transition.

In applying the materials to chemomechanical systems, it is necessary to investigate the generation of thermal shrinking stress due to the phase transition. The thermal shrinking stresses of the PVME fibrous gel at constant length and the volume changes of spongy PVME gel in water in response to temperature changes are shown in Figure 10. The thermal shrinking stresses increase rapidly at phase transition temperature (Tt=38°C), and the stress generation show a similar tendency of the volume changes of spongy PVME gel. It is proved that the shrinking of the gel respond to the generation of the stress. Thermal stress of 0.31 grams per a fibrous PVME gel is generated by changing temperature from 20°C to 40°C. The value is 98kPa in converting into stress per unit area and equivalent to a third to a tenth as much as the stress values of muscles[21).]

In application of the gels, it is little to be used only in pure water. Then it is necessary to investigate the effects of environments of the gels. The behaviors of swelling or shrinking of PVME gel are affected with kinds or concentration of salts in medium. The effects of salts on swellings of PVME gels are shown in Figure 11[22).]The swelling behavior are affected by anions of salts in medium. The gel is shrunk highly with solution containing sulfate or chloride anion in comparison with pure water, and is not affected with nitrate anion, but is swollen with iodide anion conversely. But the effects are smaller than that of ionic water absorbing polymers. In addition of methanol, the transition point of the gel is raised with increase of concentration of methanol, and disappear with concentration above 75wt%. The phenomena are due to the amphipatic properties of nonoinic thermoresponsive polymers such as PVME.

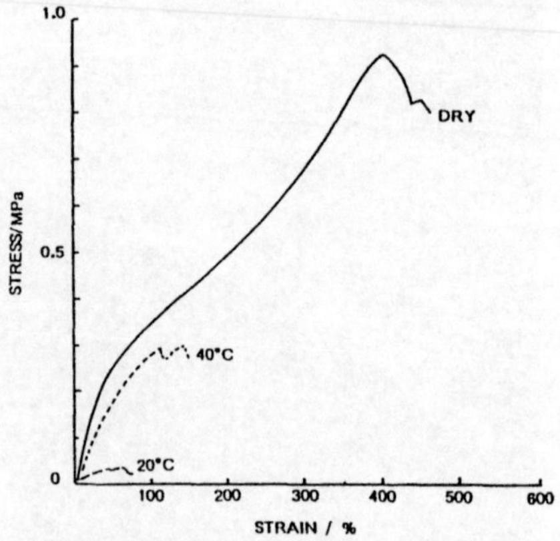

Fig.9 Stress-strain curves of PVME gel
Rate of expansion is 10mm/min.

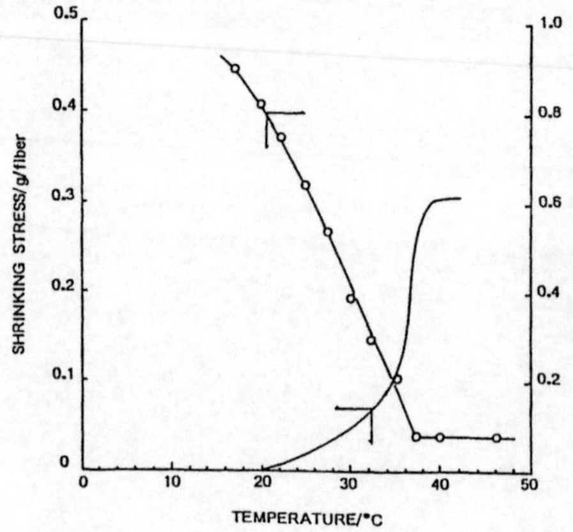

Fig.10 Thermal shrinking stress of fibrous PVME gel and Volume changes of spongy PVMEG in water

Table 2 Mechanical Properties of fibrous PVME gel(PVMEG-f)

Mechanical Property	20°C	40°C	Dry
Maximum Tensile Strength (Mpa)	0.032	0.29	0.92
Maximum Elongation (%)	44.0	114	450
Maximum Tensile Strength (g/Fib.)	0.40	0.90	2.6
Young's Modulus (MPa, 10% elongation)	0.092	0.47	0.49

PVMEG-f was prepared by means of wet spinning of mixed aqueous solution of PVME and Na-alginate into $CaCl_2$ solution at 40°C and gamma ray irradiation at 23°C.
PVME/Na-alginate; 10:1
Doses of gamma ray irradiation:100kGy

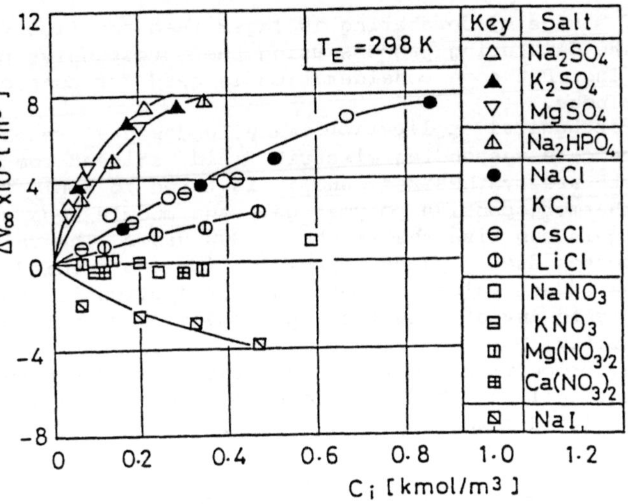

Fig.11 Difference Between equilibrium volume of PVME gel in distilled water and that in salt solutions

APPLICATION OF THERMORESPONSIVE POLYMER HYDROGEL

On stimulus responsive gels, it is expected to apply to new fields such as sensor, switch, separation membrane, adsorbent, mechano-transducer, dehydrant and materials for drug delivery system etc.

Properties of gels being changed by stimuli are as follows;

1) Chemical properties such as hydrophilicity and hydrophobicity.
2) Physical properties such as dimension, strength and elasticity.
3) Optical properties such as opacity and transparency.

On chemical properties, Thermal change of interaction between polymer gel and organic compounds is applied to thermoresponsive adsorbent [8]. On PVME gel, the hydrophilicity and hydrophobicity is changed drastically and reversibly at transition point. Adsorption isotherms of poly(oxyethylene nonylphenyl ether) on PVME gel fixed on alumina by gamma ray irradiation are shown in Figure 12. Equilibrium amounts of adsorption at 40°C is forty times as larger as the value at 5°C. It is possible to separate and concentrate the hydrophobical organic substances such as nonionic surfactant in waste water by temperature changes. The adsorption and desorption phenomena are affected with concentration of salt solution. The effects of addition of NaCl on adsorption is shown in Figure 13. The adsorption method is energy saving for having a possibility of reuse of waste heat from factory. Schematic diagram of the process is shown in Figure 14 [23].

On mechanical properties, there is application to dewatering agent using water absorption and desorption depend on volume phase transition of the gel by temperature change. Freitas and et al try to concentrate alubumin, geratin and dextran with thermoresponsive poly(N-isopropyl acrylamide) gel . On poly(vinyl methyl ether) gel, we investigated to possibility to apply to dewatering of biological sludge [24]. Relation between dewatering volume and numbers of repeat is shown in Figure 15 [25]. The PVME gels heated 48°C and shrunk are thrown in to sludge at 25°C for 10 minutes. After absorption of water, the gels are heated and desorbed water. On repeating the process, the sludges are dewatered and concentrated. In the initial step of dewatering, the gel absorbs water of 1.6 times of weight of the gel. But in the later step, the gel absorbs only water of 0.8 time of weight of the gel. The reason is due to decrease the fluidity of the sludge in later step, and the velocity of volume decreasing of sludge by dewatering is lager than the flow velocity of the sludge. The dewatering process using thermoresponsive polymer gel is energy saving. But some consideration is need for contact method of the gel and sludge.

On chemo-mechanical application many fundamental researches have been done. The gels responding electric field, solvent composition and temperature etc are synthesized, and it is tried to apply to the actuator systems. On thermoresponsive polymer gel, the module consisting 1000 fibrous PVME gels can rise the weight of 300 grams. The value of thermal shrinking stresses due to phase transition is from a third to a tenth as much as the stress values of muscles. For application to actuator, it is necessary to investigate the improvement of tensile strength and the controlability.

On optical properties, there are shielding materials for light or optical element applying the cloud point of the polymer solution. It is possible to avoid coagulation of the polymer by using the gels.

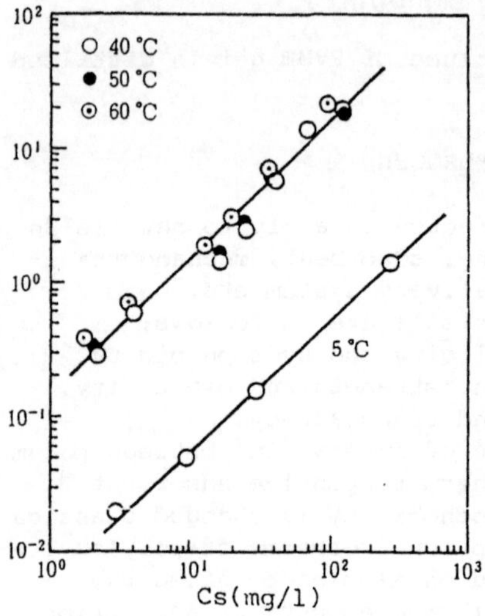

Fig.12 Absorption isotherms of nonionic surfactant on PVME gel fixed on alumina by Gamma ray irradiation

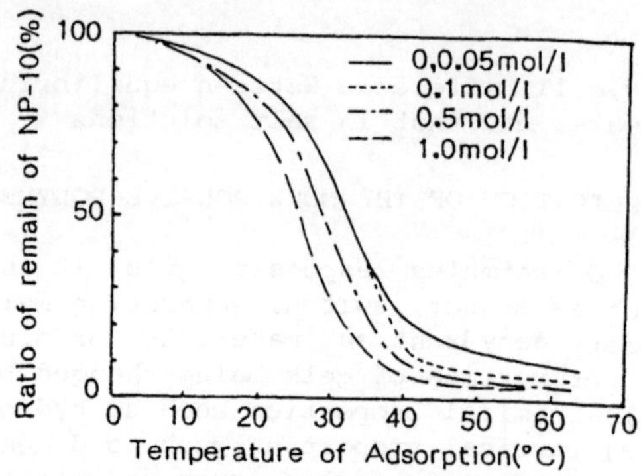

Fig.13 Effect of NaCl concentration on absorption of surfactant on PVMEG/alumina adsorbent

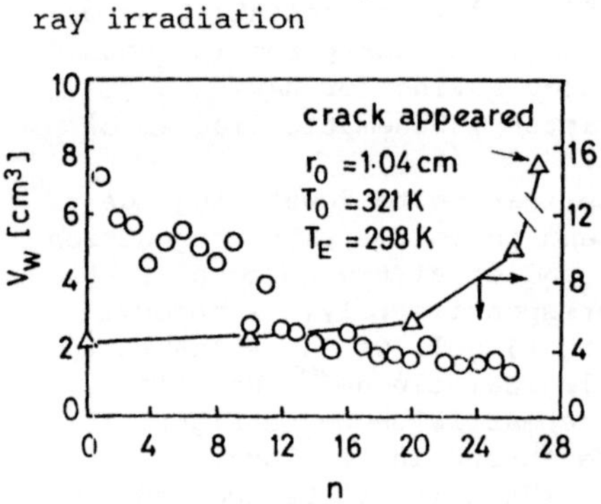

Fig.14 Changes in water amount absorbed by gel,Vw and sludge concentration C with time n, of water absorption by one spherical PVMEG

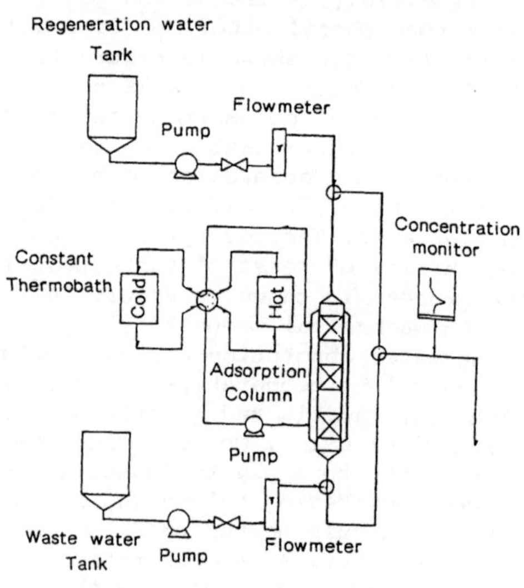

Fig.15 Schematic diagrams of thermo-reversible adsorption system

CONCLUSION

In the paper, we referred to investigation on synthesis, the properties and some applications of thermoresponsive polymers and the gels, especially, on poly(vinyl methyl ether) gel in Research Institute for Polymers and Textiles. Many gels exist near ourselves, especially in organisms. Studies on gels responding to stimuli have been practiced very often on the synthesis and the physical properties lately.

The polymer gels on soft materials are very promising both in the field of fundamental study and in the field of applied research.

ACKNOWLEDGEMENT

Research on application of PVME gel to dewatering agent for sludge has been done with cooperative study by Dr Xia Huang, Prof. Hajime Unno and Prof. Takasi Akehata in Tokyo Institute of Technology. We should like to express our grateful thanks to them. A part of this research was supported by Special Coordination Fund from Science and Technology Agency of Japanese Government.

REFERENCE

1. O. Hirasa and M. Hayashi, The effects of the addition of alcohols on the reological properties of methyl cellulose aqueous solutions, Kogyo kagaku Zasshi 73:2680(1970)
2. I. Sakurada, Y. Sakaguchi and N. Ito, Phase equilibrium of solution of partial acetylized poly(vinyl alcohol), Kobunshikagaku, 14:41(1957)
3. O. Hirasa, Thermoreversible adsorption property of poly(vinyl methyl ether) in water, Bull. Res. Inst. Polym. Text.,144:69(1984)
4. S. Ito, Phase transition of aqueous solution of poly(N-alkylacrylamide) derivatives, Kobunshi Ronbunshu, 46:437(1989)
5. S. Fujishige, K. Kubota and I. Ando, Phase transition of aqueous solution of poly(N-isopropylacrylamide) and poly(N-isopropyl-methacrylamide), J. Phys. Chem.,93:331(1989)
6. S. Ito and K. Mizoguchi, Preparation and properties of copolymers of N-isopropylacrylamide and N-isopropyl methacrylamide, Bull. Res. Inst. Polym. Text.,144:13(1984)
7. S. Ito, F. G. Gianan, E. S. Estilo and K. Mizoguchi, N-isopropylacrylamide grafted cellulose fibers and its application as an adsorbent, Bull. Res. Inst. Polym. Text., 144:93(1984)
8. O. Hirasa and S. Yamamoto, Fixation of Poly(vinylmethylether) on support by means of gamma-ray irradiation treatment, Bull. Res. Inst. Polym. Text., 144:87(1984)
9. T. Tanaka, Collapse of gel and the critical endpoint, Phys. Rev. Lett., 40:820(1978)
10. T. Amiya and T. Tanaka, Phase transitions in cross-linked gels of natural polymers, Macromolecules, 20:1162(1987)
11. S. Katayama, A. Ohata, Phase transition of cationic gel, Macromolecules,18:2781(1985)
12. S. Katayama, Y. Hirokawa and T. Tanaka, Reentrant phase transition in acrylamide-derivative copolymer gels, Macromolecules, 17:2641(1984)
13. E. S. Matsuo and T. Tanaka, Kinetics of discontinuous volume-phase transition of gels, J. Chem. Phys.,89:1695(1988)
14. I. Ohmine and T. Tanaka, Salt effects on the phase transition of ionic gels, J. Chem. Phys., 77:5725(1982)
15. Y. Hirokawa and T. Tanaka, Volume phase transition in a nonionic gel, J. Chem. Phys.,81:6379(1984)

16. Y. Hirose, T. Amiya, Y. Hirokawa and T. Tanaka, Phase transition of submicron meter gel beads, Polymer Preprints Japan for polymer gels,39(1988)
17. S. Hirotsu, Phase transition of a polymer gel in pure and mixed solvent media, J. Phys. Soc. Jpn., 56:233(1987)
18. S. Katayama, Phase transition of thermoresponsive ionic polymer gel, polymer Preprints Japan for Polymer gels, 29(1988)
19. T. Tanaka and D. J. Fillmore, Kinetics of swelling of gels, J. Chem. Phys., 70:1214(1979)
20. O. Hirasa, Thermoresponsive polymer gel, Kobunshi(High Polymers), Japan,35:1100(1986)
21. O. Hirasa, Y. Morisita, R. Onomura, H. Ichijo and A. Yamauchi, Preperation and Mechanical properties of Thermo-responsive fibrous hydrogel made from poly(vinyl methyl ether), Kobunshi Ronbunshu, 46:661(1989)
22. X. Huang, H. Unno, T. Akehata and O. Hirasa, Effect of salt solution on swelling behavior of poly(vinyl methyl ether) gel(PVMRG), J. Chem. Eng. Jpn., 21:10(1988)
23. O. Hirasa, Trial construction and performance test of thermoreversible adsorption system, Bull. Res. Inst. Polym. Text.,144:139(1984)
24. X. Huang, T. Akehata, H.Unno and O. Hirasa, Dewatering of biological slurry by using water-absorbent polymer gel, Biotech. Bioeng., 34:102(1989)
25. X. Huang, H. Unno, T. Akehata and O. Hirasa, Applicability of Temperature-sensitive water-absorbing hydrogel (polyvinylmethyl-ether gel) to dewatering of sludges, Kagaku Kogaku Ronbunshu, 13:518(1987)

Contraction Behavior of Poly(acrylonitrile) Gel Fibers

Susumu Umemoto, Norimasa Okui and Tetsuya Sakai

Department of Organic and Polymeric Materials, Tokyo

Institute of Technology, Ookayama, Meguro-ku, Tokyo, Japan

1. INTRODUCTION

The properties of volume change of gel stimulated by solvent, pH, ionic strength, temperature, electric power and so on have been dealt with the attracted subjects by many investigators[1~19]. Especially, the mechanisms of volume change in gels by solvent composition and temperature have roused many researchers from the observation of the phase transition phenomena of the gel systems [6~8]. In the case of the volume change caused by pH, ionic strength and ion concentrations, theoretical and experimental considerations on swelling of ionic gels have been represented by Flory[1], Kachalsky [2~4] and the other researchers [17~19]. It has been pointed out that the contractile force of the network, the electrostatic interactions and the interactions between polymer chain and solvent play an important role of the swelling. Recently, Tanaka and co-workers [14,15] investigated the swelling behavior of weakly charged ionic gels, such as polyacrylamide-acrylic acid copolymer gel, by the quantitative consistency of the Donnan theory. And they showed that the changes in swelling occurred at the predicted ion concentrations in the solutions.

Applications of the swell/collapse phenomena in the gels stimulated by environmental solution exchange have been attracted attention to a spontaneous motion similar to muscle contraction behavior. The pioneering works in this field were accomplished by Katchlsky[2~5] and many other researchers. However, if we apply the gel to a mechanochemical system as an artificial muscle, several problems must be overcome. For example, large swelling of the gel is very important but also the response time of elongation/contraction behavior and the contraction force must be taken into consideration. The gel contraction speed becomes more rapidly as a decrease in the gel size, since the speed is strongly related with the diffusion mechanisms such as the collective diffusion of the networks into solution, the ionic species

Polymer Gels, Edited by D. DeRossi *et al.*
Plenum Press, New York, 1991

diffusion into the networks [11,12]. Therefore, the gel must be decreased in size to achieve a high speed contraction/elongation behaviors. In order to achieve the strong contraction force of the gel, the gel must be constructed with a large number of crosslinked networks or with a rigid chemical structure.

For these purpose, fine gel fibers with strong mechanical properties were prepared from commercially available poly(acrylonitrile) (PAN) fibers. The dynamical contraction/elongation behaviors of the PAN gel fibers showed high speed response time and stronger contraction force than the human muscle. In addition, two abrupt discontinuous changes in the equilibrium swelling ratio and the contraction force, hysteresis curves on pH by successive pH changes were found [26,27].

In this paper, the structure of PAN gel fibers and contraction/elongation mechanism are presented on the basis of the hysteresis curves and the mechanical properties for the gel fibers.

2. EXPERIMENTAL

2.1 Sample preparation

PAN gel fibers were prepared from preoxidation [21] and saponification methods[23-25]. In the first step, PAN fibers (Mitsubishi Rayon Co., Ltd., Silpalon, 60 filaments, single filament diameter: 22.5 μm) were annealed at 220 ℃ in air on constant length (preoxidation step) and subsequently were saponified in boiling 1N NaOH aqueous solution for 30min (saponification step). The PAN gel fibers were purified by the distilled water to remove residual ions, impurities and non-crosslinked polymer, and then were dried in vacuum.

2.2 Structure determination of PAN gel fibers

The preoxidated PAN fibers and successively saponified PAN gel fibers are optically invisible because of their black color. So, the optical methods could not be used for determination of the chemical structure of PAN gel fibers. Chemical structures of the preoxidated PAN fibers and the saponified PAN gel fibers were measured by C^{13} NMR spectroscopy and wide angle X-ray diffraction(WAX). Aromatization index (degree of aromatic ring formation) of the samples were estimated by the WAX patterns[22].

2.3 Measurement of elongation/contraction behaviors

The dynamical and statical elongation/contraction behaviors stimulated by environmental solution change were studied by the change in length under an isotonic and the generating contraction force under an isometric state. The equilibrium state on the contraction/elongation change of the PAN gel fibers is defined as follows; changes in the gel fiber length at a given environment solution are measured repeatedly exchanging of the same solution until the saturation of the length change. This is because to avoid that pH or ion concentrations in the solution may change by the volume change in the gel. The procedure of decrease or increase of pH "step by step" is also defined that the solvent pH value changes to new pH value from previous pH value after the gel fiber length change reaches to the equilibrium state.

The dynamical elongation/contraction behaviors of PAN gel fibers in isotonic state were investigated through measurement of dynamical properties of the gel by the apparatus constructed with the cell and the chemical balance. One of the ends of the gel fibers was setted in the cell and the other end was fixed at the balance. The length change of PAN gel fibers was measured by the angle detector which is fixed in the center of the balance.

The dynamical properties of generating contraction force of PAN gel fibers in isometric state were studied through measurements of the stress change when the fibers were immersed in 1N HCl solution from 1N NaOH. The contraction/elongation response time was estimated from the value of $\Delta l_\infty/e$, where Δl_∞ is defined by $\Delta l_\infty = l_\infty - l_0$, l_∞ is the length at equilibrium state, l_0 is the initial length and e is an unit of natural logarithm.

2.4 Measurement of visco-elastic mechanical properties

Visco-elastic properties of the gel fibers in a given pH solution were evaluated through a mechanical dispersion measurement. The dynamic complex moduli (E^*) and phase shift (δ) were obtained from the results of the observed generating stress and the additional sinusoidal strain which was controlled to the value 1.0% on the frequency 0.100Hz. The dynamic storage moduli(E') and the dynamic loss moduli(E'') were also estimated by the fitting of the strain-stress Lissajous's figure data.

3. RESULTS AND DISCUSSION

3.1 Chemical structure of PAN gel fibers

In general, PAN fibers can be changed by the preoxidation step from a linear chain structure to a planar network structure constructed with cyano groups and pyridine rings as a crosslinker [21]. C^{13} NMR spectrum of preoxidated PAN fibers is shown in Fig.1. The pyridine rings and cyano groups was confirmed by the characterization of the peaks in Fig.1. The weak peak of cyano groups indicates that the degree of transformation from linear chains to planar network structure could be high.

Fig.2 shows changes in WAX pattern of oxidated PAN fiber against annealing time. As an increase in preoxidation time, the strong peak at 2θ= ca.17° and the weak peak at 2θ= ca. 28° disappeares gradually

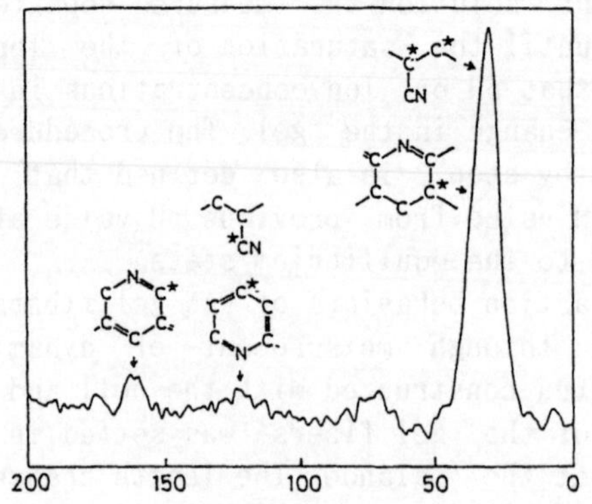

Fig.1 C^{13} NMR spectrum of pre-oxidated PAN fibers. The peaks are related to the chemical structures as shown.

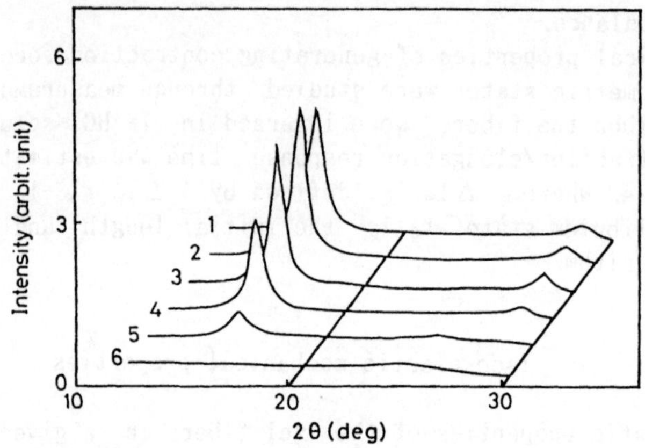

Fig.2 Changes in WAX pattern of oxidated PAN fibers against annealing time. The numbers in this figure show the sample number. The annealing time of sample No.1 is 0 min, No.2 is 50 min, No.3 is 100 min, No.4 is 200 min, No.5 is 300 min and No.6 is 1200 min.

and a new weak shallow peak at 2θ = ca. 25° appeares. It can be pointed out that the peaks at $2\theta = 17°$ and 28° are associated with the PAN crystalline spacing (100) and (010) planes, and the peak at 25° can be related to planar network spacing in preoxidated PAN fibers [22].

The degree of ring structure in the preoxidated PAN fibers is defined as "aromatization index" (AI) and estimated by the following equation[22] using the WAX peak intensities at $2\theta = 17°$:I_{17} and at $2\theta = 25°$:I_{25}.

$$AI = I_{25} / (I_{17} + I_{25})$$

The dependence of AI on preoxidation time is shown in Fig.3. The crosslinked structures due to aromatic or pyridine ring in the fibers obviously increase with the preoxidation time.

Cyano groups in the oxidated PAN fibers could be transformed to carboxyl groups by the saponification step in NaOH solution[23~25]. Fig.4 shows the C^{13} NMR spectrum of the saponified PAN gel fibers which were measured at dried state of the fibers. The peak of the cyano group disappeares and a new strong peak due to carboxyl group is found. The network of PAN gel fibers contains cationic groups and anionic groups such as the pyridine ring and the carboxyl group, respectively.

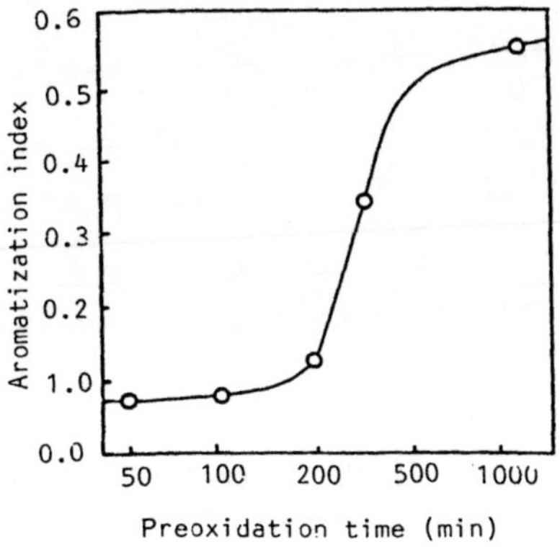

Fig.3 Dependence of aromatization index (AI) on preoxidation time.

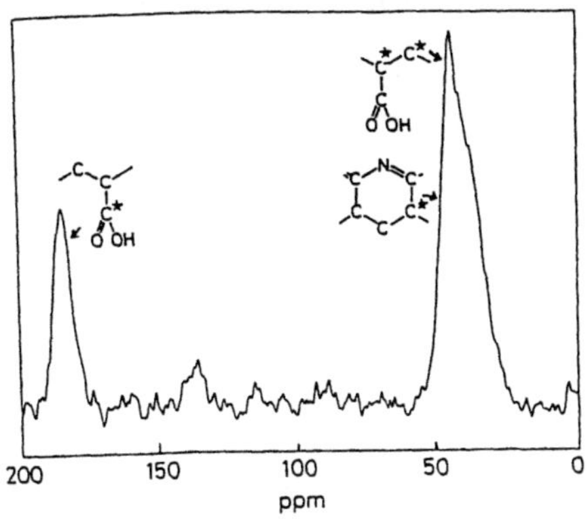

Fig.4 C^{13} NMR spectrum of saponified PAN fibers. The peaks are related to the chemical structures as shown.

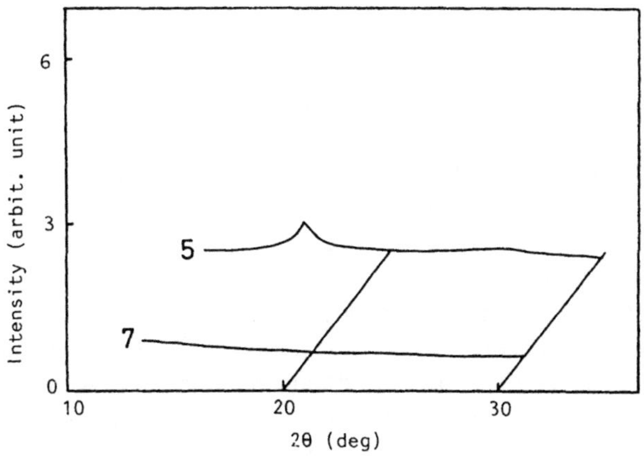

Fig.5 Changes in WAX pattern of saponified PAN gel fibers by the saponification step in NaOH solution. The sample No.5 was annealed during 300min and was not saponified.The sample No.7 was saponified No.5 during 30min.

Therefore, it can be said that the PAN gel fibers are amphoteric gel fibers.

The WAX patterns of the saponified PAN gel fibers are shown in Fig.5. Both peaks at $2\theta = 17°$ and at $2\theta = 25°$ are disappeared in the WAX patterns of saponified PAN fibers. PAN gel fibers are not soluble in aqueous solution at any pH. These results indicate that the chemical structure of PAN gel fibers prepared by the preoxidation and the saponification method could be illustrated schematically in Fig.6.

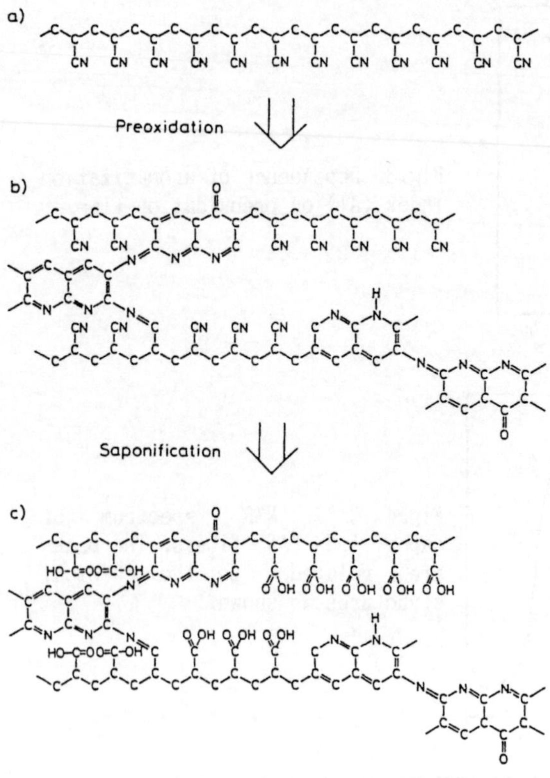

Fig.6 Schematic illustration of the structure of PAN fibers; a), preoxidated PAN fibers; b) and saponified PAN gel fibers; c).

3.2 Time dependence of elongation/contraction behavior

Thus prepared PAN gel fibers showed a large swelling in an alkaline solution and collapsed in an acidic solution. Fig.7 shows a typical dynamical elongation/contraction behavior of PAN gel fibers when the fibers are immersed in 1N NaOH solution from 1N HCl (elongation) or vice versa (contraction) in the isotonic state at 0.75Kg/cm^2 load. The collapsed PAN gel fibers are elongated rapidly and reach to the equilibrium length after ca. 20 seconds, and the swollen gel fibers contract in a similar way as the elongation. The swelling change in length: ΔL_∞ is about 80%, and the contraction/elongation response time is less than 2 seconds.

In general, the contraction/elongation behavior of a living skeletal muscle shows about 30% in length change and about $10^{-3}\sim10^{-2}$ seconds in the response time [28, 29]. From the viewpoint of application of mechanochemical materials to the artificial muscle, the length change in PAN gel fibers may be sufficient for the practical application. On the other hand, the response time is still too slow to compare with contraction speed of the living muscle. However, the response time of synthetic gel can be improved in its speed by reducing the fiber diameter, since the speed is strongly related with the diffusion mechanism.

The contraction force of the PAN gel fibers was measured at the

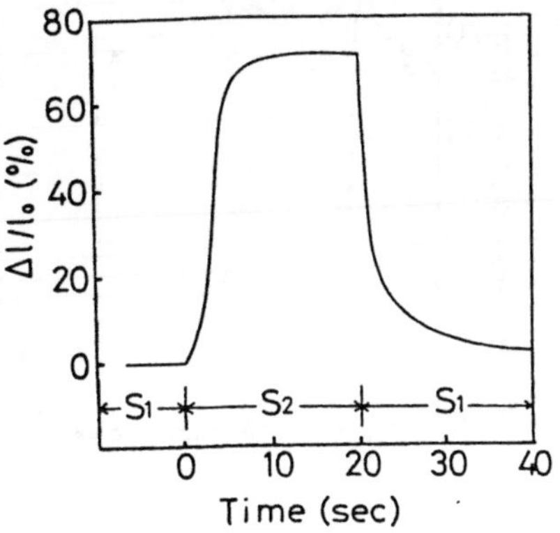

Fig.7 Dynamical changes in length of the PAN gel fibers. The solution was exchanged from S_1 to S_2 (elongation) and from S_2 to S_1 (contraction). The solution S_1 is 1.00N HCl and S_2 is 1.00N NaOH.

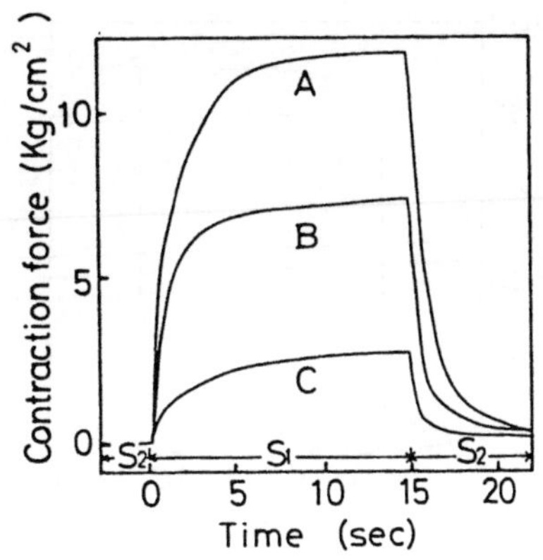

Fig.8 Dynamical changes in contraction force of the PAN gel fibers. The solution was exchanged from S_2 to S_1 (contraction) and from S_1 to S_2 (elongation). The solution S_1 is 1.00N HCl and S_2 is 1.00N NaOH. The symbols of A, B and C show the AI; 0.08, 0.12 and 0.35, respectively.

constant length. Fig.8 shows dynamical changes of contraction force of the three PAN gel fibers in the isometric state. The swollen PAN gel fibers generate the contractile force rapidly and reach to the equilibrium state after above 20 seconds. The maximum contraction force was ca. 12Kg/cm^2 . The generating force of living skeletal muscle is reported at 4~10 Kg/cm^2 [28,29] . It could be considered that the contraction force of PAN gel fibers are enough in strength for the artificial muscle. Such contraction force in PAN gel fibers strongly depends on the degree of AI. Contraction force increases with an increase in the AI, but the degree of swelling decreases with an increase in the AI. The contraction force is a function of the product of the modulus and the volume change of the gel fibers. Therefore, the rigid chemical structure might be necessary to achieve the strong contraction force.

3.3 Elongation/contraction behavior stimulated by pH

Fig.9 shows the equilibrium gel length in the isotonic state as a function of pH value which is adjusted by the concentration in HCl or NaOH. As decreasing in pH from 14 to 0 step by step, the gel fibers swell gradually in the pH ranges from 14 to 11 and show a constant length in the pH ranges from 11 to 3. At about pH 3, the swollen gel fibers contract suddenly and then show a constant length at below pH 3. On the other hand, as increasing in pH from 0 to 14 step by step, the fibers show a contracted state in the pH ranges from 0 to 11 and abruptly elongate at about pH 11. This hysteresis loop was well

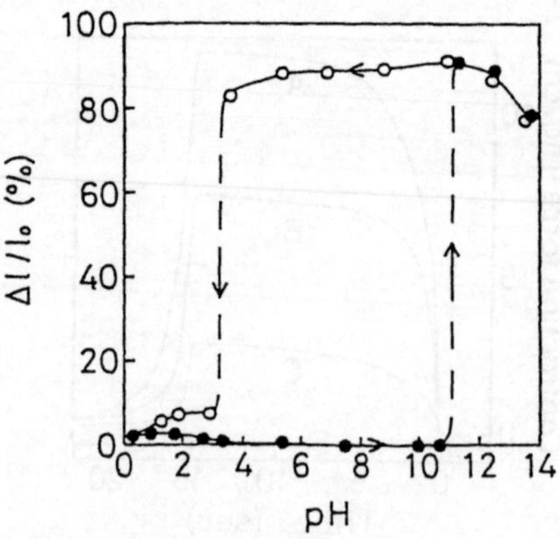

Fig.9 Equilibrium length of PAN gel fibers in the isotonic state as a function of pH (ionic strength is not controlled). The symbols show●; pH increase and ○; pH decrease.

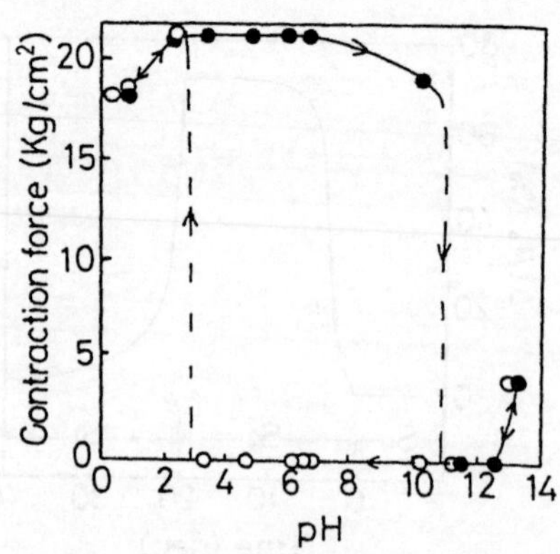

Fig.10 Equilibrium contraction force of PAN gel fibers in the isometric state as a function of pH (ionic strength is not controlled). The symbols show ●; pH increase and ○; pH decrease.

reproducible. Fig.10 shows the equilibrium contraction force of PAN gel fibers in the isometric state as a function of pH value. The contraction force behavior against pH shows the similar hysteresis loop as seen in Fig.9. Fig.9 and Fig.10 indicate that the swell/collapse behavior of the PAN gel fibers in isometric state and in isotonic state would be controlled by the same mechanism. It is interesting to point out that such hysteresis behaviors are found in the PAN gel fibers as an amphoteric gel but are not present in a monopolar cationic or anionic gel. The elongation/contraction behavior of PAN gel fibers can be divided into three pH regions; no hysteresis is found at region 1 and 3 which are in the pH range of 0 to 3 and 11 to 14, respectively, region 2 is in the pH range of 3 to 11 where the large hysteresis appeares. Further detail is the hysteresis loop will be discussed in the next section.

3.4 Hysteresis on elongation/contraction behaviors

Tanaka, et al.[14] have presented a quantitative interpretation of Donnan theory for a swelling in weakly charged gels as a function of ionic composition of the solvent. The balance of the osmotic pressure of free ions in the gel and environmental solution, the dissociation state of attached ionic groups in the gel and the contractile force of network play an important role of swelling of the gel. But the hysteresis loop in gel can not be predicted in the above theory. Here, it is necessary to take into account the effects of the osmotic pressure of free ions in the gel and in the environmental solution on the swelling behavior of PAN gel fibers. Fig.11 shows the dependence of length change of PAN gel

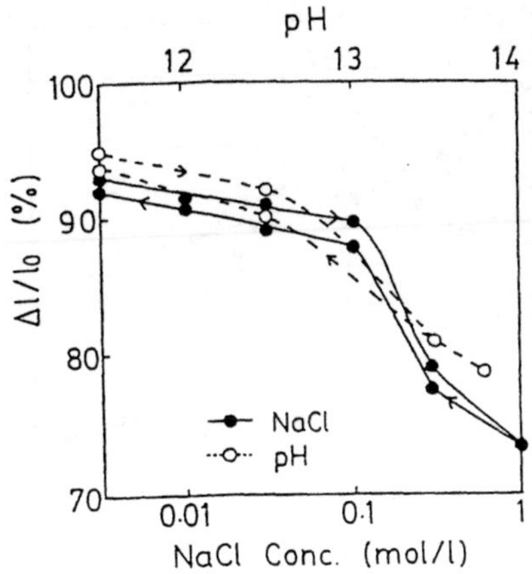

Fig.11 Dependence of length change of PAN gel fibers on the ionic strength of environmental solution controlled by NaCl concentration and on pH in the region 3.

fibers on the ionic strength of environmental solution controlled by NaCl concentration in distilled water (pH ~6) and on pH in the region 3. Changes in length by pH and NaCl concentration show almost the same characteristics. This suggests that the osmotic pressure of free ions play an important role in the swelling behavior in pH region 3. In addition, it might be thought that the osmotic pressure of free ions is one of main factor on the length change in the region 1.

It is generally thought that the amphoteric gel shows two different state corresponding to the pH history. In fact, two abrupt changes in Fig.9, 10 could be associated with phase transitions between the collapsed state and the swollen state of the PAN gel fibers. Such phase transition may be induced by the change in ionic interaction, such as association/dissociation of ionic groups in the gel network.

3.5 Mechanical properties

Fig.12 and Fig.13, respectively, show dependences of the equilibrium dynamic storage moduli(E') and dynamic loss moduli(E") of PAN gel fibers on pH value. As decreasing in pH, E' and E" values show the characteristic changes corresponding to the gel swelling behavior. E' and E" values are fairly low in the swollen state of the gel, however, they become considerably high in the contracted state of the gel. It can be thought for the swelling gel that the storage moduli(E') are associated with the rubber elasticity [1,20]. On the basis of the pure rubber elastic properties, the loss moduli(E") maintaines at almost zero and an increase in the crosslink density of the gel can not yield an increase of E". Therefore, the large increase of E" in the pH region 2 could not be explained solely by the rubber elasticity. The increase in E' and E" may be related to the formation of a molecular interaction in the amphoteric gel, such as hydrogen bonding, electrostatic interactions, hydrophobic interactions. However, an increase in E" suggests that the interactions in the gel are related to the flexible networks which is reorganized frequently in the gel.

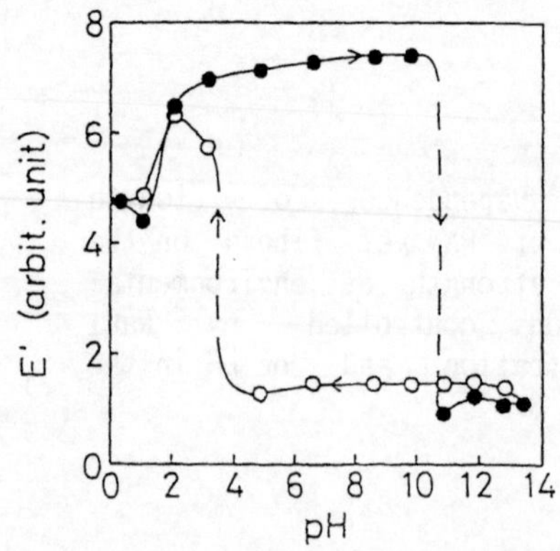

Fig.12 Dependence of equilibrium dynamic storage moduli(E') of PAN gel fibers on pH (ionic strength is not controlled). The symbols show ● ; pH increase and ○; pH decrease.

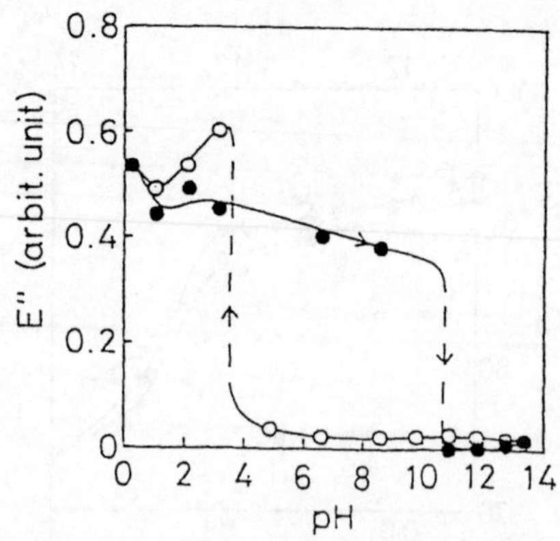

Fig.13 Dependence of equilibrium dynamic loss moduli (E") of PAN gel fibers on pH (ionic strength is not controlled). The symbols show ●; pH increase and ○; pH decrease.

3.6 Relationship between dynamical behaviors and the structure change

In order to describe the relationship between the dynamical elongation/contraction behavior and the structure change, the change of dynamic storage moduli(E'), dynamic loss moduli (E") and contraction force of the PAN gel fibers were measured simultaneously when the gel elongated and contracted corresponding to environmental solution exchange. Time dependence of the behaviors of contraction force, E' and E" at near the phase transition points were measured as shown in Fig.14 (contraction: pH 4.3 ➡ pH 3.6) and in Fig.15 (elongation: pH 9.8 ➡ pH10.4). On the contraction, the increase of contraction force was faster than that of E' and E". This result shows that the formation of molecular interactions which is related with the structure change may

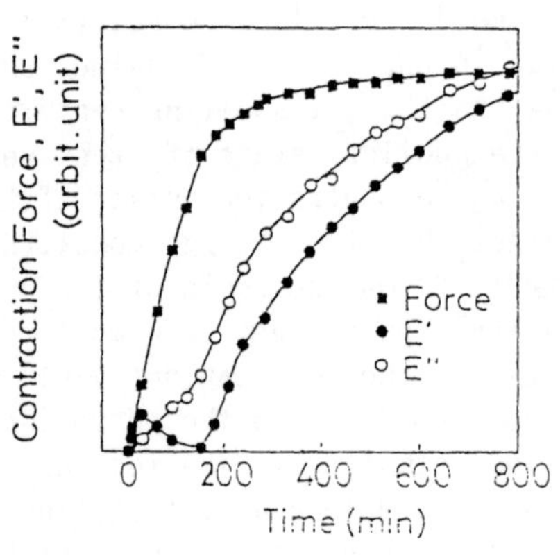

Fig.14 Time dependence of the behaviors of contraction force, E' and E" at near phase transition point on contraction(pH4.3→pH3.6).

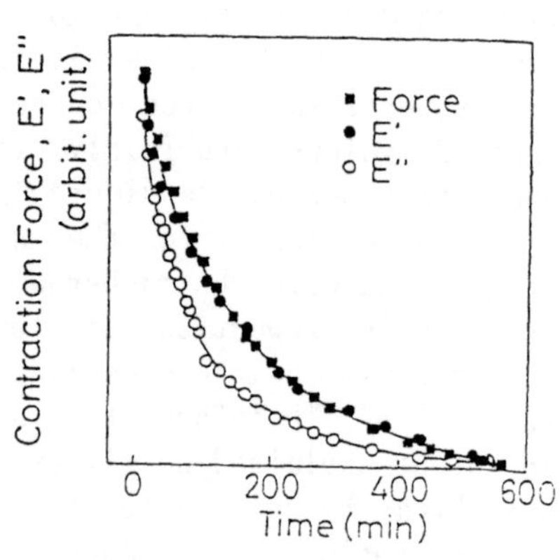

Fig.15 Time dependence of the behaviors of contraction force, E' and E" at near phase transition point on elongation(pH 9.8→pH10.4).

not be triggers on the shrinkage of the gel, but the shrinkage may bring on the formation of the molecular interactions. We considered that the shrinkage was carried out by the effect of osmotic pressure determined by the dissociation state of the gels. On the other hand, the decrease in the contraction force is similar to that in E' and E" on the elongation. This indicates that a vanishing out the molecular interactions may play an important role of swelling of the gels.

3.7 Effect of electrolytes on the elongation/contraction behavior

Simple electrolytes such as NaCl, KCl, $NaSO_4$ affect on the swelling behavior from the collapsed gel. For example, the collapsed gel swelled when the gel was immersed in 1N NaCl solution and subsequently immersed in water. This result would bring to the useful informations for the nature of collapsed state. The length of the PAN gel fibers in the collapsed state in HCl solution was measured in water as a standard length l_0. The fibers immersed in several concentration in NaCl solution were taken out from the solution and the length of the fibers was measured in water. The values of $\Delta l/l_0$ were plotted against the NaCl concentration as shown in Fig.16, where the $\Delta l = l - l_0$. No change in the gel length is found below NaCl 0.70. On the other hand, a large elongation is observed above NaCl 0.75. The values of E' and E" were also changed in a similar manner as the above length change. These results suggest that the molecular interactions in the collapsed state are strongly affected by pH and NaCl.

Fig.17 shows the equilibrium gel lengths in the isotonic state as a function of pH value which is adjusted by HCl or NaOH solution concentration, where ionic strength of solutions were controlled 0.10 by NaCl concentration. And Fig.18 also shows the equilibrium gel lengths as a function of pH value on the ionic strength controlled at 1.00. As an increase in the ionic strength, the hysteresis loop was disappeared gradually and finally the gel showed the abrupt change at only pH 3 in the ionic strength 1.00. It would be thought that NaCl could weaken the electrostatic interaction between anions and cations in the gel network. And the electrolytes may change the dissociation state of the ionic

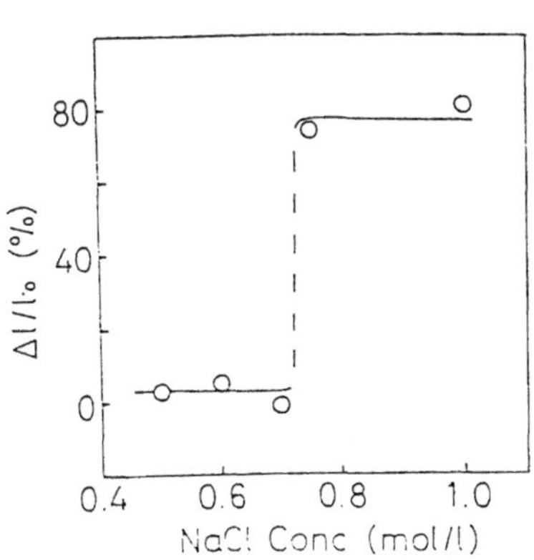

Fig.16 Plots of $\Delta l/l_0$ values against NaCl concentration. The length of collapsed PAN gel fibers by HCl solution was measured in water as a standard length; l_0. The fibers immersed in several concentration in NaCl solution were taken out from the solution and the length of the fibers; l was measured in water. The values of $\Delta l/l_0$ are $\Delta l = l - l_0$.

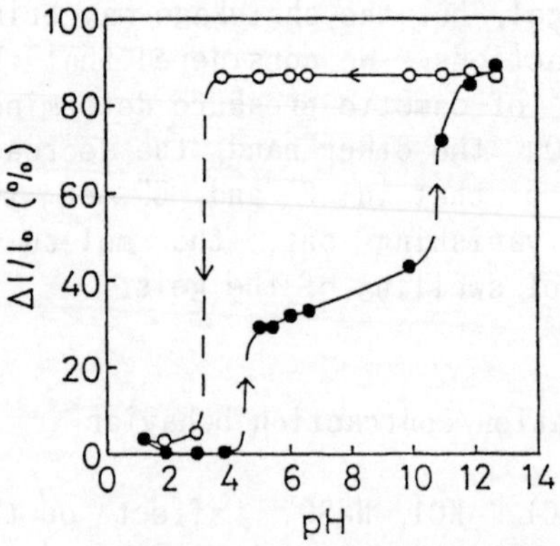

Fig.17 Equilibrium length of PAN gel fibers in the isotonic state as a function of pH (ionic strength is controlled 0.10 by NaCl). The symbols show ●; pH increase and ○; pH decrease.

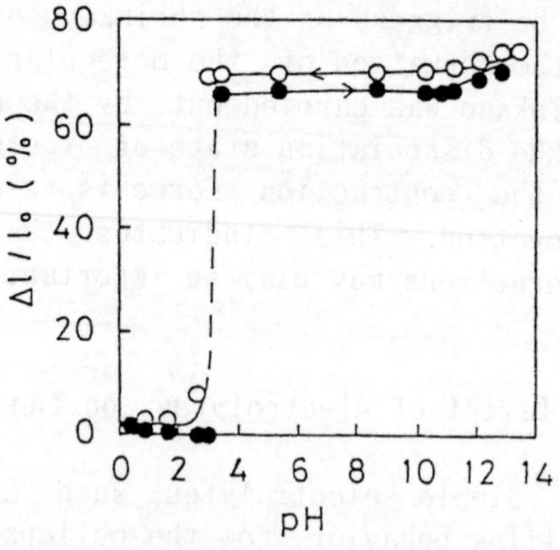

Fig.18 Equilibrium length of PAN gel fibers in the isotonic state as a function of pH (ionic strength is controlled 1.00 by NaCl). The symbols show ●; pH increase and ○; pH decrease.

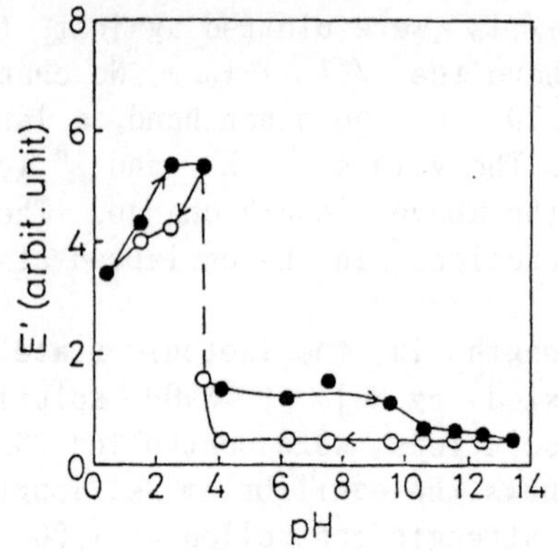

Fig.19 Dependence of equilibrium dynamic storage moduli (E') of PAN gel fibers on pH (ionic strength is controlled 1.00 by NaCl). The symbols show ●; pH increase and ○ ; pH decrease.

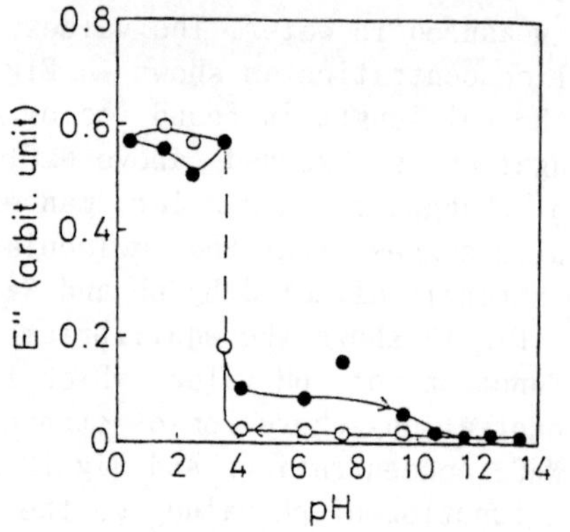

Fig.20 Dependence of equilibrium dynamic loss moduli (E") of PAN gel fibers on pH (ionic strength is controlled 1.00 by NaCl). The symbols show ●; pH increase and ○; pH decrease.

groups in the gel, the destruction of the electrostatic interaction can occur by the addition of NaCl.

It is also expected that such electrolytes affect on the dependences of equilibrium dynamic storage moduli(E') and dynamic loss moduli(E") of PAN gel fibers on pH value. Fig.19 and Fig.20 show the dependences of equilibrium E' and E" on pH value, where the ionic strength was controlled at 1.00 by NaCl concentration. As decreasing in pH from 14 to 0 step by step, the tendencies of E' and E" were also agreed with the results in Fig.12 and Fig.13, respectively. However, as increasing in pH, the E' and E" values abruptly decrease at pH 3 and large hysteresises in E' and E" disappear. Therefore, it is thought that the electrostatic interaction[30,31] formed in the collapsed state between anions and cations in the gel network are destroyed by the NaCl.

4. CONCLUSION

The fine gel fibers were prepared from commercially available poly (acrylonitrile) (PAN) fibers by the preoxidation step and the saponification step. From the measurement of C^{13} NMR and WAX, the chemical structure of PAN gel fibers are estimated to be the planar network structure with the pyridine ring and the carboxyl group, that is, the fibers could be said to be an amphoteric gel fibers. The PAN gel fibers show a large swelling in an alkaline solution and collapse in an acidic solution. The swelling change in length is about 80%, and the contraction/elongation response time is less than 2 seconds on the dynamical behavior in the isotonic state. And the maximum contraction force in the isometric state was ca. 12Kg/ cm^2 . It could be considered that the length change and the contraction force of PAN gel fibers are enough for the artificial muscle. The rigid chemical structure might be necessary to achieve the strong contraction force.

The pH hysteresis behaviors are found in the PAN gel fibers. Change in length, change in contraction force and changes of E' and E" show all similar pH hysteresis loops. The elongation/contraction behavior could be divided into three pH regions, no hysteresis is found at the region 1 and 3 which are in the pH range of 0 to 3 and 11 to 14, respectively, and the region 2 is in the pH range of 3 to 11, where the large hysteresis appears. The large increase of E" in the region 2 could not be explained solely by the rubber elasticity. The increase in E' and E" may be related to the formation of a molecular interaction and the increase in E" suggests that the interactions in the gel are related to the flexible networks which is reorganized frequently in the gel. The formation of molecular interactions may not be triggers on the shrinkage of the gel, but the shrinkage may bring on the formation of the molecular interactions and a vanishing out the molecular interactions may play an important role of swelling of the gels. The electrostatic interaction formed in the collapsed state between anions and cations in the gel network are destroyed by simple electrolytes such as NaCl.

5. REFERENCES

1. P. J. Flory, Principles of Polymer Chemistry (Cornell University, Ithaca, 1953), Chap. 13.
2. W. Kuhn and A. Katchalsky, Nature, 165, 4196 (1950)
3. A. Katchalsky and H. Eisenberg, Nature, 166, 267 (1950)
4. A. Katchalsky and M. Zwick, J. Polym. Sci., 16, 221 (1955)
5. M. V. Sussmann and A. Katchalsky, Science, 167, 45 (1970)
6. K. Dusek and D. Patterson, J. Polym. Sci., A-2, 6, 1209 (1968)
7. T. Tanaka, D. Fillmore, S. T. Sun, I. Nishio, G. swislow and A. Shah, Phys. Rev. Lett., 45, 20, 1636 (1980)
8. H. Vink, Acta Chemica Scandinavica, A-37, 187 (1983)
9. T. Tanaka, S. Ishiwata and C. Ishimoto, Phys. Rev. Lett., 38, 14, 771 (1977)

10. T. Tanaka, Phys. Rev. A, 17, 2, 763 (1978)
11. T. Tanaka and D. J. Fillmore, J. Chem. Phys., 70, 3, 1214 (1979)
12. T. Tanaka, Physica, 140A, 261 (1986)
13. I. Ohmine and T. Tanaka, J. Chem. Phys., 77, 11, 5725 (1982)
14. J. Ricka and T. Tanaka, Macromolecules, 17, 1, 2916 (1984)
15. J. Ricka and T. Tanaka, Macromolecules, 18, 83 (1985)
16. T. Tanaka, I. Nishio, S. T. Sun and S. Ueno-Nishio, Science, 218, 29, 467 (1982)
17. M. Ilavsky, Macromolecules, 15, 782 (1982)
18. M. Ilavsky and J. Hrouz, Polym. Bull., 9, 159 (1983)
19. Y. Osada, Y. Saito, Makromol. Chem., 176, 2761 (1975)
20. P. G. de Gennes, Scaling Concepts in Polymer Physics (Cornell University, Ithaca, 1979), Chap. 5.
21. T. Uchida, et al., Proc. 10th Biennial Carbon Conf., 31 (1971)
22. K. Miyachi and R. D. Andrews, Appl. Polym. Symposia, 25, 127 (1974)
23. G. F. Fanta, J. Appl. Polym. Sci., 24, 9, 2015 (1979)
24. G. F. Fanta, J. Polym. Sci., 23, 229 (1971)
25. Japan Patents, 52-9095 (1977)
26. T. Matsumura, S. Umemoto, N. Okui and T. Sakai, Reports on Progress in Polym. Phys. in Japan, 30, 323 (1987)
27. S. Umemoto, Y. Itoh, N. Okui and T. Sakai, Reports on Progress in Polym. Phys. in Japan, 31, 295 (1988)
28. S. Hatano, J.Mechanochem. and Cell Motility, 1, 75, (1972)
29. Cold Spring Harbor Symposia on Quantitative Biology, 37, The Mechanism of Muscle Contraction (Yurinsha, 1973)
30. R. M. Fuoss, H. Sadek, Science, 110, 552 (1949)
31. A. Nakajima, H. Sato, Biopolymers, 11, 1345 (1972)

MEASURING CONTACT STRESS WITH ^{19}F-NMR SPECTROSCOPY

Peter J. Basser

Biomedical Engineering and Instrumentation Program
National Institutes of Health
Bethesda, MD USA

INTRODUCTION

Increased static contact stress is associated with tissue damage *in vivo* and *in vitro*. It could be caused by edema (for example, in normal hydrocephalus) or a growing tumor which displaces surrounding tissue. Measurement of contact stress is inherently invasive, however, it can be less traumatic with a sensor having no wires or leads. One goal of this chapter is to propose that NMR spectroscopy can perform the function of a telemetry system provided that an in-dwelling transducer can convert the desired physical quantity (such as applied stress) to an NMR detectable chemical change. Previously, it was observed that compressing polyelectrolyte gels could cause a redistribution of mobile ions that is measurable with NMR spectroscopy [1]. In the sensor described below, the applied stress is inferred from a non-invasive ^{19}F-NMR spectroscopic measurement of pH.

There are many desirable attributes of this biomimetic sensor design. It exhibits tissue-like compliance. The sensor can be sterilized prior to implantation; the absence of leads improves the likelihood of maintaining a sterile field. Because the sensor measures an equilibrium mechanochemical effect, excellent d.c. response is expected. In addition, the sensor contains no cytotoxic substances [12] or non-toxic chemicals in non-physiologic concentrations whose release would present a threat to biological tissue. Since the pH indicator is not endogenous, little background noise is encountered. The operating range and sensitivity of the sensor can be tailored to suit the application because they depend upon process variables which can be controlled during synthesis of the gel. Finally, the sensor can be biodegradable, thus requiring only implantation. Its principle disadvantages include the high cost of NMR scans and low sensitivity.

Potential medical applications include: monitoring contact stress between tumor and adjacent soft tissue, tumor and bone (e.g. meningioma), two swollen tissue surfaces (e.g. hydrocephalus or brain edema), muscle and fascia (e.g. cardiac tamponade), artificial and/or normal articulated joints. Potential applications to biotechnology include measurement and monitoring of contact stresses between tissue or cells in culture.

Polymer Gels, Edited by D. DeRossi *et al.*
Plenum Press, New York, 1991

DESCRIPTION OF THE SENSOR AND ITS PRINCIPLE OF OPERATION

The sensor element consists of an impermeable but flexible pouch containing a polyelectrolyte gel bathed in an ionic solution. Figure 1 shows a schematic diagram of the bag, the external solution and the gel.

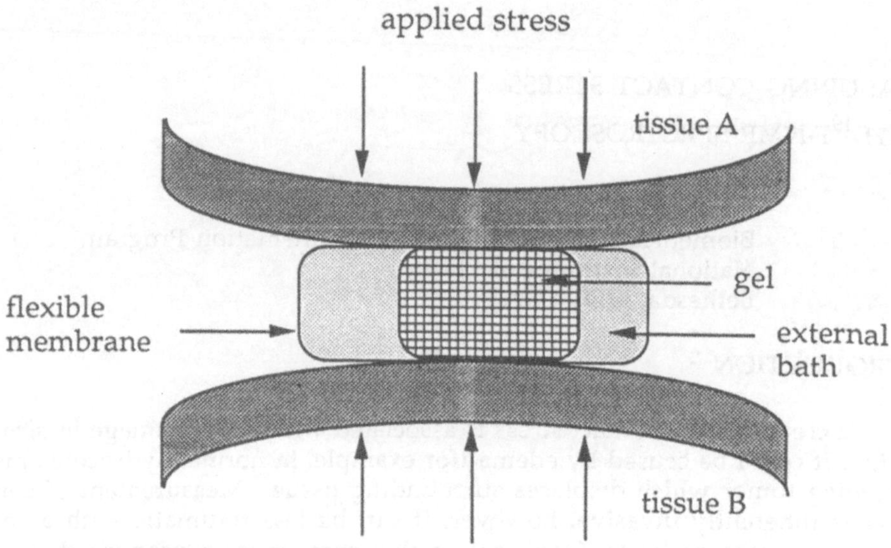

Figure 1. Schematic drawing of the contact stress sensor consisting of a flexible, impermeable membrane that contains a block of polyelectrolyte gel bathed in an ionic solution. Two representative tissue surfaces, A and B are also shown.

A polyelectrolyte gel is hydrated in an aqueous cocktail (pH 7.5) containing Na^+, Cl^- and difluoromethylalanine hydrochloride. The latter is a sensitive pH indicator originally developed to measure intra and extracellular pH using ^{19}F-NMR spectroscopy. Since the majority of the fixed charge groups on the polymer backbone of the gel are ionized at pH 7.5, the equilibrium concentration of the mobile ions ($[Na^+]$, $[OH^-]$, $[H^+]$, $[Cl^-]$, and the two charged forms of the ^{19}F-NMR indicator, $[NH_3^+C(CH_3)(CHF_2)COOH]$ and $[NH_2C(CH_3)(CHF_2)COO^-]$,) differ in the interstitial and external fluid compartments. A normal stress applied to the bag compresses the gel, squeezing fluid from its interstices into the surrounding bath. This loading decreases the volume of interstitial fluid, increasing the density of fixed charge on the gel. Mobile co- and counter-ions (including H^+) are redistributed between the interstitium and external bath. The new equilibrium interstitial and bath pH are reflected in a change in the NMR spectrum of the pH indicator molecule. A mathematical model, presented below, describes the operation of the sensor

CONTACT STRESS VS. HYDROSTATIC PRESSURE

It is important to note that this sensor measures the normal component of the mechanical contact stress, not the ambient hydrostatic pressure. The

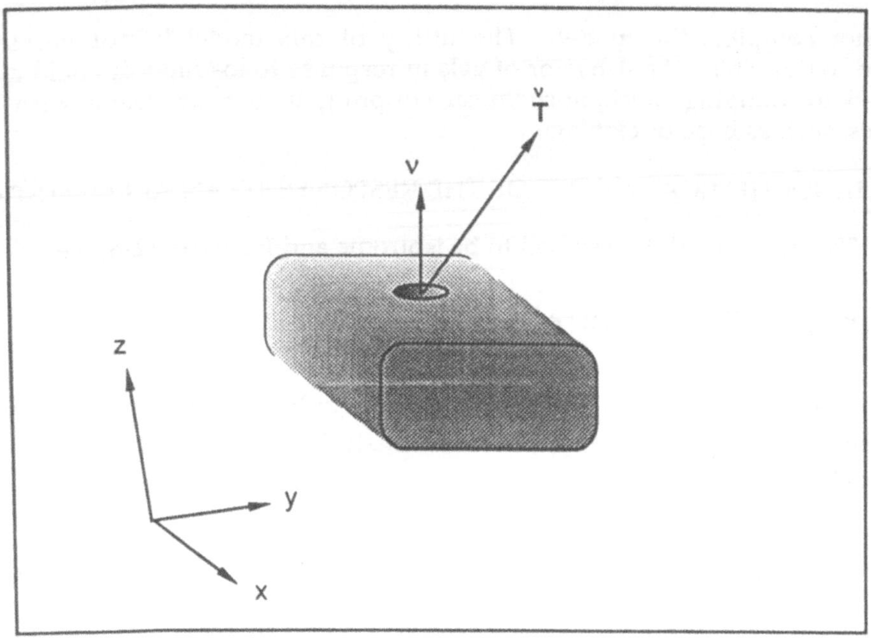

Figure 2. Schematic diagram of the traction, T, acting on a block at a point at which the unit normal vector is v.

contact stress at a point is generally described by a traction vector, **T**, which is depicted in the Figure 2 below.

The traction vector can be resolved into components normal and parallel to the surface of contact. Only the normal component can be measured by this method because, for small strains, only the normal stress changes the volume of the gel. In addition, since in the steady state, the hydrostatic pressure is uniform everywhere within the gel, changes in hydrostatic pressure also will produce no changes in gel volume.

MODELLING THE SENSOR'S RESPONSE TO A STATIC APPLIED LOAD

This mathematical model relates the applied stress to the measured shift in the ^{19}F-NMR spectrum of the sensor. Continuum mechanics is used to describe the behavior of the gel under loading while physical-chemistry is used to determine the chemical changes that these stresses induce. Finally, elementary quantum mechanics is used to determine how these chemical changes affect the frequency shifts in the ^{19}F-NMR spectra of the indicator moiety. Specifically, the equation of conservation of momentum and a constitutive law of the gel (with appropriate boundary conditions) express the relationship between the applied stress on the gel and its dilatation. The equations of conservation of mass, electroneutrality, electrochemical equilibrium and equilibrium dissociation of each chemical species describe the chemical interactions between the gel and the external bath. These expressions are simplified to a system of two non-linear algebraic equations that are solved numerically for the hydrogen ion concentrations within the gel and in the external bath. Finally, equations that relate the shift in Larmor frequencies of the observed ^{19}F-NMR spectrum to the degree of dissociation of the pH indicator complete the model. The utility of this model is not necessarily limited to describing the behavior of gels in response to loading; it could also be applied to studying mechanochemical coupling in porous tissue with fixed charges, such as bone or cartilage.

MECHANOCHEMICAL MODEL OF THE RESPONSE TO AN APPLIED STRESS

The gel network is assumed to be isotropic and Hookean [2-6], i.e.,

$$\sigma^{gel}_{ij} = K\varepsilon\,\delta_{ij} + 2\mu\left(\varepsilon_{ij} - \frac{1}{3}\varepsilon\,\delta_{ij}\right) - \sigma\,\delta_{ij} \tag{1}$$

where K and μ are the bulk and shear elastic moduli of the gel, respectively, and σ is the isotropic pore pressure [3-6]. For small deformations, the strain tensor, ε_{ij}, is defined as

$$\varepsilon_{ij} = \frac{1}{2}\left(\frac{\partial u_i}{\partial x_j} + \frac{\partial u_j}{\partial x_i}\right) \tag{2a}$$

and the dilatation, ε, is defined as

$$\varepsilon = \varepsilon_{xx} + \varepsilon_{yy} + \varepsilon_{zz} \tag{2b}$$

Since the gel is in mechanical equilibrium, the conservation of linear momentum takes the following form [3-6]:

$$\frac{\partial \sigma^{gel}_{ij}}{\partial x_j} = 0 \tag{3}$$

The gel is assumed to be caste in the shape of a rectangular prism. As shown in Figure 1, a uniform compressive contact stress, σ_0, is applied at two faces while the other four are free of mechanical traction. In the steady state, the hydrostatic pressure of the interstitial fluid equals the hydrostatic pressure of the external bath. The applied stress can be related to the dilatation of the sample by summing the diagonal components of the stress tensor in Eq. (1), i.e.,

$$-\sigma_0 = 3K\varepsilon \tag{4}$$

The importance of Eq. (4) is that if we know the bulk modulus and the dilatation of the gel, we can calculate the magnitude of the compressive stress applied to the gel. This is the strategy that is used in designing the sensor. However, it should be remembered that Eq. (4) is valid only for small deformations. The dilatation, ε can be interpreted as the fractional change in gel network volume (relative to its unstrained state),

$$\varepsilon = \frac{\delta V}{V_0} \cong \frac{V - V_0}{V_0} \tag{5}$$

where V and V_0 are the volume of the gel in the deformed and undeformed state, respectively. Using Eqs. (4) and (5), the applied stress is proportional to the gel's final volume:

$$-\sigma_0 \cong 3K\left(\frac{V - V_0}{V_0}\right) \quad \text{or} \quad V \cong V_0\left(1 - \frac{\sigma_0}{3K}\right) \tag{6}$$

THE RELATIONSHIP BETWEEN DILATATION AND pH

The gel sensor element is modelled as two compartments that are in electrochemical equilibrium. One compartment is the polymer network and its interstitial fluid; the other is the external bath. (Although we restrict our discussion to a two-compartment system, formulae below can be extended to model multi-compartment systems). Both compartments are enclosed within a flexible bag that is impermeable to water and other ions, but allows heat to be exchanged. The contents of the bag are assumed to be isothermal. Because no matter is exchanged across the impermeable barrier, the equation of the conservation of mass of each species within the bag takes the following form:

$$M_i = \sum_{j=1}^{N_c} V_j [C_i]_j \tag{7a}$$

where M_i is the total number of moles of the ith species, $[C_i]_j$ is the concentration of the ith species in the jth compartment, V_j is the volume of the jth compartment, and N_c (2 in this case) is the number of compartments. Since the number of moles of each solute is determined during synthesis, all of the M_i can be treated as constants. As such, the number of constraint equations (Eq. (7a)) equals the number of individual chemical species added to the solvent.

In addition, the contents of the bag are individually incompressible so that the total volume of the sensor, V_T, remains constant:

$$V_T = \sum_{j=1}^{N_c} V_j \tag{7b}$$

The condition of macroscopic electroneutrality in each compartment imposes another constraint. If Z_i is the valence number of the ith charged species and N_j is the total number of charged species present in compartment j, then, for all j,

$$\sum_{i=1}^{N_j} [C_i]_j Z_i = 0 \tag{8}$$

Since it is assumed that the compartments are in electrochemical equilibrium, the Donnan equilibrium equations predict the relative concentrations of *mobile* ionic species i and k in different compartments:

$$\left(\frac{[C_i]_1}{[C_i]_2}\right)^{z_i} = \left(\frac{[C_k]_1}{[C_k]_2}\right)^{z_k} \quad (9)$$

The carboxylic acid groups that are covalently linked to the backbone of the gel are modelled as a monoprotic weak acid, HP, whose dissociation is governed by the following formulae:

$$HP + H_2O \rightleftarrows H_3O^+ + P^- \quad \text{with} \quad K_a' = \frac{[P^-]\,[H_3O^+]_2}{[HP]} \quad (10)$$

where $[P^-]$ is the concentration of ionized groups in the interstitial solvent and K_a' is the acid dissociation constant. The total concentration of ionizable groups, $[P_{tot}]$, is given by,

$$[P_{tot}] = [P^-] + [HP]\ . \quad (11)$$

By combining Eq. (10) and (11), one can express the concentration of ionized or fixed-charges in terms of the hydronium ion concentration, $[H_3O^+]$ and the total concentration of ionizable groups in the network:

$$[P^-] = [P_{tot}]\,\frac{K_a'}{K_a' + [H_3O^+]_2}\ . \quad (12)$$

Moreover, since the total number of ionizable groups in the polymer is conserved, Eq. (7a) becomes:

$$M_{P_{tot}} = V_{2_0}\,[P_{tot}]_0 = V_2\,[P_{tot}] \quad (13)$$

where V_{2_0} is the volume of gel in the unstressed state and $[P_{tot}]_0$ is the total concentration of ionizable groups.

We also require that the dissociation of water obeys the familiar equation,

$$[H_3O^+]_1\,[OH^-]_1 = K_w' = [H_3O^+]_2\,[OH^-]_2 \quad (14)$$

where K_w' is the dissociation constant of water.

Because indicators are usually present in trace concentrations, they are often neglected in acid/base calculations. However, since millimolar concentrations of indicator are needed (to improve the signal/noise ratio of the spectroscopic measurement) its dissociation must also be included in this model. Once dissolved, the amphoteric ^{19}F-NMR indicator dissociates in a two-step reaction:

$$IND^+ + H_2O \rightleftarrows IND + H_3O^+\ ; \quad \text{with}\ K_1' = \frac{[IND]\,[H_3O^+]}{[IND^+]} \quad (15a)$$

and

$$IND + H_2O \rightleftarrows IND^- + H_3O^+\ ; \quad \text{with}\ K_2' = \frac{[IND^-]\,[H_3O^+]}{[IND]} \quad (15b)$$

For brevity, three forms of the indicator moieties, $NH_3^+C(CH_3)(CHF_2)COOH$, $NH_2C(CH_3)(CHF_2)COO^-$, and $NH_2C(CH_3)(CHF_2)COOH$ are denoted by IND^+, IND^-, and IND, respectively.

The total number of moles of indicator in the pouch, M_t, is given by:

$$M_t = V_1\left([IND]_1 + [IND^+]_1 + [IND^-]_1\right) + V_2\left([IND]_2 + [IND^+]_2 + [IND^-]_2\right) \quad (16)$$

The electrically neutral Zwitterion is a small molecule that can diffuse freely between compartments. Therefore, at equilibrium it is present in equal concentrations in each compartment:

$$[IND]_1 = [IND]_2 \quad (17)$$

For the particular chemical species used in these experiments, the equations of electroneutrality become,

$$[H_3O^+]_1 + [Na^+]_1 + [IND^+]_1 - [OH^-]_1 - [Cl^-]_1 - [IND^-]_1 = 0 \quad (18a)$$

in the external bath and

$$[H_3O^+]_2 + [Na^+]_2 + [IND^+]_2 - [OH^-]_2 - [Cl^-]_2 - [IND^-]_2 - [P^-] = 0 \quad (18b)$$

within the gel.

It is possible, to express Eqs. (18a,b) *only* in terms of the hydrogen ion concentrations in each compartment and the applied stress:

$$h_1(\,[H_3O^+]_1\,,\,[H_3O^+]_2\,) = 0 \quad (19a)$$

$$h_2(\,[H_3O^+]_1\,,\,[H_3O^+]_2\,,\sigma\,) = 0 \quad (19b)$$

where h_1 and h_2 are each scalar functions. These two equations contain only two unknown dependent variables, $[H_3O^+]_1$ and $[H_3O^+]_2$. Other terms which appear are either independent variables that can be dictated during gel preparation or material constants whose values can be determined by independent measurements. Since the stress, σ, can be treated as a free parameter Eqs. (19a,b) constitute a system of two non-linear, algebraic equations with two unknowns which can be solved using standard numerical techniques, such as Newton's method or the Levenberg algorithm [7]. Other mobile ionic species could serve as the dependent variables in the electroneutrality equations above.

Figure 3 shows the calculated relationship between pH in the external bath and interstitium and the normalized applied stress. The stress, σ, is normalized by the bulk modulus of the gel, K, making the ratio, σ/K, dimensionless. This figure is titled "stress titration curve" because of the close analogy that exists between mechanical and chemical titration experiments. It shows the predicted equilibrium pH in the interstitial and external solutions which would result from increasing the applied stress on the gel reversibly.

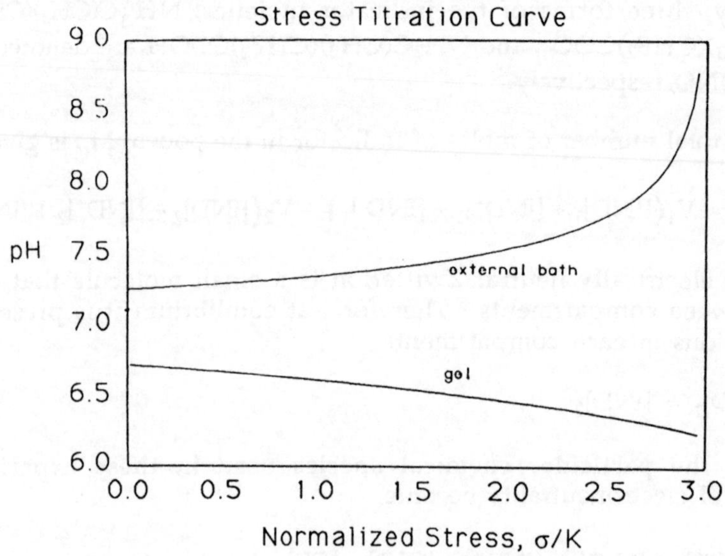

Figure 3. The predicted relationship between pH of the interstitial and external solutions and applied stress on the gel.

The difference in pH between the interstitial and external solutions at equilibrium is a consequence of the Donnan equations.

The sensitivity of the gel sensor element is given by the slope of the pH vs. σ/K curve. It can calculated numerically by differentiating the curves of pH vs. σ/K curve with respect to σ/K using centered differences. In the expected operating range of the sensor (i.e., $\sigma/K < 1$) the sensitivity is approximately 0.25 pH units per unit change in the normalized stress. However, the sensitivity and pH range are controlled by several independent variables, such as the polymer content, the degree of cross-linking, the number of ionizable groups and the bulk modulus, K.

THE RESPONSE OF THE NMR INDICATOR TO CHANGES IN pH

The hydronium ion concentration within the gel and within the external bath are measured by exploiting a relationship between the observed chemical shift in the ^{19}F-NMR spectrum and the concentration of the hydronium ions within each compartment of the sensor. The indicator molecule exists in three dissociation states; their relative populations reflect the ambient pH. Since the chemical shift is different in each state, the measured chemical shift can be related to pH.

The lifetime of a particular dissociation state of the indicator molecule is on the order of its inverse dissociation rate constant whereas the time it precesses in the magnetic field equals its inverse Larmor frequency. The rate constant is proportional to $\exp(-\Delta G^*/kT)$ where ΔG^* is the activation energy of transition between states, k is the Boltzmann constant and T is the absolute temperature. For the deprotonation reaction of an amine group, the rate constant > 10^6 Hz. However, the characteristic frequency associated with the acquisition of the spectrum is $(\nu_a - \nu_b)/2$ where ν_a and ν_b are the Larmor frequencies of one the Fluorine moieties in the protonated and deprotonated states, respectively. For this indicator, this frequency difference is only $O(10^2)$ Hz. Therefore, the indicator changes its state an average of 1000 times during the time an individual ^{19}F-NMR spectrum is acquired.

In this "rapid exchange limit", the measured chemical shift spectrum in compartment 1 can be expressed as linear combination of the chemical shift spectrum of each dissociation state, i, of the indicator molecule weighed by its probability of occurrence, $p(i)_1$ [16]. Since the indicator used in these experiments exhibits three separate dissociation states, the observed chemical shift, $E(L_j)$, is given as:

$$E(L_j) = p(0)_j\, L_j^{(0)} + p(-1)_j\, L_j^{(-1)} + p(+1)_j\, L_j^{(+1)} \tag{20}$$

The components of the vector $L_j = (l_1,l_2,l_3,l_4,l_5,l_6,l_7,l_8)_j$ represent the particular frequencies at which the chemical shift lines (Hz) occur in the observed ^{19}F-NMR spectrum of the indicator in compartment j. As the measured chemical shift is shown to be a mathematical expectation it is denoted as $E(L_j)$. Since the probability of a particular state, i, is just its relative concentration,

$$p(i)_j = \frac{[IND^i]_j}{[IND_{tot}]_j} \tag{21}$$

the measured chemical shift is the sum of the chemical shifts of the three states of the indicator weighted by their relative concentrations:

$$E(L_j) = \left(\frac{[IND^0]_j}{[IND_{tot}]_j}\right) L_j^{(0)} + \left(\frac{[IND^-]_j}{[IND_{tot}]_j}\right) L_j^{(-1)} + \left(\frac{[IND^+]_j}{[IND_{tot}]_j}\right) L_j^{(+1)} \tag{22}$$

A representative NMR spectrum of the indicator in aqueous solution is given in Figure 4.

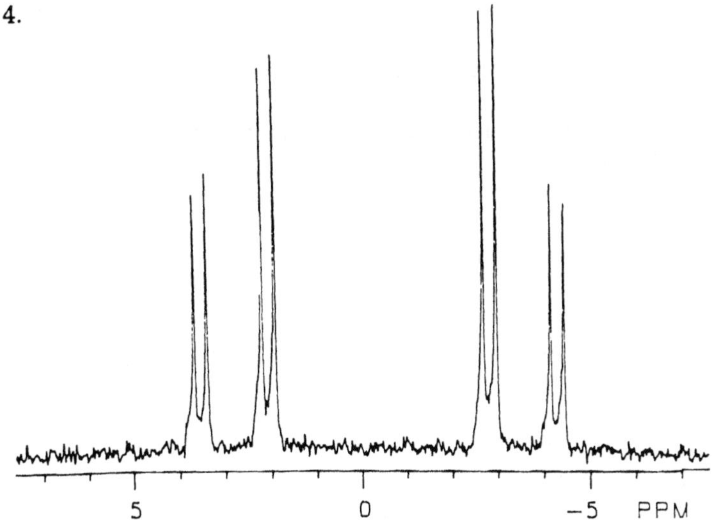

Figure 4. The NMR spectrum of the indicator in aqueous solution. The eight lines (octet) are a characteristic of an ABX spin system.

It shows eight distinct spectral lines - a characteristic of an ABX spin system [10]. Deutsch et al. observed that the difference in frequency between the two central peaks, Δ, of the spin-decoupled (AB) spectrum of difluoromethylalanine is sensitive to changes in pH [11,12]. An approximate formula for this difference signal in terms of the spectral lines that we measure without spin decoupling is:

$$\Delta \sim \frac{l_3+l_4-(l_5+l_6)}{2} \qquad (23)$$

Because the expectation operator is linear, the observed value of Δ is given by:

$$E(\Delta)_j = \left(\frac{[IND^0]_j}{[IND_{tot}]_j}\right)(\Delta)_j^{(0)} + \left(\frac{[IND^-]_j}{[IND_{tot}]_j}\right)(\Delta)_j^{(-1)} + \left(\frac{[IND^+]_j}{[IND_{tot}]_j}\right)(\Delta)_j^{(+1)} \qquad (24)$$

Figure 5 shows the calculated values of Δ(ppm) plotted against external bath and interstitial pH.

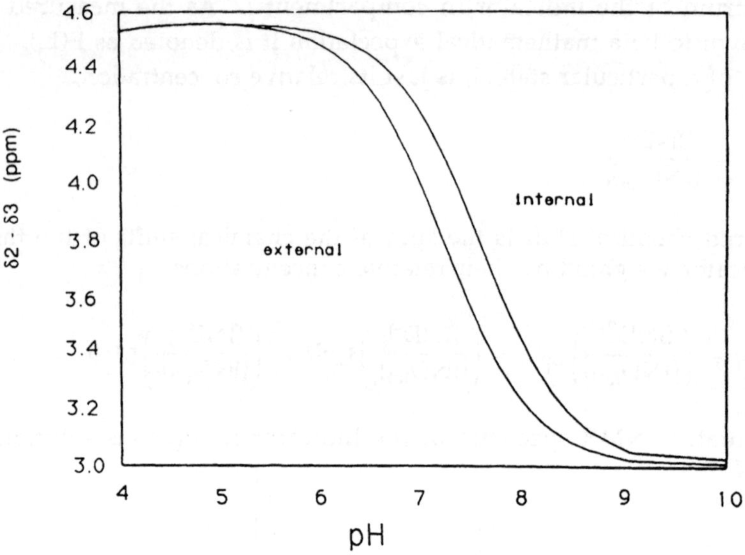

Figure 5. The predicted chemical shift in parts per million (ppm) vs. interstitial and bath pH. The shape of this curve is effectively determined by the Henderson-Hasselbach equation; the sensitivity of the Fluorine NMR indicator used in these calculations is estimated from Deutsch [12].

The relative concentrations of the indicator were obtained from Eqs. (19a,b). Once the hydronium ion concentration is known in both compartments, the concentration of each of the indicator species was determined using Eq. (15a,b,16,17). In preparing Figure 5, values of Δ in the Zwitterionic and negatively charged states were estimated from Deutsch et al. [12]. In the pH range at which the sensor is maintained, the indicator exists primarily in the Zwitterionic and negatively charged forms so that the last term in Eq. (22, 24) can be neglected.

The time it takes for the indicator to diffuse between compartments scales with δ^2/D where δ is a characteristic length and D is the diffusivity of the indicator molecule in water. For this sensor element $D \sim 0.9 \times 10^{-5}$ cm^2/sec [13] and $\delta \sim 0.1$ cm, so that $\delta^2/D \sim 1000$ sec. Since the diffusion time is much greater than $2/(\nu_a - \nu_b) \sim 0.01$ sec, it can safely be assumed that the indicator molecules will remain in their respective compartments while the spectrum is being measured. In principle, this implies that the spectrum from a two-compartment system will be the superposition of the spectrum from each compartment. Since the strength of the signal is proportional to the mass of the indicator, the measured spectrum, L_m, will be proportional to the spectra from compartments 1 and 2 weighted by their respective volumes:

$$L_m \sim [IND_{tot}]_1 V_1 L_1 + [IND_{tot}]_2 V_2 L_2 \tag{26}$$

Unfortunately, this formula is not experimentally justified. Figure 6 shows a typical measured ^{19}F-NMR spectrum of the gel and the external bath.

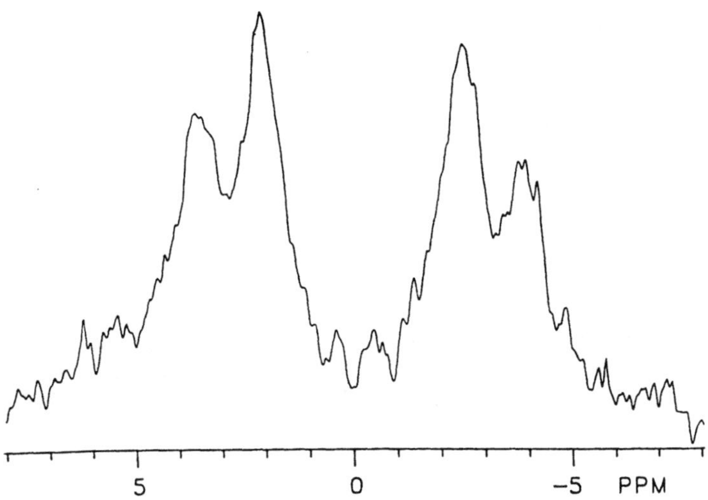

Figure 6. A typical ^{19}F-NMR spectrum of the indicator in gel.

The signal from the interstitial fluid compartment of the sensor is highly attenuated and shows line broadening. This result is attributed to dipole-dipole interactions which are negligible in most liquids but pronounced in solids and gels [10].

DETERMINATION OF MATERIAL CONSTANTS OF THE GEL

There are several parameters whose values can be prescribed *a priori* or measured directly. Since the sensor is assumed to be operating at 37 C, the value of the dissociation constant of water, K_w', is taken to be $10^{-13.68}$. The ^{19}F-NMR indicator has two dissociation constants: $K_1' = 10^{-2.34}$ and $K_2' = 10^{-7.3}$. The latter was taken from Deutsch [11] while the former was estimated from the literature values of the dissociation of the carboxylic acid residue of alanine and glycine [13]. The dissociation constant of the polyacrylamide gel is taken to be $K_a' = 10^{-7.3}$. This value is obtained from Riǒka and Tanaka [14]. The mass of polymer can be determined at the time of synthesis by a weight measurement. The polymer volume is measured using a graduated eyepiece in a microscope stage. The concentration of ionizable groups is also estimated [14]. The value of the bulk modulus is estimated from quoted values in Tanaka and Filmore [1]. The diffusivity of the indicator in water is estimated from the literature values of the diffusivities of alanine and other small amino acids [13].

THE RESPONSE OF A PROTOTYPE SENSOR

Figures 7a and b depict the system that was used to demonstrate that an applied stress can cause a measurable chemical shift during an experiment with a phantom.

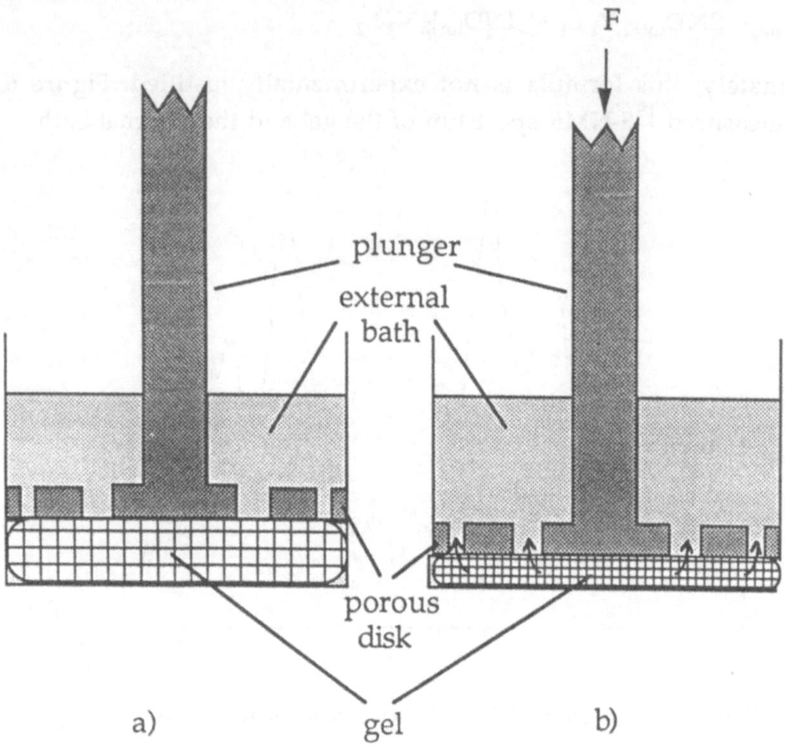

Figure 7. Schematic drawing of the plunger used to compress a polyelectrolyte gel slab immersed in an ionic solution.

The end of a 20 cm^3 plastic syringe was sealed with paraffin tape. Holes were drilled into the rubber tamper so that fluid could drain through it when the plunger was depressed. In preliminary experiments, the syringe was first filled with 3.0 ml of hydrolyzed polyacrylamide gel (15% by weight), 50.0 mg of difluoromethylalanine hydrochloride (the ^{19}F-NMR - pH indicator) and 0.1 M NaOH and distilled, de-ionized water to bring the total volume to 5.0 ml at pH 7.5. (In later tests, acrylamide-acrylic acid copolymer gels were used). The cylindrical RF transmitting coil was first tuned. Then the syringe was inserted into the coil and the entire assembly was placed into the core of a 4.7 Tesla magnet (General Electric). Initially, no compressive force was applied to the plunger (corresponding to the configuration shown in Figure 7a). Proton and

^{19}F-NMR Free Induction Decay (FID) spectra were obtained after the magnetic field gradients were shimmed with the sample in place. A normal force of approximately 1.0 x 10^5 dynes (a mean normal stress of approximately 3.2 x 10^4 dynes/cm^2) was then applied to the plunger. The system was then allowed to equilibrate for about 1 hour. Proton and ^{19}F-NMR spectra were measured again. The change in interstitial and external volume produced by the application of a compressive force are depicted in Figure 7b. Interstitial fluid from the gel flows into the reservoir above the sample, however the total volumes of polymer and solvent are equal in both experiments.

Figures 8 and 9 show the ^{19}F-NMR spectra of the uncompressed and compressed samples, respectively. The change in the relative spacing of the two sets of central peaks reflect a change in internal and external pH induced by the applied stress.

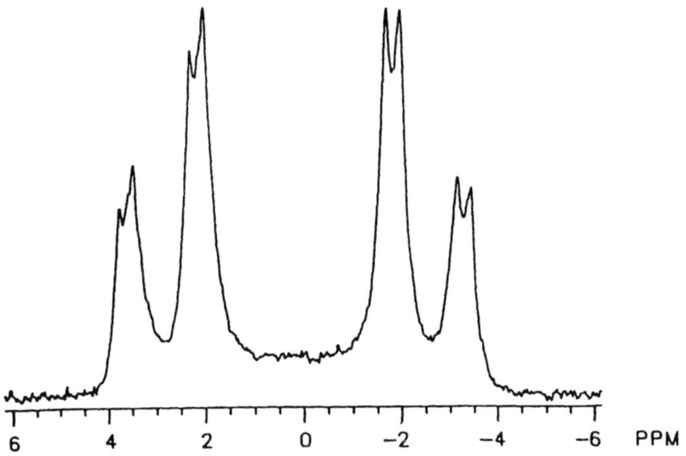

Figure 8. ^{19}F-NMR spectrum of uncompressed sample

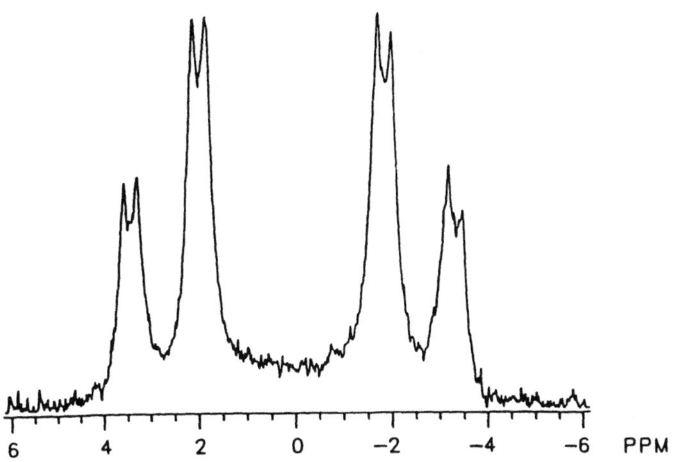

Figure 9. ^{19}F-NMR spectrum of compressed sample

NOTES ABOUT SENSOR CALIBRATION

Empirical methods can be used in conjunction with the mathematical model to determine the applied stress from the chemical shift of the NMR indicator. Before the sensor is implanted, a calibration table can be generated by placing the gel sensor in a "phantom", applying a set of known loads, and measuring the corresponding chemical shifts. Once the sensor is implanted the applied stress can be inferred from the measured chemical shifts by reading this table. An accurate method of measuring the temperature distribution *in vivo* now exists [23], which could provide a non-invasive method for temperature compensating an in-dwelling sensor.

PRESENT THEORETICAL AND EXPERIMENTAL LIMITATIONS

Several enhancements could increase the utility of this model and improve the sensitivity of the sensor. The most obvious improvement of the model is to extend its applicability to unsteady mechanical and chemical processes. Grimshaw et al. [9] recently presented a model of the dynamics of gel swelling and ionic transport, which could be adapted for this purpose. Large deformations could be encountered in practice, and should be included in the model. Now, only small deformations are described using a linear theory. To improve the sensitivity of the sensor, its charge density must be increased. However, doing so without increasing network elasticity would only cause the gel to swell to a new, larger rest volume. Nature has already solved this problem in cartilage, by trapping large molecules with fixed charges within a collagen network. This strategy could also be applied to the design of future prototype contact stress sensors.

SUMMARY

The design of this sensor is a departure from traditional solutions to measuring stress. It uses an equilibrium mechanochemical conversion mechanism exhibited in biological tissue such as cartilage to measure and monitor the stress between opposing surfaces in contact. The sensor element consists of an impermeable but flexible pouch that encapsulates a polyelectrolyte gel that is bathed in an ionic solution. The sensor requires no external connections for its continued operation since the stress applied is determined by a non-invasive ^{19}F-NMR spectroscopic measurement. The absence of leads or wires increases the probability of maintaining a sterile field. The sensor exploits an equilibrium mechanochemical effect so good d.c. response is expected. In addition, the sensor contains no cytotoxic substances or non-cytotoxic chemicals in non-physiologic concentrations whose release would present a threat to biological tissue. The operating range and sensitivity of the sensor can be tailored to suit the requirements of the specific measurement. Because the indicator used to make the spectroscopic measurement is not endogenous, little background noise is expected. Although the predicted changes in the ^{19}F-NMR spectra induced by an applied stress are described by a mathematical model empirical methods can also be used to establish a calibration curve for its operation. Potential medical applications include: measuring and/or monitoring contact stress between tumor and adjacent soft tissue, tumor and bone (as in meningioma), two swollen tissue surfaces (as in hydrocephalus or brain edema), muscle and fascia (as in cardiac tamponade), and artificial and/or normal articulated joints. Potential applications to biotechnology include measurement and monitoring of contact stresses between tissue or cells in culture.

ACKNOWLEDGEMENTS: The Society of Polymer Science, Japan should be commended for their excellent organization of the Symposium on Polymer Gels which provided a forum in which this work could be presented. Many thanks to John Eng, Bruce Berkowitz and Bob Balaban of NIH who provided expertise in the acquisition and analysis of NMR spectra and to Piero Chiarelli for informative discussions about deforming polymer gels.

REFERENCES

[1] P. J. Basser, J. Eng, & P. Chiarelli, A Polymer Based Method to Measure Contact Stress, Proceedings of the Society of Polymer Science, Japan, 20-23, (1989)

[2] T.Tanaka, D.J. Fillmore, Kinetics of Swelling of Gels, J. Chem. Phys. , 70(3), 1214-1218 (1978)

[3] M. A. Biot, Consolidation Settlement Under a Rectangular Load Distribution, J. Appl. Phys., 12(2), 426 (1941)

[4] M. A. Biot, General Theory of Three-Dimensional Consolidation, J. Appl. Phys., 12 , 155 (1941)

[5] M. A. Biot, Theory of Elasticity and Consolidation of a Porous Anisotropic Solid, J. Appl. Phys., 26 (2), 182-185, (1955)

[6] M. A. Biot, G. Willis, The Elastic Coefficients of the Theory of Consolidation, J. Appl. Mech. 24 , 594 (1957)

[7] W. Press, B. Flaherty, et al., "Numerical Recipes", Cambridge University Press

[8] E. H. Frank, A.J. Grodzinsky, Cartilage Electromechanics-II. A Continuum Model of Cartilage Electrokinetics and Correlation with Experiments, J. Biomechanics, 20 (6) , 629-639 (1987)

[9] P. E. Grimshaw, J. H. Nussbaum, A.J. Grodzinsky, and M. L. Yarmush, "Kinetics of Electrically and Chemically Induced Swelling in Polyelectrolyte Gels", Jour. Chem. Phys. (in press)

[10] E. Becker, "High Resolution NMR, Theory and Applications", 2nd Edition Academic Press, (1980)

[11] C. Deutsch, J. S. Taylor, D. F. Wilson, "Regulation of intracellular pH by human peripheral blood lymphocytes as measured by ^{19}F NMR", Proc. Natl. Acad. Sci. 79, 7944-7948, (1982)

[12] C. Deutsch, T. Kashiwagura, J. Taylor, D. Wilson, M. Erecinska, "The effect of Glucagon and Adrenergic Agonists on Intracellular pH of Isolated Rat Hepatocytes", Jour. Biol. Chem., 260 (11), (1985)

[13] R. Weast, M. Astle, ed. "Handbook of Chemistry and Physics", F-51, 1983, CRC Press, Inc.

[14] J. Rička, T. Tanaka, "Swelling of Ionic Gels: Quantitative Performance of the Donnan Theory", Macromolecules, 17 (12), 2916-2921 (1984)

[15] D. LeBihan, J. Delannoy, R. L. Levin, "Non-invasive temperature mapping using magnetic resonance imaging of molecular diffusion", Radiology, 171, 853-857, (1989)

DRUG DELIVERY SYSTEM

CONVENTIONAL AND ENVIRONMENTALLY-SENSITIVE HYDROGELS FOR MEDICAL AND INDUSTRIAL USES: A REVIEW PAPER

Professor Allan S. Hoffman

Center for Bioengineering and
Chemical Engineering Dept., FL-20
University of Washington
Seattle, Washington 98195

Abstract

Hydrogels are water-swollen polymer networks. They are of great interest both medically and commerically. Conventional hydrogels change little in swelling with environmental conditions while stimuli-responsive hydrogels may swell or deswell sharply with relatively small changes of temperature or pH. A general review of the principles and applications of these interesting materials is presented here.

Introduction

For the past twenty years many researchers have been investigating a wide range of hydrogels for biomedical and biotechnological applications (see refs. 1, 2 for reviews). Such gels are of special interest in biological environments because they have high water contents similar to body tissues and are easily washed to remove impurities. Some of these hydrogels are "conventional" in that they do not exhibit significant sensitivity to environmental changes, such as temperature and pH, while others show varying degrees of response to such stimuli.

Conventional Gels

Conventional gels have mostly been synthesized from hydroxyethyl methacrylate (HEMA), polyethyleneglycol methacrylate (PEGMA), and the methoxy-terminated derivative (MPEGMA), acrylamide (AAm), N-vinyl pyrrolidone (NVP) and polyethylene oxide (PEO). These gels have most often been prepared either as unsupported membranes or beads, or as graft copolymer coatings on more hydrophobic supports.

Because such gels are "similar" to body tissues in their high water contents, it has been assumed that they should be "biocompatible". Indeed, many conventional hydrogels appear to be biocompatible when used topically, such as in contact lenses and burn dressings (2). Animal implant studies have also suggested that they should be tissue-compatible in vivo in humans (2). It was therefore surprising to us to find that, using a baboon femoral A-V shunt model, a series of acrylic hydrogels increased in platelet destructiveness in proportion to their increase in water content (3). It was hypothesized that such behavior might actually be caused by their high water contents (or low solids contents) which leads to the observed low protein and cell adhesiveness of hydrogels.

Polymer Gels, Edited by D. DeRossi *et al.*
Plenum Press, New York, 1991

The polymer polar groups responsible for such high water contents could interact with proteins and cell surfaces via strong H-bonds or ionic bonds to disrupt the native structure of key proteins, which might then desorb in the flowing blood, causing platelet damage, and at the same time re-exposing the hydrogel surface to cause further protein and platelet destruction.

Subsequent studies with polyethylene oxide (PEO) hydrogels (4,5) which are H-bond acceptors but not H-bond donors, have suggested that the PEO hydrogels may be more blood compatible than the acrylic hydrogels containing -OH, $-CONH_2$ and $-COO^-$ groups. This cannot be simply verified by in vitro protein adsorption or cell adhesion studies. For example, our recent data on fibrinogen adsorption to a number of such gels and other substrates indicate that hydrogel-protein interactions decrease almost linearly with the increase in substrate surface water content, independent of the H-bond donating or accepting character of the hydrogel (6). We observed that the fractional decrease in fibrinogen adsorption is small in going from a PAAm gel to a P(AAm/MPEGMA) copolymer gel with a relatively long PEG (or PEO) pendant side chain. Direct comparison of the blood interactions of strong H-bond donor hydrogels with those of PEO-containing hydrogels in a widely accepted in vivo animal model, has yet to be carried out to the author's knowledge. Demonstration of low protein and cell (especially platelet) interactions in vitro is not necessarily an indication of good blood compatiblity in vivo.

Environmentally-Sensitive Gels

A new class of hydrogels whose water swelling ratios are sensitive to small changes in environmental conditions has been of great, recent interest. There are many possible environmental changes which can stimulate interesting responses in water soluble polymers and their crosslinked hydrogels (Table 1). One can incorporate many different biomolecules into these gels, such as drugs, enzymes and antibodies (7-12). We have also immobilized cells in pH and temperature-sensitive gels (9,10). These immobilized biological species may be physically entrapped or chemically bound within the gel structure.

The most commonly studied hydrogels having environmental sensitivity are either pH-or temperature-sensitive. Thermally-sensitive gels have usually been synthesized from N-substituted AAm and methacrylamide (MAAm) or related monomers (13). For example, N-isopropylacrylamide (NIPAAm) has been widely studied. We have also synthesized LCST gels from hydrophobic PVA derivatives (12). Both pH and electric field sensitivity require the addition of ionic monomers to the hydrogels, such as acrylic acid (AAc) or aminoethyl methacrylate (AEMA). Thus, one may synthesize a "simple" pH-sensitive gel from HEMA and AAc (or AEMA) or a combined pH and temperature-sensitive gel from NIPAAm and AAc (or AEMA) (14).

The temperature-sensitive gel is usually characterized by the lower critical solution temperature (LCST) of the backbone polymer or copolymer comprising the gel. The LCST of the gel polymer may be shifted up or down by copolymerization with more or less hydrophilic monomers (15,16). One may also do this by increasing the hydrophobic content of hydrophilic polymers such as PVA (12) or hydroxypropyl methylcellulose (HPMC) (17).

When an enzyme is immobilized in such gels, it may be "turned off and on" by de-swelling and re-swelling the pores as the temperature is raised and then lowered around the LCST (7,8). This is due to the closing and opening of the pathways for molecular diffusion.

Enzymes or cellular systems are usually immobilized in order to stabilize them by protecting them from the environment (and sometimes

Table 1. Examples of Possible Environmental Stimuli and Responses with Polymeric Biomaterials (Especially Hydrogels)

STIMULUS	RESPONSE
Δ (pH)	Chemical/Biochemical (stimulates or inhibits reactions or recognition processes)
Δ (temperature)	PhaseSeparation (precipitation)
Δ (chemical or biochemical agents)	Shape (shrinks or swells)
Δ (solvent or salts)	Surface (becomes non-wetting)
Δ (electrical field)	Permeability (increases or decreases sharply)
Δ (E.M. radiation)	Mechanical (hardens or softens)
Δ (mechanical stress)	Optical (clears up or opacifies; becomes colored)
	Electrical (generates signal; electrochemical reaction occurs)

Table 2. Some Polymers and Surfactants which Show LCST Behavior in Aqueous Solutions

Ether Groups
Polyethylene oxide (PEO)
Poly(EO/PO) copolymers
PEO-PPO-PEO triblock surfactants
Alkyl-PEO block surfactants
Poly(vinyl methyl ether)
Alcohol Groups
Hydroxypropyl acrylate
Hydroxypropyl methylcellulose
Hydroxypropyl cellulose
Methylcellulose
Poly (vinyl alcohol) (and derivatives)
Substituted Amide Groups
Poly (N-substituted acrylamides)
Poly (N-vinyl pyrrolidone)
Poly (ethyl oxazoline)
Other
Poly (methacrylic acid)

Table 3. Most Common Groups in pH-Sensitive Hydrogels

Anionic	Cationic
$-COO^-$	$-NH_3+$
$-OPO_3^-$	$-NRH_2+$
	$-NR_2H+$
	$-NR_3+$

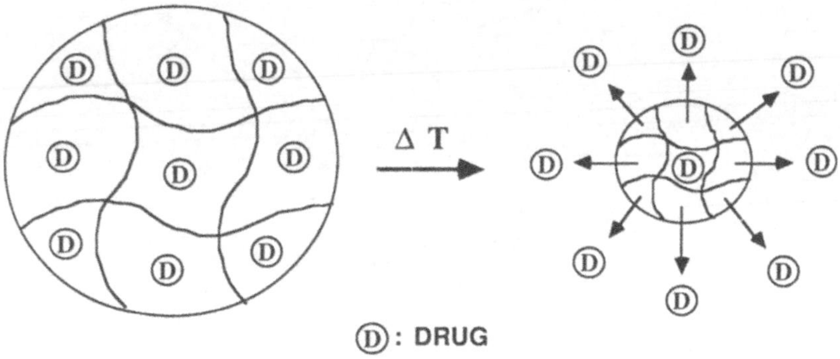

Figure 1. Physical and chemical immobilization of biomolecules in hydrogels (2).

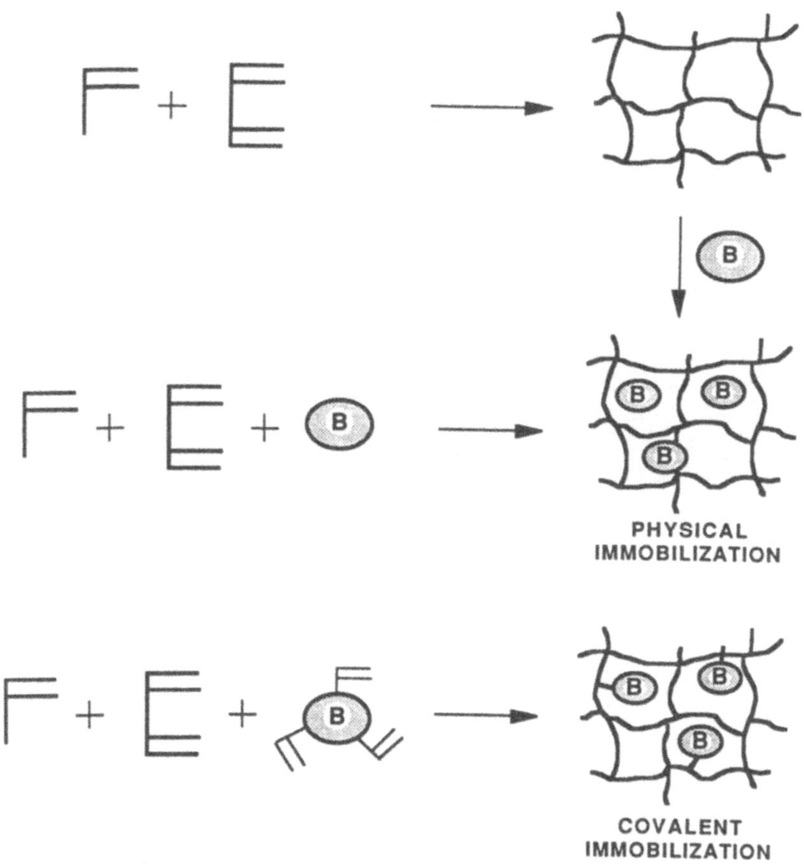

Figure 2. One mode of drug delivery from a temperature-sensitive hydrogel.

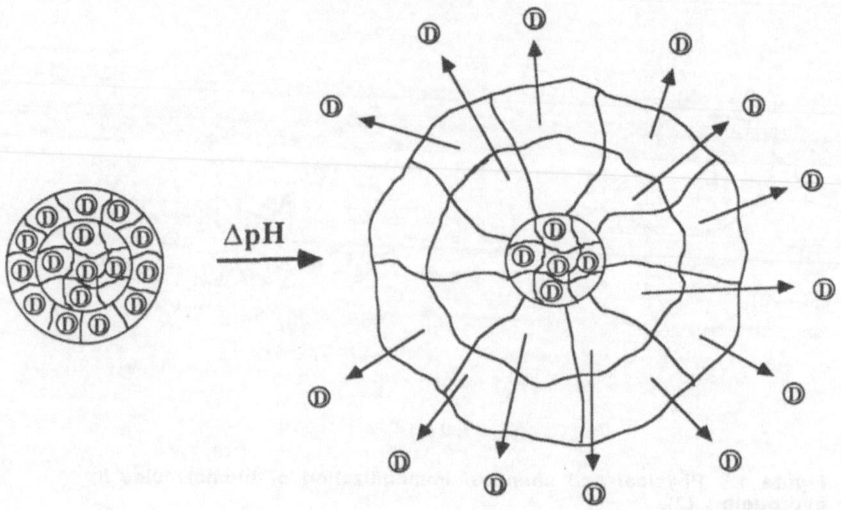

Figure 3. One mode of drug delivery from a pH-sensitive hydrogel.

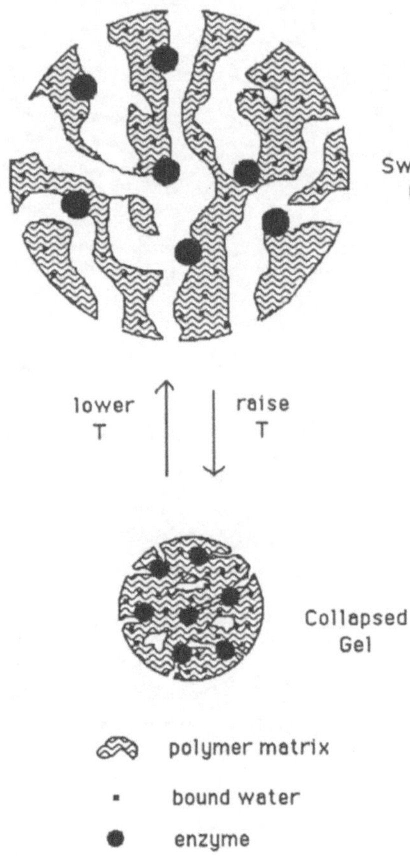

Figure 4. Schematic of collapse and reswelling of a temperature-sensitive gel containing an immobilized enzyme (7).

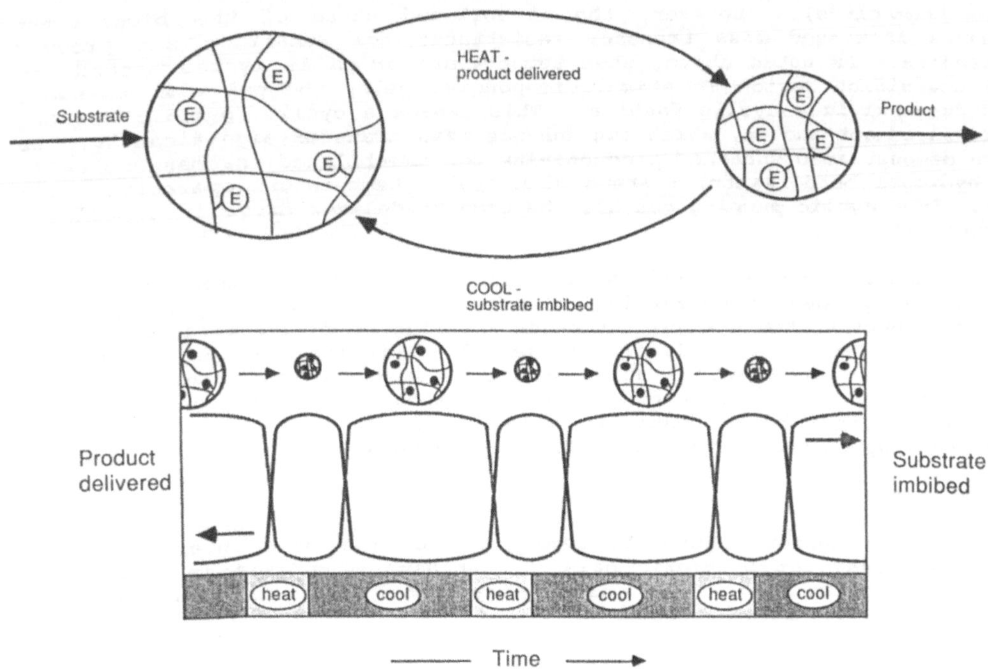

Figure 5. Thermal cycling to enhance productivity of an enzyme (or cell) immobilized in a temperature-sensitive hydrogel.

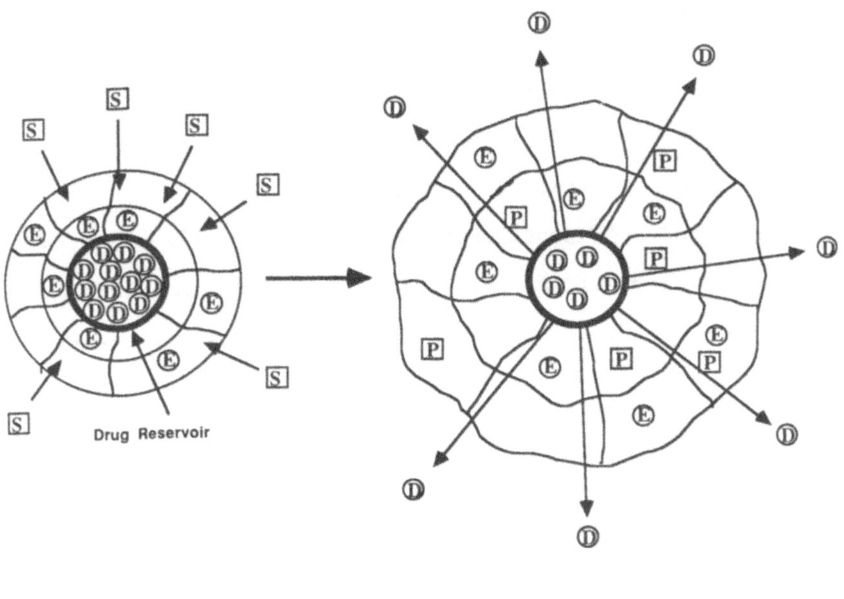

Figure 6. Drug delivery from a pH-sensitive hydrogel containing an immobilized enzyme.

from themselves). However, the immobilized state of the biocatalyst creates increased mass transfer resistances for substrate and product molecules. As noted above, when temperature or pH is cycled around the gel transition region in stimuli-responsive gels, the gel will de-swell and re-swell in a cyclic fashion. This causes a cyclic "pumping" of the pore fluid out and in, which can enhance mass transfer significantly. We have demonstrated enhanced productivity for immobilized enzymes and cells in hydrogel bead reactor systems with cyclic temperature operation. (9-11). Such cyclic pumping can also be used to deliver drugs in a periodic pattern (18, 19).

When selected enzymes are immobilized in a pH-sensitive gel, then changes in pH due to conversion of a neutral substrate to an acidic or basic product within the gel can cause swelling or de-swelling of the gel due to changes in the microenvironmental pH inside the gel. In this case, the environmental stimulus is a change in the substrate concentration. Conversion of glucose to gluconic acid by glucose oxidase immobilized in a basic, cationic hydrogel is one example (20). We are also working on such interesting and complex gels.

Conclusions

In conclusion, hydrogels remain one of the most interesting and diverse classes of polymeric gels. They have many uses in biomedical devices and implants, drug delivery systems, diagnostics, biosensors, bioreactors, and bioseparations.

REFERENCES

(1) Hoffman, A.S., & B.D. Ratner: ACS Symposium Series 31, 1 (1976).

(2) Gombotz, W.R. & Hoffman, A.S.: "Hydrogels in Medicine & Pharmacy", Vol. 1, N.A. Peppas, (Ed.), CRC Press, Boca Raton , FL (1986) 95-126.

(3) Hanson, S.R., Harker, L.A., Ratner, B.D., & Hoffman, A.S.:"Biomaterials 1980", G.D. Winter, D.F. Gibbons, & H. Plenk,(Eds.), J. Wiley & Sons, London (1982) 519-521.

(4) Mori, Y., et.al., Trans. ASAIO, 28, 459 (1982).

(5) Merrill, E.W., and Salzman, E.W., ASAIO J., 6, 60 (1983).

(6) Sheu, M. and Hoffman, A.S., unpublished data.

(7) Hoffman, A.S., J.Contr. Rel., 6, 297-305 (1987).

(8) Dong, L.C. and Hoffman, A.S., J. Contr. Rel., 4, 213-222 (1986).

(9) Park, T.G. and Hoffman, A.S., Appl. Biochem. Biotech., 19, 1-9 (1988).

(10) Park, T.G. and Hoffman, A.S., Biotech Bioeng. (1989) in press.

(11) Park, T.G. and Hoffman, A.S., J. Biomed. Mater. Res. (1989) in press.

(12) Kugo, K., Grashin, J., and Hoffman, A.S., Trans. Third World Biomtls. Congress, Kyoto, April, 1988, p. 108.

(13) Ito, S., Hirasa, O. and Yamauchi, A., Preprints, IUPAC Int'l Symposium, Seoul, June, 1989, p. 477.

(14) Dong, L.C., and Hoffman, A.S., J. Contr. Rel. (1989) in press and Proc. Controlled Rel. Soc., Aug 6-9, 1989, p. 95.

(15) Taylor, L.D. and Cerankowksi I, L.D., J. Poly. Sci., Poly. Chem., 13, 2551 (1975).

(16) Priest, J.H., Nelson, R.G., Murray, S.L., and Hoffman, A.S., ACS Symposium Series 350, 255-264 (1987).

(17) Nightengale, J.A.S., Grashin, J., and Hoffman, A.S., Trans. Soc. for Biomtls., X, N.Y.C., June 1987, p. 56.

(18) Kwon, I.C., Bae, Y.H., Okano, T., Berner, B., and Kim, S.W., Preprints, IUPAC Int'l Symposium, Seoul, June, 1989, p. 275.

(19) Osada, Y., Hasebe, M., Kishi, R., and Umezawa, K., Preprints, IUPAC Int'l Symposium, Seoul, June, 1989, p. 513.

(20) Albin, G.W., Horbett, T.A., Miller, S.R., and Ricker, N.L., J. Contr. Rel., 6, 267-291 (1987).

Thermo-Responsive Polymeric Hydrogels and Their Application to Pulsatile Drug Release

Teruo Okano*,*1, Ryo Yoshida**, Kiyotaka Sakai**, Yasuhisa Sakurai*

* Institute of Biomedical Engineering, Tokyo Women's Medical College
8-1, Kawada-cho, Shinjuku-ku, Tokyo 162, Japan
** Department of Chemical Engineering, Waseda University
3-4-1, Ohkubo, Shinjuku-ku, Tokyo 169, Japan
*1 To whom correspondence should be addressed

ABSTRACT

Poly(N-isopropyl acrylamide)(IPAAm) has characteristics to exhibit a remarkable swelling change in aqueous system in response to step-wise temperature change. Random copolymers of IPAAm with alkyl methacrylate (RMA) having different length of alkyl side chain were synthesized to achieve complete on-off drug permeation and release in response to step-wise temperature. On-off release and permeation of indomethacin were found to be due to surface shrinking layer with increasing temperature. The formation process of these surface shrinking layers were confirmed to be controlled by alkyl side chain length of RMA. From these results, the effects of chemical structure (length of alkyl side chain) on temperature modulated drug permeation and palsatile drug release pattern have been discussed.

1. INTRODUCTION

Stimuli-responsive polymers which change their structure and physical properties in response to external signals have promising potential to achieve intelligent drug delivery system[1,2]. Because, in drug delivery system, stimuli-responsive polymers may be utilized as molecular devices for self regulating drug delivery with functions of sensing, processing and acting, as shown in Figure 1. From this perspective, temperature responsive controlled drug delivery have been studied[3-6]. Normal body temperature, kept constant near 37°C, is often increased by the physiological presence of pathogens and pyrogens. Temperature change is, therefore, one of the interesting modulating stimuli corresponding to anti-hyperpyretic drugs could control drug release in response to increased body temperature in a disease state, then self regulating or auto feed-back drug delivery systems can be achieved. The concept can also be utilized in the design of drug

Polymer Gels, Edited by D. DeRossi *et al.*
Plenum Press, New York, 1991

delivery systems in which drug release rate depends on modulated external temperature. Anti-cancer drug release, for example, can be controlled by the effect of temperature on the release devices. This paper is a summary of our recent work with thermo-responsive polymeric hydrogels and their application to pulsatile drug release. Specifically, the alkyl side chain effects on thermosensitivity in swelling, temperature modulated drug permeation, and palsatile drug release pattern will be discussed.

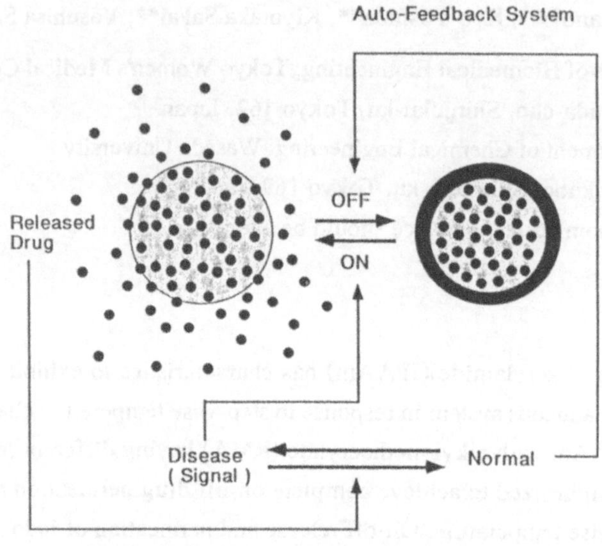

Figure 1 Self regulating drug delivery system.

Table 1 Poly(N-isopropyl acrylamide) copolymers.

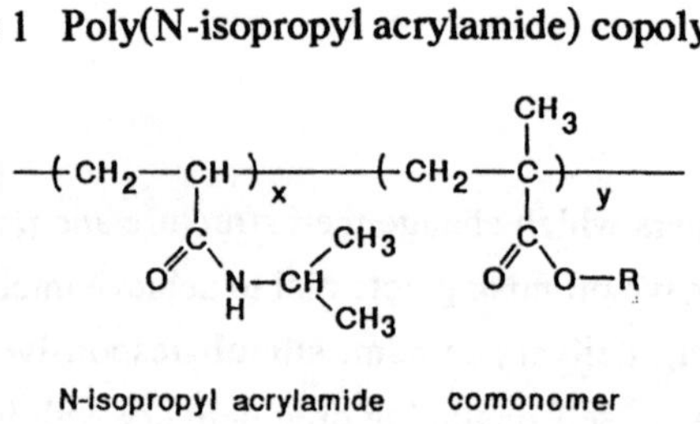

comonomer	R	Abbreviation
butyl methacrylate	C_4H_9	Poly(IPAAm-co-BMA)
hexyl methacrylate	C_6H_{13}	Poly(IPAAm-co-HMA)
lauryl methacrylate	$C_{12}H_{25}$	Poly(IPAAm-co-LMA)

2. THERMO-RESPONSIVE POLYMERIC HYDROGELS

Crosslinked random copolymers of N-isopropyl acrylamide (IPAAm) with alkyl methacrylate (RMA) (5wt% in feed composition) (Table 1) were synthesized using ethyleneglycol dimethacrylate (EGDMA) as a crosslinker, t-butylperoctanoate(BPO) as a initiator, and distilled 1,4-dioxane as a diluent. Butyl methacrylate (BMA), hexyl methacrylate (HMA), and lauryl methacrylate (LMA) were used as RMA monomer.

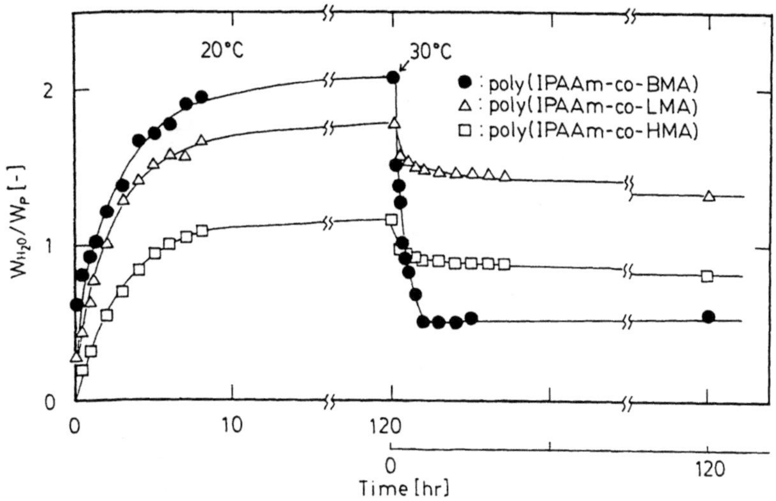

Figure 2 Swelling-deswelling kinetics of IPAAm copolymers between 20°C and 30°C.

Figure 2 shows the swelling-deswelling kinetics of poly(IPAAm-co-RMA) (disk-shaped geometry: 15mm diameter, 0.5mm thickness) in PBS. During the experiment, the temperature was stepwise changed from 20°C to 30°C. Swelling-deswelling kinetics of the thermo-responsive hydrogels can exhibit reversible swelling changes with the temperature change. Swelling rate from dried state (of deswollen state) to swollen state are much faster than the opposite deswelling process from swollen state to deswollen state. The three hydrogels having different length of alkyl side chain showed WH2O/Wp=0.2 at equilibrium state of 30°C. Changes in the deswelling process showed an initial rapid shrinking followed by a slow deswelling. This can be explained as an initial immediate shrinking of the outermost layer (skin formation) of the gel in contact with the stimulating solution, which restricts further bulk water outflow from the interior of the gel. Initial deswelling process from 20°C to 30°C showed remarkable differences among three different gels. More specifically, poly(IPAAm-co-BMA) gel revealed large deswelling change compared to poly(IPAAm-co-HMA) and poly(IPAAm-co-LMA) gels. This result suggested that the formation process of surface shrinking layer was depend on the length of alkyl side chain. Figure 3 shows the schematic illustration of the deswelling process. The thickness of skin formed on poly(IPAAm-co-BMA) gel is much bigger than that of skin formed on poly(IPAAm-co-HMA) and poly(IPAAm-co-LMA) gels. Whereas poly(IPAAm-co-LMA) having the longest alkyl side chain forms a thin, but dense and strong layer on the surface with keeping high water content inside the polymer. These quick responses of the gel surface, skin formation, to temperature changes may be utilized as an on-off switch for drug permeation and release.

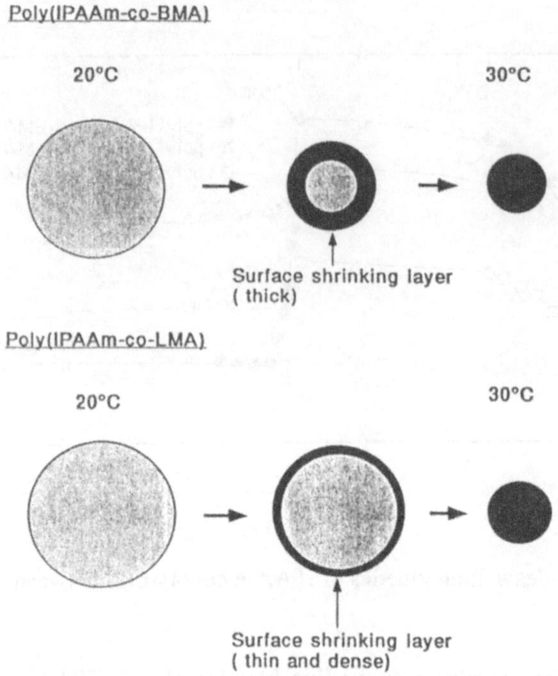

Figure 3 Process for temperature responsive swelling change mediated by surface shrinking layer.

3. TEMPERATURE MODULATED PERMEABILITY CHANGES

Figure 4 shows the permeability change of indomethacin through the poly(IPAAm-co-BMA) gel membrane in response to step-wise temperature change between 20°C and 30°C. The complete on-off control of the drug permeation was observed by the temperature change. The skin formed on the gel surface at 30°C changed from swollen state at 20°C immediately blocks the permeation or leaking of the drug from the gel membrane. By lowering the temperature to 20°C, the deswollen surface quickly recovers its equilibrium swelling and allows permeability of the drug. Same results of on-off permeation were observed for poly(IPAAm-co-HMA) and poly(IPAAm-co-LMA) membranes as shown in Figures 5 and 6. In the case of poly(IPAAm-co-BMA), small lag time was observed by temperature change from 30°C to 20°C. However, poly(IPAAm-co-HMA) and poly(IPAAm-co-LMA) showed no lag time by lowering the temperature to 20°C. In the case of poly(IPAAm-co-LMA), the initial permeation rate during the second 20°C period was greater than the constant permeation rate at steady state during the initial 20°C period. These results suggested that the drug concentration profile in the membrane changed during the "off" period. As noted in the preceding section, poly(IPAAm-co-LMA) formed outermost thin layer on the surface with keeping high water content inside the polymer. Therefore, the drug can diffuse inside the polymer membrane even when the drug permeation was stopped by the surface shrinking layer. During the 30°C period, the linear concentration profile (donor-side concentration > receptor-side concentration) formed in the membrane during the initial 20°C period approached uniform concentration profile by diffusion from the donor side to the receptor side in the membrane. This redistribution of drug gave greater concentration gradient at the receptor-side surface of membrane when the temperature was lowered to 20°C again. Concequently, rapid permeation occured at the initial state of the second 20°C period in the case of poly(IPAAm-co-LMA). It was suggested that the length of alkyl side chain was an important parameter to control the thickness and density of the surface shrinking layer.

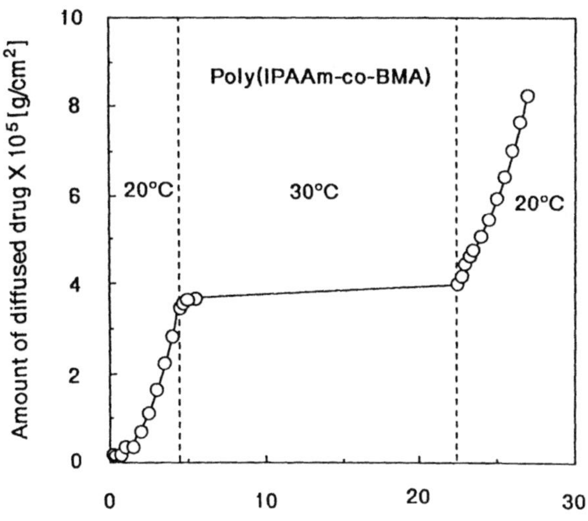

Figure 4 Diffused amount of indomethacin through poly(IPAAm-co-BMA) membrane in response to step-wise temperature change between 20°C and 30°C.

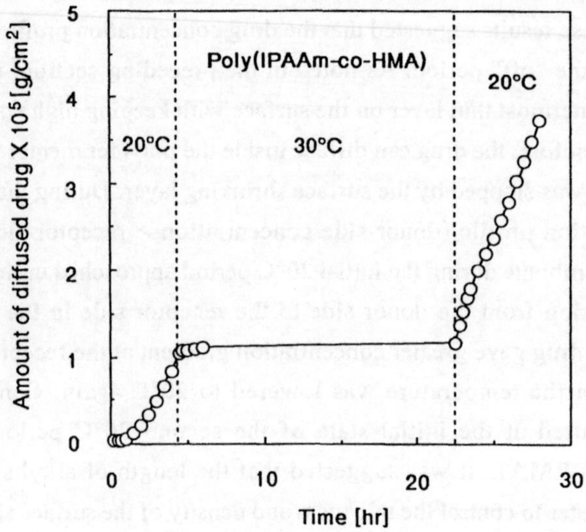

Figure 5 Diffused amount of indomethacin through poly(IPAAm-co-HMA) membrane in response to step-wise temperature change between 20°C and 30°C.

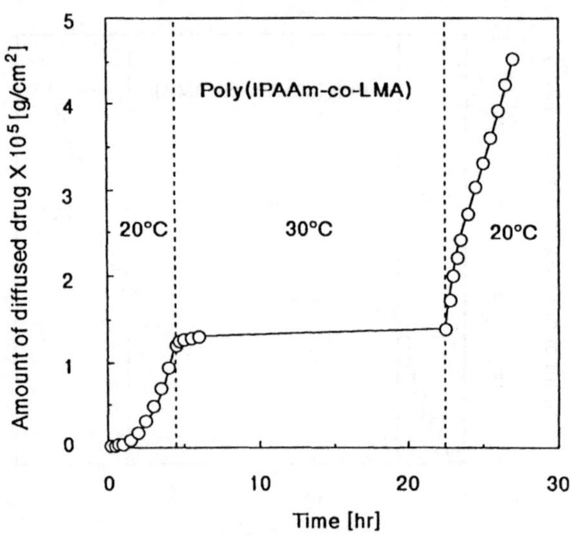

Figure 6 Diffused amount of indomethacin through poly(IPAAm-co-LMA) membrane in response to step-wise temperature change between 20°C and 30°C.

4. ON-OFF RELEASE CONTROL

Figure 7 shows the fractional release amount of indomethacin from poly(IPAAm-co-RMA) gels in response to a step-wise temperature change between 20°C and 30°C. Complete on-off drug release was found to be regulated by the temperature changes. A possible mechanism for on-off drug release was schematically represented in our previous paper [2]. During the "off" state (increased temperature), the surface of polymer may shrink before the bulk material. This action may cause the drug partitioned near the

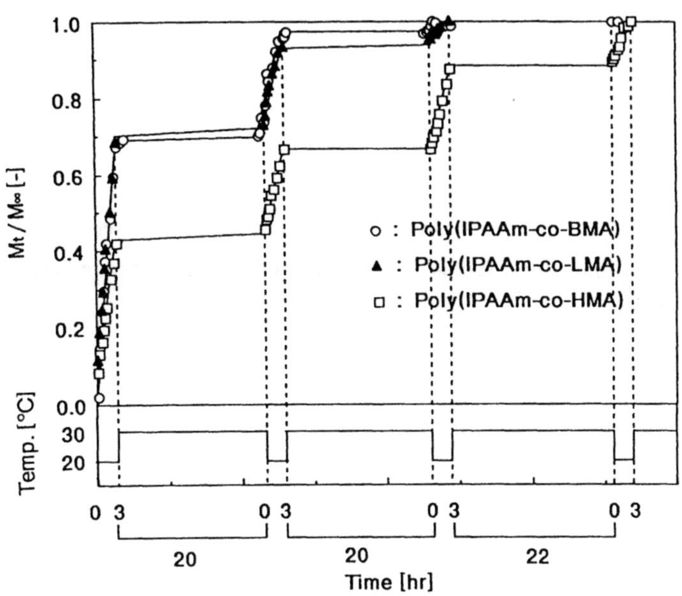

Figure 7 Release profiles of indomethacin from poly(IPAAm-co-RMA) (RMA5wt%) copolymers in response to step-wise temperature change between 20°C and 30°C.

surface to be squeezed out during the surface shrinking to show the initial burst effects. This dense surface structure will then slow or prevent further drug release. During the "on" state (low temperature), the shrunken surface layer starts to reswell and can allow drug release after a certain period, depending on the time for minimum reswelling of the surface. Another possible aspect affecting the drug release pattern during "on" state may be drug solubility at different temperatures and redistribution of drug and water concentration profiles during the "off" period.

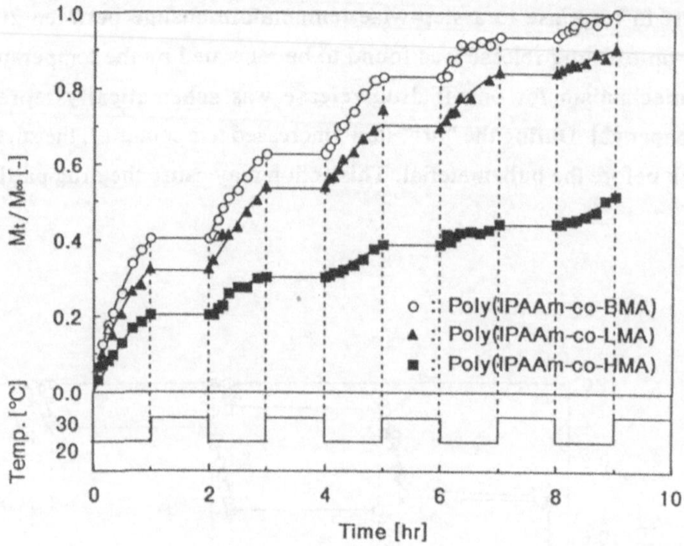

Figure 8 Release profiles of indomethacin from poly(IPAAm-co-RMA) (RMA5wt%) copolymers in response to step-wise temperature change between 20°C and 30°C.

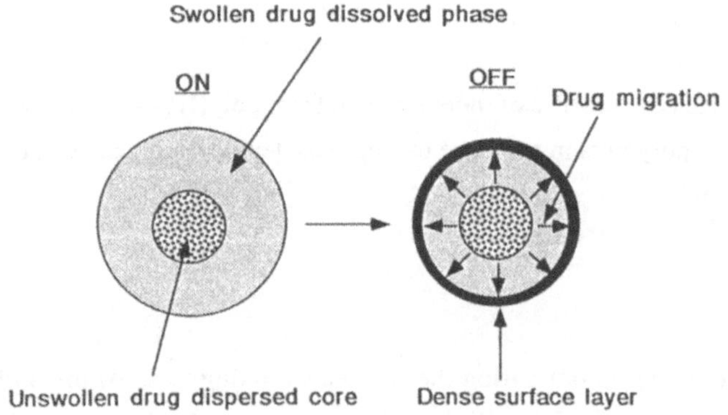

Figure 9 Drug migration in the thermo-responsive hydrogel for "off" state of drug release.

Figure 8 shows the pulsatile release profiles of indomethacin from poly(IPAAm-co-RMA) when "on-off" process was repeated every one hour. The release rate during each "on" state decreased as "on-off" process was much repeated. This result suggested that the drug loaded in the polymer gel migrated from the inside to the surface by diffusion during "off" state even if the drug release was completely stopped. Figure 9 shows the schematic illustration of this drug migration. The lowering of drug concentration gradient near the surface, that is, the decrease in the drug release rate was depressed by the drug migration during "off" state. This redistribution of drug was so time-consuming that the release rate decreased as repeating "on-off" process when it was repeated at short time interval as shown in Figure 8. When the period of "off" state was enlarged, the release rate was maintained constant before and after "off" state as shown Figure 7. Concequently, the polymer forming the thin and dense surface shrinking layer is favorable for on-off drug release and keeping constant pulsatile release rate. The formation process of surface shrinking layer can be regulated by changing the length of alkyl side chain of RMA.

REFERENCES

A significant observation in this study was the formation of dense surface layer during rapid surface deswelling, which was shown to affect subsequent drug release. "On-off" drug permeation and release through poly(IPAAm-co-RMA) were regulated by the swelling/deswelling aspects of the surface, rather than the bulk matrix. An "on-off" temperature regulation of drug release was demonstrated in this experiment, and drug release pattern may be affected by the formation process of surface shrinking layer of polymer gel. The length of alkyl side chain was found to be an important parameter to control the thickness and density of surface shrinking layer in order to achieve complete on-off drug release and constant palsatile release rate.

5. SUMMARY

1) T.Okano, Y.H.Bae and S.W.Kim : "Temperature-responsive controlled drug delivery", in Modulated Control Release System, Ed.by J.Kost, CRC Press, in press.
2) T.Okano, T.H.Bae, H.Jacobs and S.W.Kim : "Thermally on-off switching polymers for drug permeation and release", J.Controlled Release, 11, 255-265 (1990)
3) Y.H.Bae, T.Okano and S.W.Kim : "A new approach for thermo-sensitive hydrogels:Interpenetrating polymer networks of N-acryloyl pyrrolidine and poly(oxyethlene)", Makromol.Chem.Rapid Commun., 9, 185-189 (1988)
4) Y.H.Bae, T.Okano and S.W.Kim : "Thermo-sensitive polymers as on- off switches for drug release", Makromol.Chem.Rapid Commun., 8, 481-485 (1987)

5) Y.H.Bae, T.Okano, and S.W.Kim : "Effect of temperature in aqueous swelling of crosslinked poly(N,N-alkyl substituted acrylamides)", J.Polymer Sci.: Polymer phys. Ed., in press
6) Y.H.Bae, T.Okano, and S.W.Kim : "Insulin permeation through thermo-sensitive hydrogels", J.Controlled Release, 9, 271-279 (1989)

HYDROPHOBIC WEAK POLYBASIC GELS:

FACTORS CONTROLLING SWELLING EQUILIBRIA

Ronald A. Siegel, Bruce A. Firestone, Jose Cornejo Bravo and Barry Schwarz

Departments of Pharmacy and Pharmaceutical Chemistry
University of California, San Francisco, California 94143-0446, USA

INTRODUCTION

The equilibrium and kinetic swelling properties of polymer gels are of considerable interest in the pharmaceutical field. Gels have been proposed as size selective extraction solvents[1] and environmentally sensitive permeability barriers for a variety of pharmacologic agents.[2,3] The use of gel mechanochemistry to drive an insulin pump is also under investigation.[4]

For the past few years, we have been studying the effects of pH, ionic strength, and buffer environment on the swelling properties of a certain class of hydrophobic, polybasic gels. These gels are copolymers of the hydrophobic monomer methyl methacrylate (MMA: 70 mol %) and the tertiary amine-bearing monomer N,N-dimethylaminoethyl methacrylate (DMA: 30 mol %), lightly crosslinked with divinyl benzene (DVB: 0.1% w/w). They are formed by bulk polymerization at 60 °C between narrowly spaced, silanized glass plates, using azobisisobutyronitrile as initiator.[5] Upon formation the gels are glassy. At neutral pH and above, the gels imbibe very small amounts of water, but below a critical pH they swell to an appreciable extent. This property may make these gels useful as drug carriers for oral delivery, since they can be designed to release drugs in specific pH environments.[6]

To utilize these gels optimally in drug delivery and other applications, a thorough understanding of their physical chemistry is necessary. In this article, we present some experimental results concerning swelling equilibria of MMA/DMA gels, and assess the ability of swelling models, developed primarily for hydrophilic systems, to account for the data. We argue that a new class of models will be necessary to describe the swelling behavior of hydrophobic polyelectrolyte gels.

EXPERIMENTAL RESULTS

Swelling equilibria of MMA/DMA copolymer gels were measured as a function of pH and ionic strength in both buffered and unbuffered aqueous media. Figure 1a displays the pH dependence for citrate buffered saline solutions (0.01 M citrate, adjusted to ionic strength I=0.1 M with NaCl), and for unbuffered HCl/NaCl solutions (I=0.1 M). Results are displayed as the <u>equilibrium water fraction</u> (EWF), defined by

Polymer Gels, Edited by D. DeRossi *et al.*
Plenum Press, New York, 1991

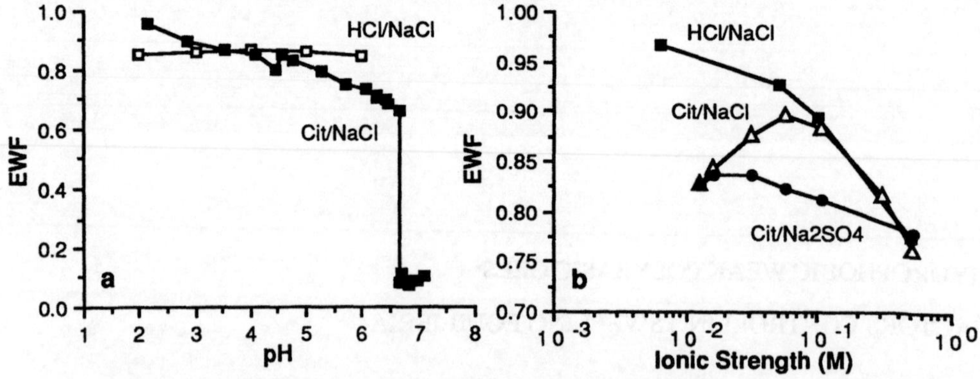

Figure 1. Equilibrium water fraction (EWF) for swelling of MMA/DMA gels. a) Swelling as a function of pH in unbuffered HCl/NaCl and citrate/NaCl solutions at ionic strength 0.1 M. b) Swelling as a function of ionic strength in unbuffered HCl, and citrate buffered solutions with ionic strength adjusted with NaCl or Na_2SO_4.

$$EWF = \frac{\text{weight of water in gel}}{\text{total weight of gel}}$$

The citrate data was gathered between pH 2.0 and 7.2. Above pH 6.6 swelling is limited to EWF=0.08. However, a pronounced swelling phase transition occurs at pH 6.6, where EWF increases abruptly to about 0.65. As pH is decreased further, swelling increases. Data for unbuffered HCl/NaCl was gathered between pH 2 and 6. Within this range swelling is approximately constant at EWF=0.85. The constancy in EWF in the unbuffered solutions in this pH range contrasts with the citrate data.

The effect of ionic strength (I) on swelling equilibrium at constant pH=4 is shown in Figure 1b. Unbuffered HCl/NaCl solutions and citrate (0.01 M) buffered solutions were studied. The ionic strength of the buffered solutions was adjusted with either NaCl or Na_2SO_4. HCl/NaCl yields the highest swelling at a given ionic strength, with EWF decreasing monotonically over the range of ionic strength that was studied. Citrate/NaCl buffers yield intermediate swelling, with a pronounced peak near I=0.05 M. The citrate/Na_2SO_4 gels show the lowest degree of swelling and the weakest dependence on ionic strength.

Specific ion effects on swelling equilibria were investigated using a series of four chloride salts (representing different gel coions) and five sodium salts (representing different

Table I. Equilibrium water fractions for MMA/DMA gels at pH 4.0, ionic strength 0.1 M, with various sodium salts (representing different gel counterions) and chloride salts (representing different gel coions). Measurements were made in duplicate, and variability in measured EWF was always <0.01.

Sodium Salts	EWF	Chloride Salts	EWF
NaSCN	0.34	NaCl	0.88
NaI	0.51	KCl	0.89
$NaNO_3$	0.81	LiCl	0.89
NaBr	0.85	CsCl	0.89
NaCl	0.88		

gel counterions). In all cases pH=4.0 and I=0.1 M. Results are presented in Table I. The coion has virtually no effect on swelling, but different counterions lead to substantially different degrees of swelling. Swelling increase can be correlated with the counterion's position within the Hofmeister, or lyotropic series,[7] with greater swelling occurring with water structure makers (kosmotropes) than with structure breakers (chaotropes).

THEORETICAL CONSIDERATIONS

We have attempted to apply Flory's theory of the swelling of ionic networks[8] to our data. The volume fraction of polymer in the gel at equilibrium, ϕ, can be related to EWF by

$$\phi = \frac{1 - \text{EWF}}{1 + (\rho_p - 1)\,\text{EWF}}$$

where ρ_p is the density of the polymer relative to water ($\rho_p \cong 1.1$ for MMA/DMA). The Flory theory then predicts:[8]

$$\left[\ln(1-\phi) + \phi\right] + \chi\phi^2 + \bar{v}\rho_0\left(\phi^{1/3} - \frac{\phi}{2}\right) = \bar{v}\sum\left(c_i - c_i'\right) \tag{1}$$

In Eq. (1), v is the partial molar volume of water (0.018 L/mol), χ is the Flory interaction parameter, ρ_0 is the molar crosslink density in the initial, dry state, and c_i and c_i' are the molar concentrations of the i'th ion inside the gel and in the outer solution, respectively. The first, bracketed term on the left hand side of Eq. (1) is due to the configurational entropy of mixing of polymer with solvent, the second term accounts for polymer/solvent contacts formed during swelling, and the third term on the left hand side represents polymer elasticity. The right hand side of Eq. (1) accounts for the reduction in chemical potential of water due to mobile ion concentrations inside and outside the gel. This term is analogous to an osmotic pressure, and the form of Eq. (1) is analogous to van 't Hoff's law.

The outer solution ion concentrations c_i' are set by the experimental conditions. In order to calculate the concentrations c_i inside the gel, and ideal Donnan equilibrium is assumed, wherein[9,10]

$$c_i = \lambda^{z_i} c_i' \tag{2}$$

The term λ is the Donnan ratio, calculated as the root of the following equation[5]:

$$\sum z_i c_i' \lambda^{z_i} + \frac{\phi\,\sigma_0}{1 + \lambda 10^{\text{pH-p}K_a}} = 0 \tag{3}$$

In Eq. (3), σ_0 is the density of gel amines in the dry state and pK_a is the ionization constant of the amines, which for DMA is taken to be 7.7.[11] Equation (3) is a statement of electroneutrality inside the gel, which takes into account both mobile ions and protonated amines, where the protonation of amine groups is assumed to follow Langmuir behavior.

Before evaluating this theory, it is useful to point out several simplifications that have been made. First, a rigorous statement of Donnan equilibrium would require concentrations in Eq. (2) to be replaced by activities.[10] Similarly, the right hand side of Eq. (1) should include an osmotic coefficient. The osmotic and activity coefficients are assumed to be unity in the ideal theory. Second, the ionization constant of the amines, K_a, is assumed to be independent of ϕ. This seems unlikely, since the electrostatic environment of the amine groups should change with ϕ, as will be described below.

An immediate discrepancy between the ideal theory and our data can be identified. Ideal Donnan theory prescribes swelling only as a function of the concentration and valences of the ions in the outer solution. However, strong specific counterion effects exist (Table I) which cannot be described by the ideal theory. On the other hand, the nature of the coion does not affect the degree of swelling. The latter observation is in accord with Donnan theory, which predicts that relatively small amounts of coion will be present inside the gel when the gel is highly charged.[8,10]

We now consider applying Eqs. (1)-(3) to the pH-dependent swelling results of Figure 1a. At neutral and high pH there is virtually no charge on the gel. In this case the ion concentrations $\{c_i\}$ inside the gel are expected to lie between zero and $\{c_i'\}$, the corresponding concentrations in the outer solution. Since the crosslink density of these gels is extremely low ($\rho_0 \approx 0.1$ M), the third, elastic term on the left hand side of Eq. (1) can be ignored. Solving Eq. (1) for χ obtains the value 1.7-1.8. This represents extreme hydrophobicity. For such high values of χ, calculations using Eqs. (1)-(3) show that even at very low pH values, where the gels are fully ionized, the Donnan "osmotic," or ionic term on the right hand side of Eq. (3) is at least an order of magnitude smaller than the "mixing entropy" and "contact" terms in that equation, so that gel ionization will have virtually no effect on swelling equilibrium. This prediction is clearly at variance with the data of Figure 1 and Table I.

There are a number of possible explanations for the failure of Eqs. (1)-(3) to account for the experimental observations. One distinct possibility is, that the interaction parameter, χ, is an increasing function of ϕ and hence a decreasing function of EWF. Erman and Flory[12] have demonstrated that critical behavior in gels can occur when this is the case, particularly when the gels are ionized.

A second possible explanation pertains to the dielectric behavior of the gels, which varies with hydration (EWF). At a given hydration, an ion in the gel will see a complex dielectric environment which affects the solvation energy of the ion. At low hydrations, the effective dielectric constant of the gel, ε_{gel}, will approximate that of the pure polymer, i.e., $\varepsilon_{gel} \approx 5$. At higher hydrations, the effective dielectric constant will approach that of pure water, i.e., $\varepsilon_{gel} \approx 80$. When ε_{gel} is low, there will be a large solvation energy for ions entering the gel; hence, they will tend to be excluded. However, when the pH of the outer solution is sufficiently low, mixing entropy will demand that some protons and counterions enter into the gel, with concomitant ionization of the pendant amines. In this case, the large solvation energies of the ions and charged amine groups inside the gel can be reduced by hydration of the gel, which increases ε_{gel}. (Note: this description is purely thermodynamic and does not presume a temporal relation between the ionization and swelling processes.) A simple model, to be presented elsewhere,[13] incorporates a hydration dependent dielectric coefficient with the Born model[14] for ion solvation energies. This model predicts a first order swelling transition without requiring the polymer/water interaction parameter χ to change with hydration. In addition, the model predicts substantial swelling at low pH, despite the high degree of hydrophobicity of the polymer. It should be pointed out that the accurate specification of ε_{gel} as a function of hydration is extremely difficult, so that models of the type described here should not be expected to provide quantitative predictions.

The ideal Donnan theory should be least tenable at low hydrations, since the high polymer volume fraction should lead to significant effects on ion activity coefficients, for reasons discussed in the previous paragraphs. We have tested ideal Donnan theory in the pH range below the swelling transition, where the polymer volume fraction is lower and activity coefficient effects are expected to be less significant.[15] To do so, we chose an arbitrary hydration (EWF=0.7) and calculated the reduced ideal Donnan ion pressure,

$$\Pi_{ion} /RT = \Sigma (c_i - c_i') \tag{4}$$

which can be regarded as the pressure (in excess of the ambient solution pressure) that would be required to maintain the gel at that hydration, divided by RT. Equations (2) and (3) were used to calculate c_i, with ϕ in Eq. (3) corresponding to EWF=0.7. We argue that conditions that predict a higher Π_{ion} at this hydration should lead to a higher ultimate degree of swelling. [Note: this argument does not require a specific model relating hydration to Π_{ion}, such as is prescribed by Eq. (1). One needs only to assume that the opposing "network pressure," which includes effects of mixing and network elasticity, depends only on degree of hydration, and not on the ionization state of the network.[9]] The pK_a of the amines in the gels was assumed to be 7.7, in accord with the data of Shatkay and Michaeli.[11]

Figure 2a shows calculations for pH-dependent conditions corresponding to the data of Figure 1a. Π_{ion} is predicted to be nearly constant between pH 2 and 6 when the outer solution is HCl/NaCl, but Π_{ion} increases with decreasing pH for citrate/NaCl solutions. These observations are in accord with the data of Figure 1a, although the observed citrate gel swelling surpasses that of the unbuffered gels at low pH. This discrepancy might be attributed to slow hydrolysis of the gel at low pH in citrate buffer.

The trends predicted in the theoretical curves of Figure 2a are readily explained in terms of Donnan equilibrium and mass action concepts. Considering first the HCl/NaCl curve, we note that as pH decreases, the charge density on the gel will increase, which will increase the ionic swelling pressure. However, as pH is lowered below 6 the gel becomes fully ionized, and since the NaCl concentration is virtually constant, swelling should be nearly constant below pH 6. The gel ionization behavior will be approximately the same for

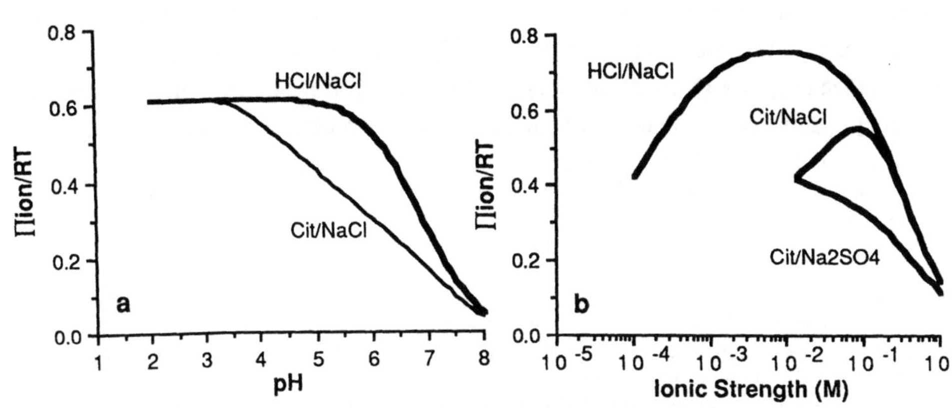

Figure 2. Predicted values of the reduced ion osmotic pressure, Π_{ion}/RT, for conditions corresponding to the swelling data in Figure 1. Concentrations of ions inside gel calculated according to Eqs. (2) and (3), with ϕ corresponding to EWF=0.7 and the pK_a of the amines set to 7.7.[11]

the Citrate/NaCl buffered system. However, citrate exists in three states of ionization, with pK_a's 3.15, 4.78, and 6.40. Therefore, at high pH the divalent and trivalent species will predominate, and these ions will be preferentially absorbed into the gel to neutralize the protonated amines. This will lead to a relatively small ion swelling pressure, since a multivalent ion will neutralize more than one amine. As pH is lowered, the citrate ions shift to lower degrees of ionization. This will lead to an increase in ion swelling pressure, since more ions must be absorbed into the gel to maintain electroneutrality.[16]

Figure 2b shows predictions of Π_{ion} for conditions corresponding to the ionic strength data of Figure 1b. Again, the behavior of the theoretical curves in Figure 2b is qualitatively similar to that of the corresponding experimental curves of Figure 1b, at least over the range where experimental points were recorded. The HCl/NaCl data does not extend to a sufficiently low ionic strength to observe the predicted maximum in the curve.

The theoretical predictions of Figure 2b have straightforward explanations.[5] Consider first the predictions for HCl/NaCl. At very low ionic strength, little ionization of the amines can occur since electroneutrality requires transfer of a chloride ion into the gel for every ionized amine. The localization of many chloride ions into the gel from a solution that is sparse in such ions would lead to a large loss in entropy, so such a process will not occur. However, as ionic strength is increased, more chloride ions are available to enter the gel, so that gel ionization can increase. During this phase swelling pressure will increase with ionic strength. Eventually the gel will become fully ionized, and further increases in ionic strength only cause ion concentrations inside and outside the gel to equalize, leading to a decrease in swelling pressure.

The peak in the Citrate/NaCl predictions and data can be explained by recalling that at pH 4 citrate will exist in a partly divalent form, and the divalent citrate ions will be preferentially absorbed compared to the monovalent citrate ions. With added NaCl, some of the divalent citrate ions will leave the gel, to be replaced by twice as many chloride ions. This leads to an increase in ion swelling pressure. Eventually this exchange process becomes complete, however, and further increase in NaCl concentration simply equalizes ion concentrations inside and outside the gel, as before. When Na_2SO_4 is added to citrate, divalent sulfate ions now exchange with divalent citrate, so no swelling pressure increase is predicted at low ionic strength. In fact, the sulfate ions will also exchange with monovalent citrate ions, thus lowering the swelling pressure. (In this regard the slight but noticeable peak for the Citrate/Na_2SO_4 data in Fig. 1b is not expected.)

A more stringent test of ideal Donnan theory is to use the measured EWF to calculate Π_{ion}/RT, and then plot Π_{ion}/RT versus EWF.[9] The ideal Donnan theory might be considered plausible if the relationship between Π_{ion} and EWF does not depend on the ionic conditions leading to a given value of Π_{ion}. In Figure 3 it is shown that the relationship between Π_{ion} and EWF depends on whether pH or ionic strength is varied, at least for values of EWF below about 0.85. For swelling above this value, Π_{ion} appears to be a reasonably unique function of EWF. From this data it appears that the ideal Donnan theory is invalid at lower degrees of swelling, but becomes more appropriate when the gels are more highly swollen.

As noted previously, the ideal Donnan theory does not take into account nonidealities in ionic interactions which would give rise to nonunity activity coefficients. These effects would be most expected when the volume fraction of polymer is appreciable. Thus it is not surprising that the ideal Donnan theory becomes more valid when conditions lead to high swelling. However, it should be noted that ideal Donnan theory is expected to break down again for gels that are highly charged and more highly swollen than the gels considered in the present work. Katchalsky and Michaeli[17-19] showed that the swelling behavior of highly swollen methacrylic acid gels cannot be accounted for by the ideal theory. An improved theoretical description incorporated an additional electrostatic term into the free energy of the

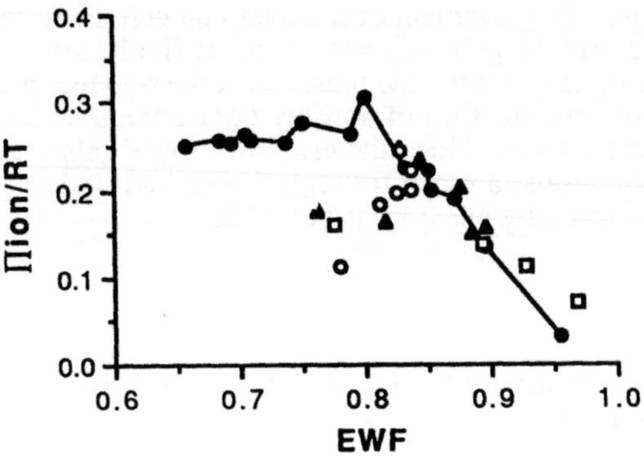

Figure 3. Plots of Π_{ion} versus EWF. Π_{ion} is calculated using ideal Donnan theory based on pH, ion composition in the outer solution, and the measured EWF using Eqs. (2) and (3). (●) pH varied in citrate/NaCl buffered solutions and unbuffered HCl/NaCl solutions at ionic strength 0.1 M; (□) pH 4 HCl/NaCl solutions at various ionic strengths; (▲) pH 4 citrate/NaCl solutions at various ionic strengths; (○) pH 4 HCl/Na_2SO_4 solutions at various ionic strengths.

gel. This term, which corresponds to the "polyelectrolyte effect," engenders two opposing contributions to the force balance that determines swelling equilibrium. The first effect is an enhanced swelling due to electrostatic repulsions between charge groups fixed to the gel. The second effect is a decreased swelling due to the localization of counterions near the fixed charges, which reduces the osmotic activity of the counterions. Both effects are associated with a nonuniformity in the electrostatic potential inside the gel which arises when highly charged chains are separated by distances comparable to or larger than the Debye screening length.[10,17-20] These effects probably contribute negligibly when the charge density is low, such as in highly swollen, lightly charged acrylamide gels,[9,16] but they can make significant contributions when the gel has a high charge density and is very highly swollen.[17-19,21] Interestingly, it appears that at high charge densities and intermediate degrees of swelling (the case corresponding to the present set of data) the polyelectrolyte effect can probably be discounted, since the distances between charged polymer chains may be smaller than the Debye screening distance, so variations in the electrostatic potential in the gel may be quite small.[10,20] This could explain the consistency of the ideal Donnan model for the data we observed with EWF>0.85.

It has been observed above that the specific counterion effects displayed in Table I also argue strongly against the ideal Donnan theory. A tentative explanation for these effects is as follows. Chaotropic ions will tend to adsorb to the gel network. This will lower their activities substantially, and reduce their contribution to Π_{ion}. On the other hand, kosmotropes will adsorb to a lesser degree, and hence their contribution to the ion osmotic pressure will be closer to ideal. This mechanism is consistent with the results of column chromatographic studies, in which the elution times of chaotropic ions were seen to be smaller than the elution times for kosmotropes.[22]

SUMMARY

To summarize, theories developed to describe swelling behavior of hydrophilic, lightly ionized gels are less successful in predicting the behavior of hydrophobic, highly ionized gels, although some qualitative trends are correctly predicted when the gels are in a relatively swollen state. Development of a model that can quantitatively predict swelling in hydrophobic polyelectrolyte gels appears to be difficult, since the degrees of swelling achieved are sufficiently low that the the presence of the polymer itself affects the activities of the ions. The precise specification of activity coefficients of ions in such heterogeneous phases is an unsolved problem. Nevertheless, since these gels show unusual properties and have promising applications, a more through theoretical understanding of the balance of forces governing their swelling behavior is desirable.

ACKNOWLEDGMENT

This work was funded in part by N.I.H. grant DK 38035 and a predoctoral N.I.H. training grant to BAF (GM 07175).

REFERENCES

1. E. L. Cussler, M. R. Stokar and J. E. Varberg, Gels as Size Selective Extraction Solvents, A.I.Ch.E. J., 30:578 (1984).
2. J. H. Kou, G. L. Amidon and P. I. Lee, pH-Dependent Swelling and Solute Diffusion Characteristics of Poly(Hydroxyethyl Methacrylate-co-Methacrylic Acid) Hydrogels, Pharm. Res., 5:592 (1988).
3. K. Ishihara, M. Kobayashi and I. Shinohara, Insulin Permeation through Amphiphilic Polymer Membranes Having 2-Hydroxyethyl Methacrylate Moiety, Polym. J., 16:647 (1984).
4. R. A. Siegel and B. A. Firestone, Mechanochemical Approaches to Self-Regulated Insulin Delivery, J. Controlled Release, in press (1990).
5. R. A. Siegel and B. A. Firestone, pH-Dependent Equilibrium Swelling Properties of Hydrophobic Polyelectrolyte Copolymer Gels, Macromolecules, 21:3254 (1988).
6. R. A. Siegel, M. Falamarzian, B. A. Firestone, and B. C. Moxley, pH-Controlled Release from Hydrophobic/Polyelectrolyte Hydrogels, J. Controlled Release 8:179 (1988).
7. K. D. Collins and M.W. Washabaugh, The Hofmeister Effect and the Behavior of Water at Interfaces, Q. Rev. Biophys., 18:323 (1985).
8. P. J. Flory, "Principles of Polymer Chemistry," Cornell, Ithaca (1953).
9. J. Ricka and T. Tanaka, Swelling of Ionic Gels: Quantitative Performance of the Donnan Theory, Macromolecules, 17:2916 (1984).
10. J. Th. G. Overbeek, The Donnan Equilibrium, Prog. Biophys. Biophys. Chem., 6:57 (1956).
11. A. Shatkay and I. Michaeli, Potentiometric Titrations of Polyelectrolytes with Separation of Phases, J. Phys. Chem., 70:3777 (1966).
12. B. Erman and P.J. Flory, Critical Phenomena and Transitions in Swollen Polymer Networks and in Linear Macromolecules, Macromolecules, 19:2342 (1986).
13. B. Schwarz and R. A. Siegel, in preparation.
14. A. A. Rashin and B. Honig, Reevaluation of the Born Model of Ion Hydration, J. Phys. Chem., 89:5588 (1985).
15. B. A. Firestone, Swelling and Transport Properties of Hydrophobic Polyelectrolyte Gels. Ph.D. Thesis, University of California, San Francisco (1989).
16. I. Ohmine and T. Tanaka, Salt Effects on the Phase Transition of Ionic Gels, J. Chem. Phys., 77,5725 (1982).
17. A. Katchalsky, Polyelectrolyte Gels," Prog. Biophys., 4:1 (1954).
18. A. Katchalsky and I. Michaeli, Polyelectrolyte gels in salt solutions, J. Polym. Sci., 15:69 (1955).
19. I. Michaeli and A. Katchalsky, Potentiometric Titration of Polyelectrolyte Gels, J. Polym. Sci., 23:683 (1957).
20. A. J. Grodzinsky. Electromechanical and Physicochemical Properties of Connective Tissue, CRC Crit. Rev. Bioeng. 9:133 (1984).

21. J. Hasa, M. Ilavsky, and K. Dusek, Deformational, Swelling, and Potentiometric Behavior of Ionized Poly (methacrylic Acid) Gels. I. Theory, J. Polym. Sci., 13:253 (1975).
22. M. W. Washabaugh and K. D. Collins, The Systematic Characterization by Aqueous Column Chromatography of Solutes Which Affect Protein Stability, J. Biolog. Chem. 261:12477 (1986).

Diffusion-reaction through Coated Gel Containing Enzyme
Effects of Asymmetry and Sol-Gel Transition

Masakatsu Yonese, Kazuhiko Baba, Hideki Murabayashi

Faculty of Pharmaceutical Sciences, Nagoya City University
Tanabe-dori, Mizuho-ku, Nagoya 467 Japan

ABSTRACT

Diffusion-reactions through coated gels containing an enzyme were studied from basical point of views by using a multimembrane, consisting of an enzyme layer sandwiched between support films. An urease and a glucoamylase were used as enzymes, and alginates (Alg) gelled by adding Ca^{2+} ions were used as an adding polymer. By measuring membrane potentials of the multimembranes, the changes of the charge densities due to the sol-gel transition of Alg were discussed and the network effects were found to enhance the Ca ion bindings. Total effluxes and efflux ratioes of products through the multimembrane were studied, and effects of adding Alg, its sol-gel transition and of the asymmetries on the diffusion-reactions were discussed.

INTRODUCTION

Hydrogels have been applied widely to pharmaceutical field, such as drug delivery systems and matrices of immobilized enzymes. Recently, they have attracted special interest[1] as promising materials due to the developments of studies in the gel properties, such as phase transition[2] and mechanochemical reactions.[3] We have paid attention to coated gels in which enzymes are immobilized. They can be expected to possess multifunctions resulting from the characteristics of gels, the coated films and the enzyme reaction, such as the laminated glucose-oxidase membrane which is an insulin delivery system sensitive to glucose[4], and the arranged multienzyme system (hexokinase and phosphatase) which concentrates glucose.[5]

Polymer Gels, Edited by D. DeRossi *et al.*
Plenum Press, New York, 1991

This report discusses diffusion-reaction through the coated gel membranes containing enzyme from basical point of views to make possible wider applications to pharmaceutical fields. The multimembrane, as shown in Fig.1, consisting of an enzyme layer sandwiched between supporting films, Solution I|Film 1|Enzyme Layer|Film 2|Solution II, is very convenient to basic studies of the diffusion-reaction, and has following characteristics; i.e., (1) the microenvironments around the enzyme can be altered arbitrarily by adding polymers or liposome etc, and especially the effects of sol-gel trandition around the enzyme can be studied. (2) By using various supporting films, the asymmetries of the

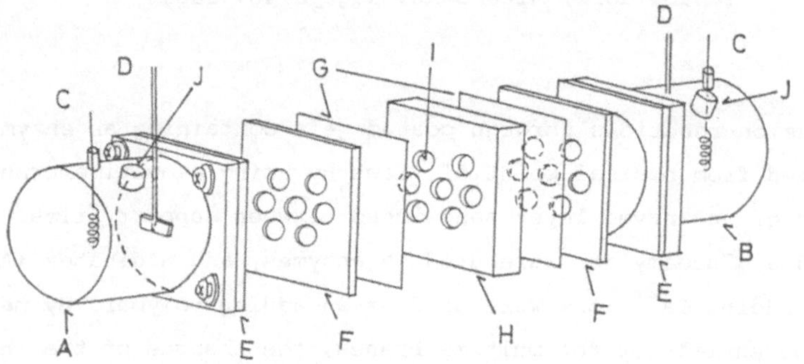

Fig.1 Scheme of Multimembrane

A,B; compartment I and II, C; Ag/AgCl electrode, D; stirring bar, E; silicone rubber, F; perforated support plate, G; support film(first and third layer), H; perforated plate, I; sample injection hole (second layer), J; inlet for solution injection

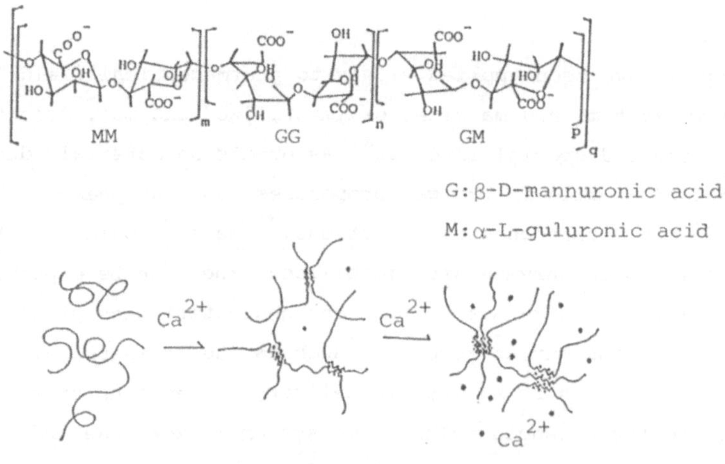

Fig.2 Schematic Chemical Structure and Sol-Gel Transition of Alg

multimembrane can be controlled easily. In the study of the diffusion-reaction of urea through the multimembrane containing urease, the effects of adding alginate and the asymmetries were reported.[6] Together with these results, this report discusses the effects of sol and gel states around enzymes on diffusion-reaction. Alginates used as an adding polymer are gelled, as shown in Fig.2, by the presence of divalent cation such as Ca^{2+} and are convenient to study the effects of sol-gel transition of the matrices[7].

EXPERIMENTALS

Materials. Urease was of commercial origin (Cooper Biochemical, 82U/mg, Lot No. G4P6952) extracted from soy beans. Glucoamylase, GA, was also of commercial origin (Funakoshi Yakuhin Co. Ltd., Japan, 30 Units/mg) produced by Rhizopus. They were used without any purification. Purified sodium alginate(NaAlg) used as an adding polymer was prepared from commercial NaAlg (Tokyo Kasei Kogyo Co. LTD.) by dializing against distilled water and by filtrating to remove insoluble substances. The purified NaAlg was stored in refrigerator after freeze drying. The weight average molar mass of NaAlg was determined to be M_w=1.32 10^5 g mol^{-1} by using a light scattering photometer LS-8 (TOYO SODA MANUFACTURING Co., Ltd., Japan). The gluronate fraction F_G and the consecutive gluronate one F_{GG} were determined to be 0.37 and 0.28 by CD[8]) and H-NMR[9]) methods. Urea, ammonium chloride and all other reagents were of special grade (Katayama Kogyo Co. Ltd.). Distilled and deionized water was used for preparation of aqueous solutions. The pH values of solutions were adjusted by using Maleic acid- Tris buffer or acetate buffer solutions. The former was ranged from pH=6.0 to pH=8.0 and the latter was ranged from pH=3.5 to pH=6.0. All other reagents were special grade.

As the support films sandwiching the enzyme layer of the multimembrane, three kinds of films were used, i.e., Spectrapor 2 (SP) (0.0033 cm in thickness, Spectrum Medical Industries, Inc.) which is a kind of cellulosic dialysis membrane, dialysis membrane CM (0.0033 cm in thickness, Seamless Cellulose Tubing, Union Carbide Corp.) and Selemion DMV (SDMV) (Asahi Garasu Co. Ltd.) which is an anion exchange membrane.

<u>Multimembrane for Measuring Membrane Potential and Diffusion Reaction through Alg Layer</u>

The multimembrane cell shown in Fig.1 was used for measuring membrane potentials of Alg layer and diffusion-reaction through Alg layer containing the enzyme. The cell was made of acrylic resins, and consisted of a polymer solution layer sandwiched between support films: Solution I |Film 1|Polymer Layer|Film 2|Solution II. Alg solutions were put into holes (I) (Radius = 0.25 cm) in a perforated acrylic resin plate (H) (0.049, 0.30 or 0.50 cm thick , with 7 holes in it). To prevent the deformations of the support films due to the osmotic pressure of the polymer layer and its vibration due to the stirring of the solution I and II, the support films on both sides were fixed by means of perforated acrylic resin plates (F) (thickness:0.049 or 0.5 cm). The volumes of compartment I and II were 51-52 cm^3, in which the solutions I and II (50cm^3) were poured. Total permeable area of the multimembrane was 1.374 cm^2.

Measurements of Membrane Potentials

The membrane potentials E between the solution I and II were measured by Ag/AgCl electrodes whose electrode potentials were obtained by using Kieland's individual activity coefficients.[10] Concentration ratios of electrolyte solutions in the compartment I and II, R ($=C^{I}/C^{II}$) were 10 or 2. Both solutions were stirred at 500 r.p.m. by stirring bars (D) so as to attain a constant thickness of stagnant layer (= 40 μm), which was obtained from measuring the permeability coefficients of NaCl through the single and dual films. To obtain the effective charge density of the support films CM, the membrane potentials were measured in the following systems; Solution I|CM|Solution II. All membrane potentials were measured at 25°C.

Measurement of Diffusion Reaction

Urease, GA solution (0.06 w/v%) or the mixed solution composed of one of these enzymes (0.06 w/v%) and NaAlg (0.6 w/v%) were put into holes (I) as the enzyme layer. The compartment I and II were usually a donor and an acceptor part respectively. The initial concentration of the urea or maltose as a substrate in the compartment I was usually 0.01 mol dm^{-3}. Time courses of concentrations of products C_p in the compartments I and II flowed out from the enzyme layer were measured by using ammonium electrode or the glucostat method. The linear relations between C_p and time t were obtained under steady state and from their slopes the fluxes of the products into the compartment I, J_p^{I}, and II, J_p^{II} mol cm^{-2} s^{-1}, were determined. The direction of flux from the compartment I to II is positive.

The gelation of NaAlg in GA layer was performed for 1 day by adding $CaCl_2$ (0.1 M) to buffer solution in the compartment I and II. In the experiments of the diffusion-reaction, $CaCl_2$ was added to the solution I (maltose solution) and II (buffer solution) to keep the conditions of the gel constant. Their concentrations of $CaCl_2$ were 0.1 mol dm^{-3}. The effluxes of the glucose into the both compartments were measured by the same method as in the case in which the GA layer was the sol state.

Theory of Membrane Potential through Multimembrane

The membrane potential E through the multimembrane is the summation of interfacial potentials generated at each interface and the diffusion potential in each layer. According to the notations in Fig.3, E is expressed by the following equation,

$$E = \sum_{j=0}^{3} E_B^{j,j+1} + \sum_{j=1}^{3} E^{*j} \tag{1}$$

where $E_B^{j,j+1}$ and E^{*j} are Donnan potential at each interface and diffusion potential in each layer, and superscripts j and (j,j+1) denote the jth layer and the interface between j and (j+1)th layer. The superscripts 0 and 4 denote the solution I and II, and first and third layers are support films and second one is the polyelectrolyte solution layer. Denoting the concentration of jth layer at the interface j and (j+1)th one by $\bar{C}^{j,j+1}$, and the effective charge density of the jth layer by θ^{*j}, E can be expressed by

$$E = \frac{RT}{F}\left[\sum_{j=0}^{3} \ln \frac{\bar{C}_A^{j+1,j}}{\bar{C}_A^{j,j+1}} + \left\{\left(\frac{1}{z_M} + \frac{1}{z_A}\right)T_A^{*j} - \frac{1}{z_M}\right\} \times \sum_{j=1}^{3} \frac{\bar{C}_A^{j,j-1} + \frac{1}{z_A}(1-T_A^{*j})\theta^{*j}}{\bar{C}_A^{j,j+1} + \frac{1}{z_A}(1-T_A^{*j})\theta^{*j}}\right] \tag{2}$$

where T^{*j} is transference number of anion A in jth layer. Donnan equilibrium and the electroneutrality at each interface are expressed by the following equations,

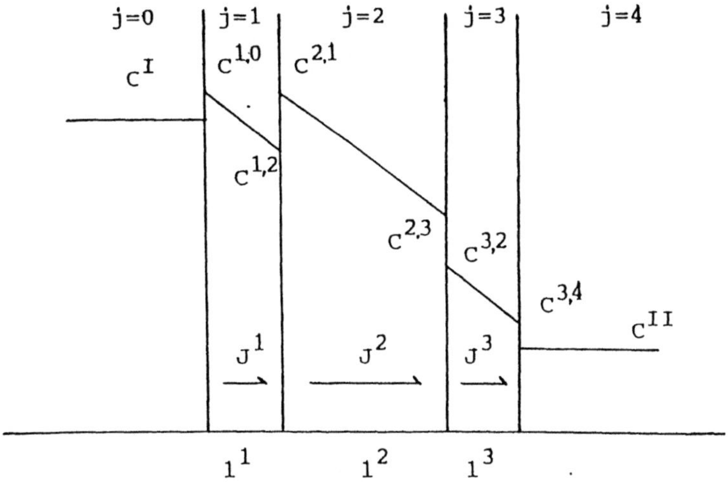

Fig.3 Notations of Concentrations for Multimembrane System

$$(\overline{C}_A^{j+1,j}/\overline{C}_A^{j,j+1})^{1/z_A}=(\overline{C}_M^{j+1,j}/\overline{C}_M^{j,j+1})^{1/z_M} \quad (3)$$

$$z_M\overline{C}_M^{j,j+1}+z_A\overline{C}_A^{j,j+1}+\theta^{*j}=0 \quad (4)$$

$$z_M\overline{C}_M^{j+1,j}+z_A\overline{C}_A^{j+1,j}+\theta^{*j+1}=0 \quad (5)$$

$$j=0,\ 1,\ 2,\ \text{and}\ 3$$

where θ^{*0} and θ^{*4} denoting the values of solution I and II are equal to 0, and the subscript M denotes a cation. When θ^{*1} and θ^{*3} are known, $\overline{C}_A^{1,0}$, $\overline{C}_M^{1,0}$, $\overline{C}_A^{3,4}$ and $\overline{C}_M^{3,4}$ can be obtained by Eqs.(3), (4) and (5) as a function of C^I or C^{II}. Under a steady state condition, the fluxes of ions J_i^j are equal in each layer according to the continuity law,

$$J_i^1 = J_i^2 = J_i^3 \quad (6)$$

Approximating that concentrations of ions in each layer vary linearly,[11] the fluxes of anion A are expressed by[12]

$$J_A^j = -P_A^j\,\frac{\overline{C}_A^{j,j+1}-\overline{C}_A^{j,j-1}}{l^j}\,(1+X) \quad (7)$$

$$X=\frac{\dfrac{z_M}{z_M-z_A}\,\theta^{*j}\left(\dfrac{1}{z_M}-\dfrac{1}{z_A}\right)T_A^{*j}}{\overline{C}_A^{j,j+1}-\overline{C}_A^{j,j-1}}\times\ln\frac{\dfrac{z_M-z_A}{z_M}\,\overline{C}_A^{j,j+1}-\left(\dfrac{1}{z_M}-\dfrac{1}{z_A}\right)(1-T_A^{*j})\theta^{*j}}{\dfrac{z_M-z_A}{z_M}\,\overline{C}_A^{j,j-1}-\left(\dfrac{1}{z_M}-\dfrac{1}{z_A}\right)(1-T_A^{*j})\theta^{*j}}$$

$$j=1,\ 2,\ \text{and}\ 3 \quad (8)$$

where P_A^j is a permeability coefficient of anion in j layer, and l^j is the thickness of j layer. So as to satisfy Eq.(6), $\overline{C}_A^{1,2}$, $\overline{C}_A^{2,1}$, $\overline{C}_A^{2,3}$ and $\overline{C}_A^{3,2}$ can be obtained from Eq.(7) by numerical analysis. Then, The membrane potentials E can be obtained from Eq.(2).

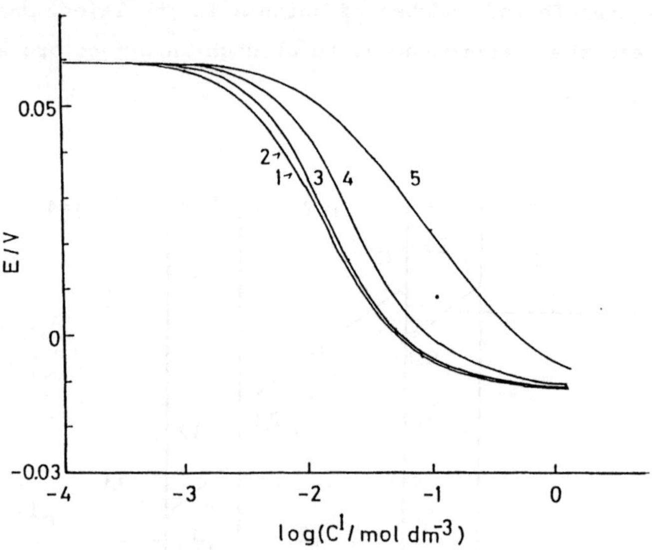

Fig.4 Effects of Effective Charge Densuties of Support Films on Membrane Potentials of Multimembrane for NaCl Solutions (R=10)

Effective charge density of the second layer θ^{*2}:-0.01, Those of first and second layers $\theta^{*1}=\theta^{*3}$: 1;0, 2;-10^{-4}, 3;-10^{-3}, 4;-0.01, 5;-0.1 mol.dm^3.

The effects of support films on the membrane potential of the multimembrane can be discussed theoretically[13)] Figures 4 shows one of the calculated membrane potentials, i.e., the effects of θ^{*2} on E. The effective charge densities of support films are $\theta^{*1}=\theta^{*3}=-0.01$ M and that of the second layer θ^{*2} is varied from 0 to -0.1. In the It should be noted that the membrane potentials are affected by not only effective charge densities of these three layers but also diffusivities of ions through them. As decreasing the effective charge density of a support film $\theta^{*1}(=\theta^{*3})$ under constant θ^{*2}, membrane potentials E decrease abruptly in lower concentration region. Even in the cases that θ^{*2} are smaller than $\theta^{*1}(=-0.01)$, such as $\theta^{*2}=-0.001$ and -0.005, the differences between their membrane potentials are found to be clear, and then even in these cases the values θ^{*2} can be obtained by the curve fitting method.

RESULTS

Effective Charge Densities of Alginates

Membrane potentials of the multimembranes (|CM|Alg Solution|CM|) were measured for NaCl and $MgCl_2$ solutions whose concentration ratio R was 10. Alg solutions in the membrane were in the range of $W_{Alg}=$ 0.00250 to 0.00626 g cm^{-3}. As shown in Fig.5, the results E showed abrupt decreases in the higher concentration regions as increasing W_{Alg}. Solid curves show theoretical values obtained by selecting the most suitable values for the effective charge densities of alginate solutions, θ^{*2}, so as to agree with experimental results. The results θ^{*2} are shown in Table 1, in which the effective charge densities of support films CM, $\theta^{*1}(=\theta^{*3})$, and transference numbers of anion T^{*j} used in calculations are also shown in Table 1. The absolute values θ^{*2} or NaCl solution were bigger than those for $MgCl_2$, and they increased as increasing W_{Alg}.

In the cases of NaCl and $MgCl_2$ solutions, Alg were in sol states. The membrane potentials of Alg in gel state were measured as follows. Alg sols in the second layer were transited to gel by putting $CaCl_2$ solutions (1.0 mol dm^{-3}) into both compartments I and II, and before measuring the membrane potentials free ions in CaAlg gel layer were removed by dialysis against water. Figure 6 shows the membrane potentials of the multimembranes containing CaAlg gels for NaCl and $CaCl_2$ solutions (R=10) and the theoretical curves obtained by using the most suitable effective charge densities θ^{*2} which are shown in Table 1. Absolute values of the effective charge density of CaAlg were

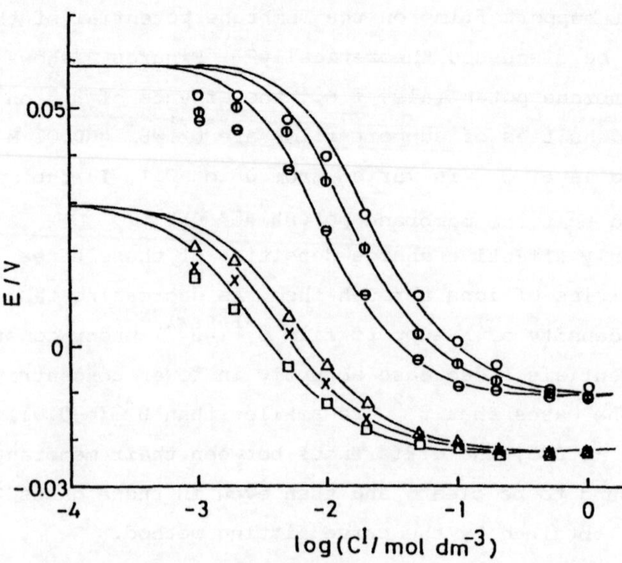

Fig.5 Concentration Effects of Alginates on Membrane Potentials of Multimembrane for NaCl and $MgCl_2$ Solution (R=10)

W_{Alg} (for NaCl Solution): ○ ;0.00625, ⦶ ;0.00420, ⊖;0.00250
W_{Alg} (for $MgCl_2$ Solution): △ ;0.00625, X ;0.00420, □ ;0.00250
Solid curves show theoretical values obtained by selecting the most suitable effective charge densities as well as in Fig.6.

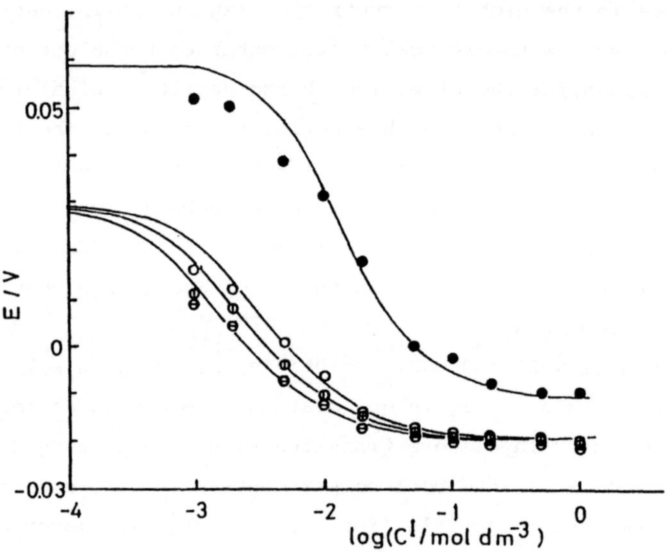

Fig.6 Membrane Potentials of Multimembranes Composed of Alginate Gel Layer for NaCl and $CaCl_2$ Solution (R=10)

W_{Alg} for $CaCl_2$ solution: ○ ;0.00626, ⊖;0.00420, ⦶ ;0.00250.
W_{Alg} for NaCl solution: ● ;0.00626.

found to be less than those of MgAlg. This is due to the capture of Ca ion in the junctions. Absolute values of θ^{*2} of CaAlg in NaCl solution systems were bigger than those for $CaCl_2$ solutions. This is considered to result from the partial ionexchange of Ca^{2+} with Na^{+} ions.

Counterion Binding Characteristics of Alginates

The ratios of the effective charge density of Alg to the concentration of carboxylic groups, i.e., the concentration of uronate units θ_0, θ^{*2}/θ_0, show the fractions of the ionized carboxylic groups. W_{Alg} in the second layer might decrease due to osmotic pressures. However, we approximated the decrease in W_{Alg} could be neglected by a help of the support plates F shown in Fig.1. As shown in Table

Table 1. Characteristic Values of Alginate Sol and Gel

Solute	$-\theta^{*1}$/mol dm^{-3}	R	W_{Alg}/g cm^{-3}	$T_\star^{b)}$	$-\theta^{2*}$/mol dm^{-3}	θ^{2*}/θ_0
NaCl	0.0028	10	0.00626	0.601	0.014	0.44
NaCl	0.0028	10	0.00420	0.601	0.010	0.47
NaCl	0.0028	2	0.00420	0.601	0.010	0.47
NaCl	0.0028	10	0.00250	0.601	0.0050	0.40
$MgCl_2$	0.0025	10	0.00626	0.605	0.0080	0.25
$MgCl_2$	0.0025	10	0.00420	0.605	0.0065	0.31
$MgCl_2$	0.0025	2	0.00420	0.605	0.0060	0.28
$MgCl_2$	0.0025	10	0.00250	0.605	0.0035	0.28
$CaCl_2$[a)]	0.0026	10	0.00626	0.562	0.0035	0.11
$CaCl_2$[a)]	0.0026	10	0.00420	0.562	0.0025	0.12
$CaCl_2$[a)]	0.0026	10	0.00250	0.562	0.0015	0.12
NaCl[a)]	0.0028	10	0.00626	0.601	0.0070	0.22
NaCl[a)]	0.0028	2	0.00420	0.601	0.0060	0.28

a) CaAlg gelled by 1.0 mol dm^{-3}. b): $T_\star = T_\star^1 = T_\star^2 = T_\star^3$. θ^{*1}: charge density of supporting film C.M.(=θ^{*3}). θ^{*2}: charge density of Alg. θ_0: concentration of Alg on the basis of uronate units.

1, the values θ^{*2}/θ_0 were found not to depend on W_{Alg} in the experimental region. The mean values θ^{*2}/θ_0 for Na-, Mg- and CaAlg were 0.45, 0.28 and 0.12. The interactions of Na^{+} and Mg^{2+} with Alg are dominantly electrostatical. Then, the differences of their value of θ^{*2}/θ_0 can be approximated to result from their valence effects. Ca ions mediate the stackings of Alg molecules to form the junctions.[14)] Then, the much smaller value of CaAlg than MgAlg results from the constrains of Ca ions in the junctions and in the networks.

Effects of gellation on Ca ion bindings

The electrostatical interactions of Mg and Ca ions with Alg are approximated to be equal. Then, the difference of θ^{*2}/θ_0 between Mg-

and CaAlg, $|(\theta^{*2}(\text{MgAlg})-\theta^{*2}(\text{CaAlg}))/\theta_0|=0.16$, shows the ratio of enhanced Ca ion bindings accompanied by the gelation of Alg. Whether this increase results from primarily the entrap of Ca ion in the junctions can be discussed as follows. When the fraction of the junctions in a Alg molecule is F_j on the basis of the concentration of the uronate unit and Ca ion binding ratio in the junctions and in Alg chains between the junctions are $(\sigma_j)_{Ca}$ and $(\sigma_i)_{Ca}$, the fraction of ionized carboxylic groups in CaAlg gels, θ^{*2}/θ_0, can be expressed by following equation,

$$\theta^{*2}/\theta_0 = F_j(1-(\sigma_j)_{Ca}) + (1-F_j)(1-(\sigma_i)_{Ca}). \qquad (9)$$

Assuming that Ca ions entrapped in the junctions bind completely to Alg ($(\sigma_j)_{Ca}=1$) and all GG blocks contribute to the formation of the junctions ($F_j=F_{GG}=0.28$), the value $(\sigma_i)_{Ca}$ of CaAlg calculated from the result $\theta^{*2}/\theta_0=0.12$ is 0.83 which is bigger than the binding ratio of Mg, σ_{Mg} $(=1-(\theta^{*2}/\theta_0)_{MgAlg}=0.72)$. Actually, F_j is considered to be smaller than F_{GG}. Then, $(\sigma_i)_{Ca}$ should be greater than 0.83. From these considerations, it can be concluded that Ca ion bindings are enhanced by the gelation not only in the junction but even in the regions between the junctions, i.e., in the networks.

Diffusion-Reaction of Urea through Multimembrane Containing Urease Layer

Symmetric Multimembrane (Urea Solution (C_s=0.01 mol dm^{-3})|SP| Urease Layer|SP|Buffer Solution (C_s=0)): Diffusion-reactions of urea through the symmetric multimembrane were measured in the range of pH=6.5-8.6. The urease solution (0.06 w/v%) or the mixed solution composed of urease and NaAlg (0.06 and 0.6 w/v%) were used as the urease layer, and the fluxes of the product NH_4^+ into the compartment I and II, J_p^I and J_p^{II}, were obtained under the steady state condition. The total fluxes of NH_4^+, J_p^T $(=-J_p^I+J_p^{II})$, and the flux ratios $-J_p^I/J_p^{II}$, are shown in Fig.7 and 8 as a function of pH. By adding NaAlg, the optimum pH, i.e., the value of pH showing the maximum J_p^T, shifted to alkaline pH region by 0.6 pH units from pH=6.9 to 7.5 and the maximum J_p^T increased slightly. The values of pH showing the maximum of $-J_p^I/J_p^{II}$ were almost as same as that of J_p^T, and the values of $-J_p^I/J_p^{II}$ were greater than unity, i.e., the fluxes of the product into the compartment I (donor part) were larger than those into the compartment II (acceptor part) in all experimental conditions.

Asymmetric Multimembrane (Urea Solution (C_s=0.01 mol dm^{-3}) |SDMV|Urease layer|SP| Buffer Solution (C_s=0)): The diffusion-

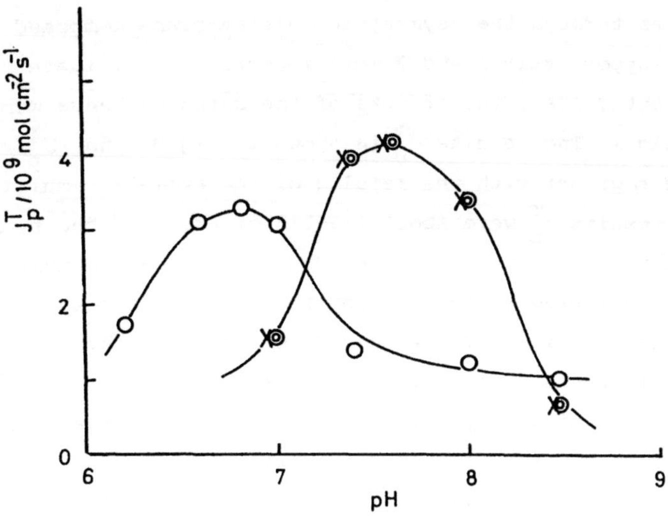

Fig.7 pH of Solutions *vs*. Total Effluxes of Product J_p^T through Symmetri Multimembrane

Urease in Enzyme layer: ○ . Urease and NaAlg in enzyme layer: ◎ and X. Marks ◎ show J_p^T *vs*. pH of the solutions in the compartment I and II. Marks X show J_p^T *vs*. pH in the enzyme layer calculated by asuuming Donnan equilibrium.

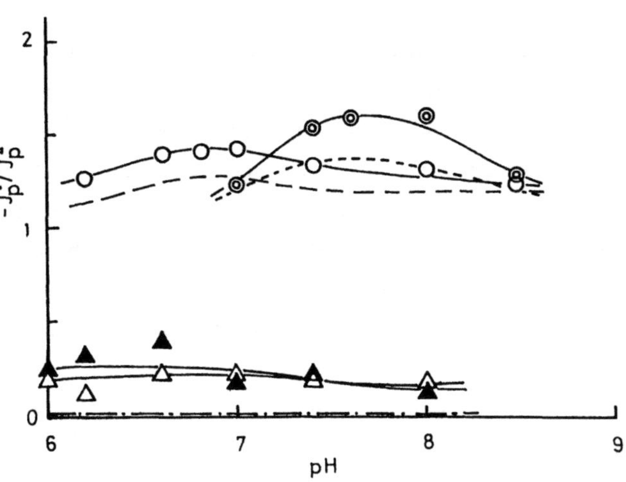

Fig.8 pH of Solution *vs*. Efflux Ratios of Products $-J_p^I/J_p^{II}$ through Symmetric and Asymmetric Multimembrane

Symmetric multimembrane: Urease in enzyme layer; ○ , Urease and NaAlg in enzyme layer; ◎ . Asymmetric multimembrane: Urease; , Urease and NaAlg; ▲ . Dotted and broken curves show theoretical values of the symmetric multimembrane containing urease, and urease and NaAlg in the enzyme layer. A line (—·—·—) shows the theoretical values of the asymmetric ones containing urease, and urease and NaAlg.

reaction of urea through the asymmetric multimembrane composed of SDMV and SP as the support film 1 and 2 were measured to elucidate the asymmetric effects, i.e., the effects of the anion exchange membrane used as the film 1. The results J_p^T is shown in Fig.9, and $-J_p^I/J_p^{II}$ are shown in Fig.8 together with the results of the symmetric multimembrane. The results J_p^T were about 1/3-1/5 of those of the symmetric multimembranes. Lower values of J_p^T through the asymmetric multimembranes are considered to result from the reduced permeability of urea through the film 1, SDMV. It should be noted that by adding Alg the maximum J_p^T increased slightly. The optimum pH was 6.3 for the multimembrane containing only urease and by adding Alg the value of pH shifted to alkaline pH region by 0.3 pH unit as well as the symmetric ones. However, the pH shift was smaller than that of the symmetric one. The flux ratios $-J_p^I/J_p^{II}$ were much less than unity in the range 0.2-0.3. As expected by the effects of the positively charged SDMV used as the support film 1, the fluxes J_p^{II} were much greater than the absolute value of J_p^I.

Diffusion-Reaction of Maltose through Multimembrane Containing GA Layer

Effluxes of glucose through the multimembrane (Maltose solution (C_s=0.01 mol dm^{-3}) |SP| GA layer|SP| C_s=0) were measured in the range of pH=3.5-6.0. The GA solution (0.06 w/v%) or the mixed solution composed of GA and NaAlg (0.06 and 0.6 w/v%) were used as the enzyme layer. The fluxes of the product (glucose) into the compartment I and II, J_p^I and J_p^{II}, were obtained under the steady state condition. The total fluxes of glucose, J_p^T, and the flux ratios $-J_p^I/J_p^{II}$, are shown in Fig.10 and 11 as a function of pH. In the case containing only GA in the enzyme layer, J_p^T showed the maximum at the optimum pH of GA. By adding NaAlg, the value of pH showing the maximum J_p^T shifted to alkaline pH region by 0.7 pH units from pH=4.6 to 5.3 as well as the case of the urease. However, the maximum J_p^T decreased contrary to the case. The values of pH showing the maximum of $-J_p^I/J_p^{II}$ were almost as same as that of J_p^T, and the values of $-J_p^I/J_p^{II}$ were greater than unity.

Effects of Gelation of Alg

To elucidate the effects of the gelation in the microenvironments around GA on the diffusion-reaction, Alginates in the enzyme layer was gelled by adding Ca^{2+} ions. The effluxes of glucose

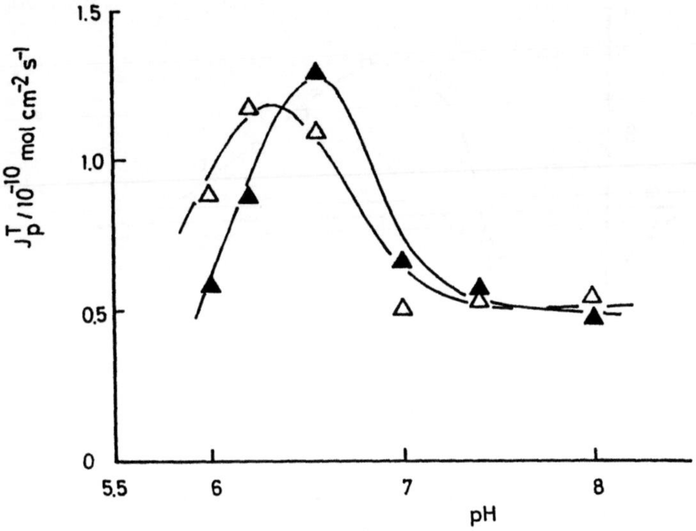

Fig.9 pH of Solutions *vs*. Total Effluxes of Product J_p^T through Asymmetric Multimembrane

Urease in enzyme layer: △ , Urease and NaAlg in enzyme layer: ▲ .

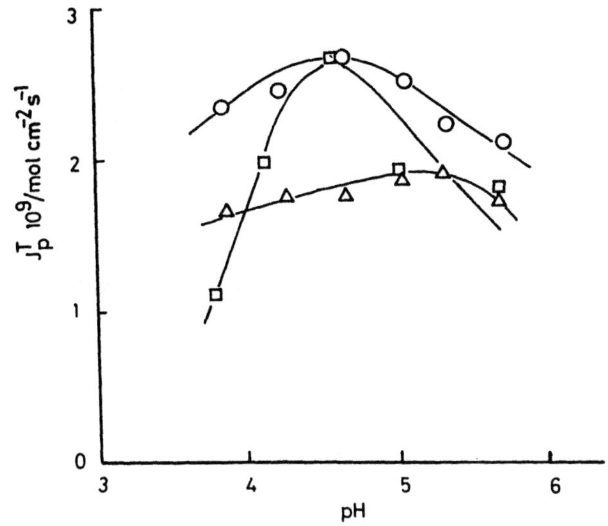

Fig.10 pH of Solution *vs*. Total Effluxes of Product J_p^T through Symmetric Multimembrane Containing Glucoamylase (GA)

GA only: ○ , GA and NaAlg: △ , GA and CAaLg: □ .

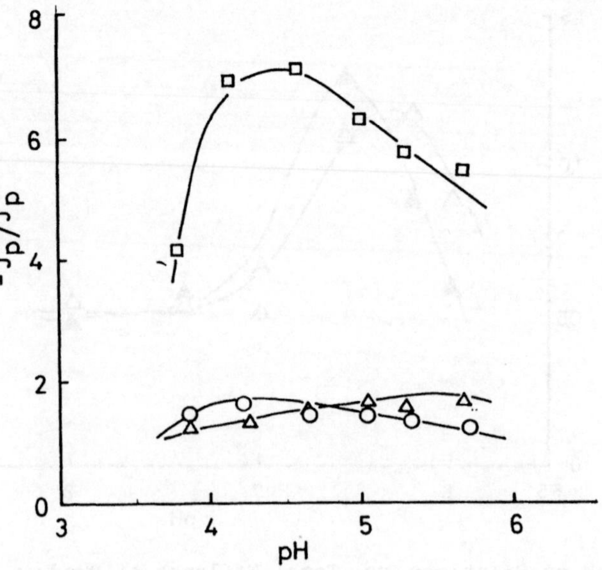

Fig.11 pH of Solution *vs.* Efflux Ratio of Product $-J_p^I/J_p^{II}$ through Symmetric Multimembrane Containing Glucoamylase

through the multimembranes (Maltose (0.01 mol/dm^3)+$CaCl_2$(0.1 mol/dm^3) | GA +CaAlg | $CaCl_2$(0.1 mol/dm^3)) were measured. The results J_p^T and $-J_p^I/J_p^{II}$ are shown in Fig.10 and 11 together with the results in the case that the enzyme layers were sol states. By gelling Alg, the value of pH showing the maximum J_p^T were found not to shift from that of GA only, and the maximum J_p^T increased almost to the same value as that of GA only. Furthermore, the flux ratios $-J_p^I/J_p^{II}$ increased significantly by the gellation.

DISCUSSION

Effects of Adding NaAlg to Enzyme Layer of Symmetric Multimembrane on Optimum pH

For the diffusion-reaction through the multimembrane, the optimum pH which shows the maximum J_p^T shifted to more alkalin pH by adding NaAlg to the enzyme layers. As Alginate is an anionic polysaccharide having a carboxylate group per a uronate unit, the pH shifts are considered to result from the pH decrease in the micro-environment around the active sites of the enzymes and the changes of H^+ ion distributions between the enzyme layer and the bulk solution. As the concentrations of the enzymes were very small comparing with that of Alg, the effects of the enzymes on the effective charge density can be neglected. The H^+ ion concentrations in the enzyme layer containing Alg can be approximated to be the same as that in the solution layer containing only Alg. From the effective charge density of the Alg layer shown in Table 1, the H^+ion distributions between the enzyme layer and the bulk solution can be obtained by assuming Donnan equilibrium and electroneutrality.[16] Taking the ions from the buffer into consideration, the pH differences between the enzyme layer and the bulk solution were calculated to be in the range of -0 - -0.02 in the experimental region. In Fig.7, J_p^T are also shown as a function of pH in the urease layer obtained as above. The deviations of H ion concentrations due to the Donnan equilibrium were found to be too small to explain the optimum pH shifts. Therefore, the optimum pH shift to alkaline region must result predominantly from an additional pH decrease in the microenvironment around the active sites of the enzymes. Their active sites are surrounded by the carboxylic groups of Alg, and the negative electrostatic potential ψ^α prevails in the region. Under equilibrium conditions, the electrochemical potential μ^α of the H^+ions in the region α around the active site equals to that of H ions μ^β in the region β apart from the region α. Then the pH difference between these regions can be expressed by

$$pH^\beta - pH^\alpha = -0.434\ e\psi^\alpha/kT \qquad (10)$$

The optimum pH shift ΔpH corrected by the contributions of the Donnan equilibrium due to the addition of NaAlg is correspond to $pH^\beta - pH^\alpha$. The local electrostatic potential ψ^α induced by Alg can be obtained from ΔpH. For the case of urease and GA, ΔpHs were -0.58 and -0.66. From these results, ψ^α can be calculated to be -0.0343 and -0.0296 V respectively, which were expressed in values relative to the β region. The results were almost equal to that obtained from the optimum pH shift in the solution.[6] The local electrostatic potential ψ^α induced by Alg under I=0.12-0.15 mol dm^{-3} was found to be almost equal to the surface potential of Chs-C[17]

By the gelation of Alg, the optimum pH shift of GA multi-membrane, as shown in Fig.10, was almost negligible. This results from the decrease of the charge densities due to the gellation as shown in Table 1.

The value J_p^T at the optimum pH through the urease multimembrane increased slightly by adding NaAlg as shown in Fig.7. The result coincides with those of Nichol et al.[18,19] They studied the adding effects of polymers to urease solutions on the hydrolysis of urea, and showed V_{max} increased slightly by adding polymers such as bovine serum albumin, ovalbumin and dextran. They attributed these effects to the change of the conformations due to the excluded volume of the adding polymers. However, for the case of GA, the value J_p^T decreased by adding NaAlg contraly to the case of urease and by the gelation increased. The difference in the adding effects of NaAlg and the effects of the gelation should be studied further.

<u>Theoretical Analysis of Diffusion-Reaction through Multimembrane</u>

Urease catalyze the hydrolysis of urea,

$$NH_2CONH_2 + 2H_2O \rightleftharpoons 2NH_4^+ + CO_3^{2-}, \qquad (11)$$

and GA catalyze the hydrolysis of maltose,

$$\text{Maltose} \rightleftharpoons 2\ \text{Glucose}. \qquad (12)$$

In the former reaction, we payed attention to NH_4^+ which is one of the products P. Then, the reactions of substrates S in the enzyme layer are expressed by $S \rightleftharpoons 2P$. In the enzyme layer, if the concentrations of substrate are much less than Michaelis constants K_M of the enzymes, the enzymic reactions can be approximated to obey to the first order reaction.[20,21] Under steady state conditions, the diffusion-reactions can be expressed by the following equations in the enzyme layer

$$D_s^2\ (d^2C_s/dx^2)\ -k\ C_s\ = 0 \qquad (13)$$

$$D_p^2\ (d^2C_s/dx^2)\ +\ 2k\ C_s\ = 0 \qquad (14)$$

where k is a first order reaction rate constant, D_s^2 and D_p^2 are diffusion coefficients of S and P, and C_s and C_p are concentrations of S and P at the position x of the enzyme layer. Concentrations of S and P in the enzyme layer are obtained by solving Eqs.(13) and (14) under the following boundary conditions, i.e., the interfacial concentrations of S and P between the support film 1 and the enzyme layer are $\bar{C}_s^{1,2}$ and $\bar{C}_p^{1,2}$ and those between the film 2 and the enzyme layer are $\bar{C}_s^{2,3}$ and $\bar{C}_p^{2,3}$ respectively. From continuities of the fluxes of S and P at each interface, the concentration profiles of S and P can be obtained, and the total fluxes of product J_p^T and their flux ratios $-J_p^I/J_p^{II}$ are expressed as a function $\alpha(=(k\ D_s^2)^{1/2}$. Their theoretical values through the symmetric and the asymmetric multimembrane containing the urease are shown in Fig.12. In the asymmetri multimembrane, due to less permeability of urea and NH_4^+ ions through the support film 1 SDMV, the theoretical results of J_p^T and $-J_p^I/J_p^{II}$ are lower than those of the symmetric ones. In the calculations, the permeability coefficients of the substrates and the products obtained experimentally were used. The diffusion coefficents in the enzyme

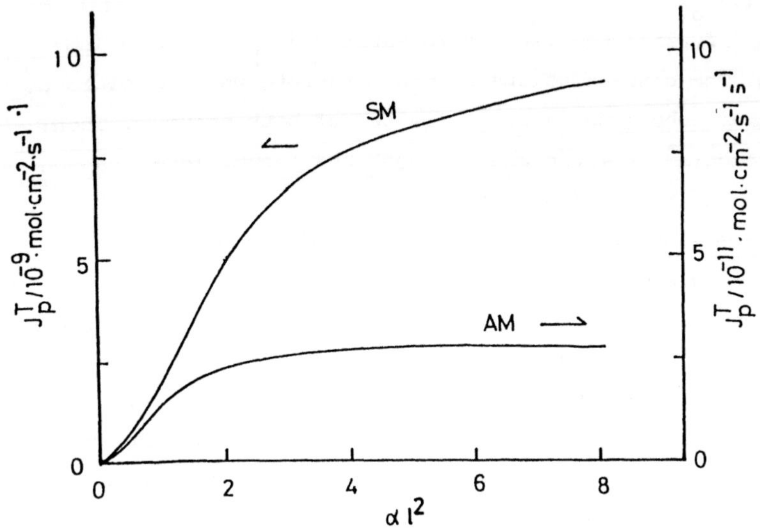

(a) J_p^T *vs.* αl^2

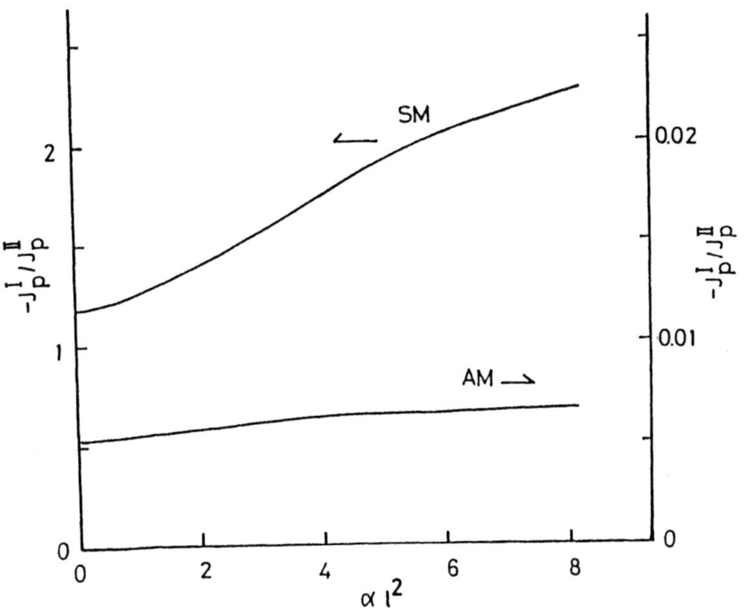

(b) $-J_p^I/J_p^{II}$ *vs.* αl^2

Fig.12 Variation with αl^2 $(=(k/D_s^2)^{1/2}l^2)$ on Calculated Total Efflux J_p^T and of Efflux Ratio of Product $-J_p^I/J_p^{II}$ for Symmetric and Asymmetric Multimembrane

SM: Symmetric multimembrane, AM: Asymmetric multimembrane

layers are obtained by Wang's equation[22,23)]

$$D = D_0(1-1.5v_p) \qquad (15)$$

where D_0 is the diffusion coefficient in water and v_p is the volume fraction of Alg. The concentrations of the substrate were found to be less than or almost the same as the K_M values of both enzymes. Then, the diffusion-reaction of substrates through the enzyme layer can be approximated to obey to Eqs.(13) and (14).

Diffusion-reaction through Multimembrane Containing Urease (Adding Effects of NaAlg)

By fitting the experimental results J_p^T shown in Fig.7 and 9, with the theoretical values, the reaction rate constants k in the urease were obtained. The efflux ratio of the product $-J_p^I/J_p^{II}$ expected from these k values are also shown in Fig.8. The expected values of the symmetric multimembrane were slightly smaller than the experimental results. However, we can consider that the theoretical values are coincide semiquantitatively with the experimental results. However. in the asymmetric multimembranes, the expected values of $-J_p^I/J_p^{II}$ were bout 0.006, which were much less than those of the experimental results. This difference can be explained by the presence of carbamate ions NH_2COO^- being an intermediate in the hydrolysis of urea as pointed by G.B.Tanny and coworkers.[24)]

$$NH_2CONH_2 + H_2O \rightleftharpoons NH_2COO^- + NH_4^+ \qquad (16)$$

They studied precisely ionic flows in a membrane composed of an anion exchanger and a urease phase, and showed that the carbamate ions produced near the interface between the anion exchanger and the urease phase exchange with anions in a donor solution through the anion exchanger, in which the carbamates decompose spontaneously with a half life of several msec.[18)]

$$NH_2COO^- + H_2O \rightleftharpoons NH_4^+ + CO_3^{2-} \qquad (17)$$

As pointed out by Tanny, the carbamate ions diffuse into the film 1 SDMV and the products NH_4^+ and CO_3^{2-} diffuse out to the donor compartment due to their concentration gradients. Consequently, the experimental results of $-J_p^I/J_p^{II}$ became greater than the theoretical ones which did not take the presence of the carbamates into consideration.

Effects of Sol-Gel Transition of Alg on Diffusion-reaction through Multimembrane Containing Glucoamylase

When the environment around GA changes from sol to gel state, J_p^T and $-J_p^I/J_p^{II}$ of the glucose increased significantly. The formmer increrased to the extent of the case of GA only, and the latter increased significantly comparing with the cases of the sol state.

This might be the increase of the reaction rate constant k by the gelation of Alg. Furthermore, the reaction might occur at nearer part to the interface the film I and the GA layer due to slower diffusivity of substrates.

CONCLUSION

(1) By using the multimembrane systems, the basic properties of coated gels can be studied, such as the effects of sol-gel transitions on their charge densities and on diffusion-reactions, and the effects of the asymmetries on them.

(2) By the sol-gel transition of Alg, Ca ion bindings are enhanced not only in the junctions but even in the regions between the junctions, i.e., in the networks, and the charge densities decrease.

(3) By Adding NaAlg to the urease or the gulucoamylase layer, the optimum pH shifts to alkaline pH region, which results from the negative electrostatic potential of Alg. The adding effects of NaAlg on the maximum J_p^T depend on the properties of the enzymes, i.e., in the case of the urease increases but in the case of the glucoamylase decreases.

(4) For the symmetric systems, the flux ratioes of product $-J_p^I/J_p^{II}$ are greater than unity, and are not affected significantly by adding NaAlg. However, in the asymmetric systems, they are affecte by the prpperties of the coated films.

(5) By the gelation in the glucoamylase layer, the optimum pH shift, tha maxumum J_p^T and $-J_p^I/J_p^{II}$ are affected significantly and the network effects on the diffusion-reactions are very important from not only basical but also practical point of views.

REFERENCES

(1) M.Yonese, *Pharm. Tech. Japan*, in press.

(2) T.Tanaka, D.Fillmore, S.-T.Sun, I.Nishino, G.Swislow, A.Shah, *Phys. Review Letter*, 45, 1636(1980).

(3) Y.Osada, M.Hasebe, *Chemistry Letter*, 1285(1985).

(4) K.Ishihara, M.Kobayashi, I.Ishihara, *Macromol. Chem. Rapid. Commun.*, 4,327(1983).

(5) G.Brown, D.Thomas, E.Selegny, *J. Memb. Biol.*, 8, 313(1972).

(6) M.Yonese, H.Murabayashi, H.Kishimoto, *J. Memb. Sci.*, in press.

(7) M.Yonese, K.Baba, H.Kishimoto, *Bull. Chem. Soc. Jpn.*, 61,1857(1988).

(8) E.R.Morris, D.A.Rees, D.Thom, *Carbohydr. Res.*, *81*, 305(1980).

(9) H.Grasdalen, B.Larsen, O.Smidsrod, *Carbohydr. Res.*, 68, 23(1979).
(10) J.Kiellnd, *J. Ame. Chem. Soc.*, 59, 1675(1937).
(11) P.Henderson, *Z. Physik. Chem.*, 59,118(1907).
(12) M.Yonese, M.Nakagaki, *YAKUGAKU ZASSHI*, 101, 493(1981).
(13) M.Yonese, K.Baba, H.Kishimoto, *Bull. Chem. Soc. Jpn.*, 61,1107 (1988).
(14) D.A.Morris, D.A.Rees, D.Thom., *Carbohydr.Res.*, 66,143(1978).
(15) A.Katchalski, R.E.Cooper, J.Upadhyaym A.Wassermann, *J.Chem.Soc.*, 1961.5198.
(16) P.C.Hiemenz, Principles of Colloid and Surface Chemistry, Marcel Delker, Inc.. New York(1977),p145.
(17) M.B.Mathews, *Biochem. Biophys. Acta.*, 35,9(1959).
(18) L.W.Nichol, M.J.Sculley, L.D.Ward, D.J.Windzor, *Arch. Biochem. Biophys.* 225,574(1983).
(19) L.W.Nichol, M.J.Sculley, L.D.Ward, D.J.Windzor, *Arch. Biochem. Biophys.*, 239155(1985).
(20) R.Goldman, O.Kedem, I,H,Silman, S.R.Carplan, E.Katchalski, *Biochemistry*, 7,486(1968).
(21) R.Goldman, O.Kedem, E.Katchalski, *Biochemistry*, 7,4518(1968).
(22) J.H.Wang, *J. Ame. Chem. Soc.*, 76,4755(1954).
(23) M.Nakagaki,M.Yonese, *YAKUGAKU ZASSHI*, 94,1170(1974).
(24) G.B.Tanny, O.Kedem, E.Selegny, *J. Memb. Sci.*, 4,363(1979).

INDEX

GPSR Compliance

The European Union's (EU) General Product Safety Regulation (GPSR) is a set of rules that requires consumer products to be safe and our obligations to ensure this.

If you have any concerns about our products, you can contact us on ProductSafety@springernature.com

In case Publisher is established outside the EU, the EU authorized representative is:

Springer Nature Customer Service Center GmbH
Europaplatz 3
69115 Heidelberg, Germany

Batch number: 10684222

Printed by Printforce, the Netherlands